SEMIOCHEMICALS
THEIR ROLE IN PEST CONTROL

Edited by

DONALD A. NORDLUND

AR-SEA-USDA
Southern Grain Insects Research Laboratory
Tifton, Georgia

RICHARD L. JONES

Department of Entomology, Fisheries, and Wildlife
University of Minnesota
St. Paul, Minnesota

W. JOE LEWIS

AR-SEA-USDA
Southern Grain Insects Research Laboratory
Tifton, Georgia

A Wiley-Interscience Publication
JOHN WILEY & SONS, New York • Chichester • Brisbane • Toronto

Library of Congress Catalog Card Number: 81-40184

ISBN 0-471-05803-3

Printed in the United States of America

10 9 8 7 6 5 4 3 2 1

SEMIOCHEMICALS

THEIR ROLE IN PEST CONTROL

To our wives;
Sheron, Ann, and Dianne,
for their ever-present
support

CONTRIBUTORS

ALFRED P. ARTHUR, Agriculture Canada, Research Station Research Branch, Saskatoon, Saskatchewan, Canada

PATRICK D. GREANY AR–SEA–USDA, Insect Attractants, Behavior and Basic Biology Research Laboratory, Gainesville, Florida

HARRY R. GROSS, Jr., AR–SEA–USDA, Southern Grain Insects Research Laboratory, Tifton, Georgia

KENNETH S. HAGEN Division of Biological Control, Department of Entomology, University of California at Berkeley, Albany, California

ROBERT D. JACKSON, AR–SEA–USDA, National Program Staff, Beltsville, Maryland

RICHARD L. JONES, Department of Entomology, Fisheries, and Wildlife, University of Minnesota, St. Paul, Minnesota

W. JOE LEWIS, AR–SEA–USDA, Southern Grain Insects Research Laboratory, Tifton, Geogia

DONALD A. NORDLUND, AR–SEA–USDA, Soutnern Grain Insects Research Laboratory, Tifton, Georgia

PETER W. PRICE, Department of Biological Sciences, Northern Arizona University, Flagstaff, Arizona

RONALD J. PROKOPY, Department of Entomology, University of Massachusetts, Amherst, Massachusetts

WENDELL L. ROELOFS, New York State Agricultural Experiment Station, Geneva, New York

LOUIS M. SCHOONHOVEN, Department of Animal Physiology, Agricultural University, Wageningen, The Netherlands

JOOP C. VAN LENTEREN, Department of Ecology, Zoology Laboratory, University of Leiden, Leiden, The Netherlands

S. BRADLEIGH VINSON, Department of Entomology, Texas A&M University, College Station, Texas

RONALD M. WESELOH, Connecticut Agricultural Experiment Station, New Haven, Connecticut

FOREWORD

Virtual reliance on chemicals for controlling many important insect pests has led to intensive investigations of alternative methods that avoid or minimize the environmental hazards often associated with broad-spectrum insecticides. Substantial research on various alternative methods of control has been under way for more than two decades and has involved research on such methods as selective insecticides, host plant resistance, insect attractants, various biological organisms, autocidal techniques, and the integration of complementary techniques. Research on insect attractants and biological organisms has revealed the existence of a diversity of naturally produced chemicals that influence the behavior of both destructive and beneficial insects in the environment.

These chemicals, grouped under the broad term "semiochemicals," include pheromones, allomones, kairomones, synomones, and apneumones; which were hardly known to exist a few years ago. In a somewhat restricted sense, however, they can be called insect attractants, repellents, and deterrents. The nature and functions of these chemicals are discussed and described in detail in this book. The substances considered include insect sex pheromones, which are vital to successful mate finding and reproduction by insects; kairomones, which are cues for parasitoids and predators in their search for hosts and prey; and allomones, which minimize the competition between coexisting organisms in natural habitats. The function of these highly active chemical substances makes it apparent that the behavior, or physiology, of insects is profoundly influenced and regulated by naturally produced substances in the environment. Current information on insect sex pheromones has been compiled in several earlier books. However, this book is the first to include extensive discussions on other types of semiochemicals that govern or regulate the behavior of insects in the environment. The knowledge that there are certain interactions between the different naturally produced chemicals makes it important that scientists keep abreast of new information on the broad complex of chemicals involved.

The pioneering investigators in this new field have already made outstanding contributions to a better understanding of the relationship of insects to their environment, and this knowledge, in turn, may lead to more effective and more desirable solutions to insect problems. Like any publication that deals with a relatively new field of science, however, this book presents a knowledge of the nature, function, actions, and interactions of these new chemical substances that is far from complete. Nevertheless, sufficient information to warrant a progress report has already been obtained by the increasing number of scientists engaged in research on semiochemicals. This report should lay a firm foundation for even more productive research in the years ahead.

The identification and synthesis of sex pheromone complexes produced by many of the more important insects have already given pest managers highly sensitive tools for the detection of insects. This alone represents an outstanding contribution to more effective and efficient insect suppression systems. More important, we now have prospects for new and desirable ways of suppressing and managing some insects by employing confusion or mass trapping techniques. The discovery of the existence and nature of kairomones produced by insects offers the possibility of manipulating natural or augmented biological organisms in ways never before considered. Plants also produce chemicals to which both destructive and beneficial insects respond. Plants produce chemicals that deter attack by pests, and destructive and beneficial insects produce chemicals that minimize the intense competition for survival that exists in a natural environment. As some of the authors point out, there are also certain interactions between semiochemicals of different kinds. Such interactions may have to be understood before we can make optimum use of these materials in integrated systems.

The outstanding advances in chemical technology during the past two decades has made possible the discovery, identification, and synthesis of many semiochemicals not heretofore known to exist. However, this is only the first step in the development and eventual practical use of these chemicals for pest control. It is necessary for entomologists, or, more specifically, insect behaviorists, to investigate how and under what conditions these substances influence the behavior, physiology, and dynamics of insects under investigation. Such research will require careful planning, innovative techniques, and keen observations, to obtain necessary data for critical analysis and interpretation. The efficiency of semiochemicals is likely to be strongly influenced by such factors as the density and distribution of the target pest. Also, the degree of control achieved is likely to depend on the distance and extent of movement of the insects into the areas under suppression. It should be emphasized also that practical use of certain semiochemicals may require the appropriate integration of two or more suppressive measures at strategic times and places in the pest ecosystems. Therefore, suppression experiments and special field investigations designed to manipulate the behavior of entomophagous insects may have to be conducted in large or well isolated areas. Many recent experiments using some of the newer methods of control have yielded less than expected results—not because of a failure of the suppression technique but because of a lack of adequate isolation of the experimental areas.

The prevailing concept of placing primary reliance on natural control factors and applying control measures only where and when the pest reaches economic threshold levels may have to be abandoned for many of our major insect pests if optimum advantage is to be taken of the potential of semiochemicals in insect pest management systems. Not only will the efficiency of these substances hinge on the density and distribution of the pest to be controlled, but they may have to be used in an organized way, in areas large enough to eliminate the factor of excessive pest movement into or out of the management areas. Unlike applications of chemical insecticides, which result in almost immediate control of the treated pests, the action of the semiochemicals or the manipulated biological organisms may be too slow to achieve adequate control of pests that have reached damaging or threatening levels. Therefore, satisfactory insect management by the use of semiochemicals will likely require the implementation of management programs on an areawide basis well in advance of the occurrence of damaging numbers. Fortunately, semiochemicals are generally highly pest specific and should cause little or no adverse effect on nontarget organisms. Thus the prospects of managing many of our major pests by taking full advantage of semiochemicals should be welcomed by environmentalists and the public who are increasingly concerned about the maintenance of environmental quality.

E. F. Knipling
Expert (Pest Management)
AR–SEA–USDA
National Program Staff
Beltsville, Maryland

PREFACE

The chapters in this book were selected from papers presented at two symposia that addressed the potential importance and role of semiochemicals in the control of pest insects. One, "Recent Advances in Biological Control Technology: Interactions of Entomophages and Semiochemicals," was held in conjunction with the 1978 National Meeting of the Entomological Society of America. The other, "Behavioral Chemicals: Role and Employment in Plant Protection," was held in conjunction with the Ninth International Congress of Plant Protection in 1979.

The papers are grouped into five topic groups: introduction, role and significance of allelochemics, role and significance of pheromones, chemistry and evolution of semiochemicals, and conclusion.

The two chapters in the introduction review the various changes that are taking place in pest control, the pressing need for control techniques with limited environmental impact, the advantages of the uses of various biological agents in pest control, and semiochemical terminology. They provide a foundation and framework for the remaining papers.

The section on the role and significance of allelochemics contains six chapters that discuss the involvement of allelochemics in various interspecific interactions that are of importance to pest control, and the potential uses of these chemicals in the pest control strategies. Interactions between plants and phytophagous insects, interactions between entomophagous insects and the food plants of their hosts or prey, and interactions between entomophagous insects and their hosts or prey are all examined.

The section on the role and significance of pheromones contains three chapters dealing with intraspecific interactions and also relates to the interspecific interactions discussed in the preceding section. It discusses the potential uses of pheromones that inhibit oviposition or feeding or encourage emigration, or pheromones that can be used to attract insects to traps or to disrupt their mating behavior.

The chemistry and the evolutionary history of semiochemicals are covered in the fourth section. The two chapters here deal with the identity and functions of kairomones and demonstrate how the various groups of semiochemicals are interrelated.

The concluding chapter brings together and expands on what has been presented with respect to techniques for utilization and significance of semiochemicals to future pest control strategies.

This book is a concise presentation of a wide variety of semiochemicals and many approaches to their use in the control of pest insects. The chapters reveal that although we know a great deal about semiochemicals, we have only begun to analyze all the subtle interactions involved, the modes of action, and the use of this important group of chemical-releasing stimuli. Because of the rather limited state of our knowledge, the authors were asked to be speculative—to present what is known about the topic, and then, as if alone in an easy chair, to consider what it means, to examine its potentials, and to consider possible methods of its use. We hope that this approach will help to increase the dimensions of research in this area.

We express our sincere appreciation and gratitude to the authors for their efforts, cooperation, and understanding; to our respective support staffs, whose help and cooperation made editing this book a pleasant task and for their many contributions to our research efforts; and to the many colleagues with whom we have had numerous inspiring exchanges of information and without whose research efforts this book would have been impossible.

Tifton, Georgia
St. Paul, Minnesota
Tifton, Georgia
February 1981

DONALD A. NORDLUND
RICHARD L. JONES
W. JOE LEWIS

CONTENTS

SEMIOCHEMICALS

THEIR ROLE IN PEST CONTROL

SECTION I

INTRODUCTION

CHAPTER ONE

SEMIOCHEMICALS: THEIR ROLE WITH CHANGING APPROACHES TO PEST CONTROL

W. JOE LEWIS

AR–SEA–USDA,
Southern Grain Insects Research Laboratory
Tifton, Georgia

Our search for ways to increase our control over the environment has been a fundamental component of human history. At the forefront of these struggles is the quest for ways to deal with the plagues and ravages of insects and similar pests. We have competed directly with insects for crops since the beginning of agricultural activities in early settlements near rivers and streams. As these agricultural practices progressed in sophistication, various cultural, physical, and even biological procedures for dealing with insect pests were discovered. They included such cultural practices as proper field sanitation, crop rotation, and choice of the time of planting. It would also be expected that these early practitioners selected strains giving greater and more desirable yield, along with noting and selecting varieties that exhibited less disease and insect infestation. In this we find the early roots of agricultural pest control.

These practices, though far from totally effective, were generally adequate during the early years. However, after the Dark Ages, pest problems became more severe because of the intensification of agriculture to provide for an increasing human population, and the introduction of pests into new areas (Smith et al. 1976). Scientific knowledge grew, and the thirst for new and better ways of dealing with these pest problems increased. A desire for a magical "cure-all" for our pest problems and the expectation of finding one seemed to grow with the expansion of scientific knowledge. The principal interest began focusing on the use of chemical compounds. Though their initial use can be traced much earlier to the Orient, suggestions that various chemical concoctions be used for the control of insect and disease pests became pronounced in the eighteenth century (Lodeman 1903).

Numerous scattered discoveries and propositions of biological pest control arose throughout the eighteenth and nineteenth centuries (Weiss 1936). Despite these findings, and cautions by some that our pest control programs should be based on ecological principles (Metcalf 1930, Smith 1975), we moved persistently toward increasing use of chemical compounds at the expense of an appreciation for desirable cultural practices and furthering of biological concepts (Stevens 1960).

CONVENTIONAL PESTICIDE ERA

Initial Success

The advent of DDT during World War II, the flood of subsequent synthetic organic pesticides, and the seemingly unqualified success of these broad-spectrum pesticides plunged pest control into an almost total preoccupation with this approach from the mid-1940s until the mid-1960s. To be sure, these conventional pesticides have brought inestimable benefits to humanity by relieving human suffering and providing economic gains (Metcalf 1968, Smith and van den Bosch 1967). Not only did they produce spectacular results, but they were cheap, effective in small quantities against a wide array of pests, and easily deployed. The use of DDT alone resulted in the elimination of malaria-carrying mosquitoes and other health pests from entire countries (Wright et al. 1972). Consequently, these pesticides rapidly became the pest control tool for most insect pests, almost to the complete exclusion of interest in other methods, and their manufacture is a gigantic economic activity (van den Bosch 1979). In 1971 chemical pesticides worth $3.4 billion (retail) were applied worldwide for control of agricultural, industrial, and household pests (Muir 1978).

Complications

The conventional chemical pesticides are of immense value, and their ready and continuing availability is indispensable to agriculture as well as to public health. Yet the shortcomings of such a heavy dependence on these materials have become all too apparent and have been discussed in detail by numerous authors (van den Bosch 1979, Newsom 1971, Klassen 1976).

Four basic problems have emerged from the heavy use of pesticides: (1) development of resistance to the chemicals by various pests; (2) sudden resurgence of some target pests to even higher levels after the initial drop; (3) induced outbreaks of secondary pests that previously had remained below damaging levels; and (4) environmental contamination. The first three complications resulted in use of even larger quantities of these highly toxic chemicals, thereby intensifying the fourth problem.

Rachel Carson's *Silent Spring,* published in 1962, contained shocking evidence of damage to the environment that had resulted from the use of pesticides, and our society was aroused by the findings that many toxic chemicals or their metabolites were infiltrating all aspects of the biosphere, with some accumulating in our own bodies. Such knowledge has precipitated strong public demand for more limited and discriminate use of these pesticides. Compounding these difficulties, an expanding world population imposes a growing demand for agricultural and forest products, while the land available for such production is diminishing.

SEARCH FOR ALTERNATIVES

Initial Redirections

Pest control scientists, with few if any exceptions, have come to agree that there is an urgent need to develop more effective, alternative procedures for pest control based on sound biological principles—Rachel Carson's "other road." In fact, some of our scientific leaders recognized the complications years earlier and began initiating redirection that is now almost universal (Knipling 1979). This redirection has resulted in a strong revival in the search for biological solutions for our pest problems and an extensive resurgence in the use of the term "biological control."

There is considerable controversy over the proper use of the term "biological control," since it is employed to embrace two basic concepts. The traditional, narrower view holds that use of the term should be limited to regulation resulting from natural enemies; in the wider view, biological control includes such factors as host plant resistance, autosterilization, and manipulation with sex attractants (van den Bosch and Messenger 1973). Regardless of the concept to which one subscribes, the alternative efforts have the common objectives of seeking more natural ways of treating pest problems, to circumvent the hazards of toxic residues and to provide more permanent and ecologically harmonious pest management procedures (Price 1975).

Various Alternatives

Research on pest control approaches that are ecologically rational, yet effective has received high priority in most institutions and agencies for the past 10–15 years. The scientific community has responded admirably, and there have been many innovative and intriguing developments that offer alternatives to conventional chemical pesticides. These developments include improved techniques for breeding resistant plant varieties; applications of various genetic manipulations such as the sterile-male release, reducing the genetic fitness of the pest species, and improving the genetic quality of natural enemies; improved and innovative approaches for the use of natural enemies by classical importation, augmentation-manipulation, and resource management techniques; and the use of natural chemicals to regulate growth and behavior.

The last two alternatives include the study of semiochemicals, their role in mediating behavioral interactions among pests and associated organisms, and approaches for their employment in pest control (Figure 1.1). Research developments with these behavioral chemicals are among the most significant advances in pest control technology. Many recent developments that are discussed throughout this book accentuate the great strategic value of chemically mediated conversation in the interrelationships of various components of our crop ecosystem.

SEMIOCHEMICALS: STATUS AND ROLE

Semiochemicals, especially sex pheromones, have been intensively investigated by numerous pest control researchers for more than a decade. As a result, many exciting avenues for their employment have been discovered, and a number of programs using them have been implemented. Furthermore, these mediators, properly understood and exploited, offer unlimited potential and remain one of our most strategic tactics in designing pest control strategies. However, successful management with these semiochemicals stands far below the hopes and expectations anticipated 10–15 years ago.

Figure 1.1. *Microplitis croceipes*, a key larval parasitoid of *Heliothis zea*, responding to 13-methylhentriacontane. This kairomone was isolated from the frass of *H. zea* larvae and is an important cue in mediating the host-seeking behavior of this entomophage.

Constraints to Semiochemical Employment

Let us examine some of the possible reasons for our failure thus far to capitalize more heavily on semiochemicals, which obviously play such an extensive and vital role in the fundamental ecology of pests competing for our food and fiber.

Organizational and Economic Constraints

Part of the difficulty is organizational and general economics. Programs utilizing semiochemicals, as well as other biological methods, do not lend themselves as readily to development, adoption, and implementation by our agribusiness systems as do conventional pesticides. Broad-spectrum pesticides can be protected by patents, manufactured in bulk, stored readily, and sold for use against a wide array of pests. They can be applied easily and reliably with standard spray equipment, and they produce immediate and dramatic effects. Furthermore, they can be screened and evaluated in a regimented system (Klassen 1976).

In contrast, the development and employment of each behavioral chemical technique may require certain unique technologies involving input from varying combinations of specialists. Such programs may require very basic long-term studies, with subsequent intense evaluations in large-scale field trials. They may require high initial monetary outlays, and even if success is achieved, only a fairly limited market is assured. These organizational and economic complications offer a strong challenge to the pest management profession and will require close cooperation among government, growers, and agribusiness. However, successful meeting of this challenge will provide the long-term, safe, effective, and economically feasible pest control systems that are essential to the future of our society.

Lack of Appreciation for Fundamental Biology

Our greatest constraint seems to be a much more subtle one, more basic even than the economic and organizational factors noted above. Our period of almost total dependence on conventional pesticides has resulted in a loss of appreciation for and understanding of fundamental biological principles and their incorporation into our applied programs. This influence seems to have had a carryover effect on our basic approaches to research programs in general. This effect perhaps can be observed most clearly in the design and execution of mating pheromone research programs for insect pests. Most of us can probably agree that we have generally approached mating pheromone studies with an insufficient appreciation of ethological complexities, particularly in light of some recent findings with several insect species, that the mating process consists of a compound sequence of behavioral activities involving a number of chemical mediators, acting individually or perhaps in various separate combinations, at various points in the sequence.

For perhaps a decade, the typical insect-mating pheromone explorations consisted of exposing the male to various candidate extracts from the females under laboratory conditions until some sexual excitation was elicited. When this occurred, the inducing chemical was purified, identified, and declared to be *the* pheromone for the subject species. The newly discovered ''pheromone'' was produced synthetically and extensively evaluated with field trapping procedures, with the notion that it should capture the males at equivalent or better rates than virgin females. In most cases, the results have not met expectations, and we have proceeded with extensive testing of various analogues and time-release formulations on the idea that the problem must be a chemical one.

The scientists in the chemical disciplines have responded admirably to the challenge of such an approach. It is to their credit that they have been able to isolate, identify, and synthesize these natural chemicals. The progress that has been made would not have been possible without rapid advances in chemical technology. However, we must now match the chemical expertise with greater incorporation of biological input.

The recognition that the mating process involves a sequence of behavioral acts brought about by a variety of chemicals acting at separate levels should emphasize the necessity of deciphering the context of various interactions of chemical stimuli and the resulting behavioral activities prior to choosing a particular pheromone and the desired response for attempted field manipulations.

The importance of these points applies in the same manner to approaches for the employment of behavioral chemicals to manipulate the host- or prey-searching behavior of entomophages, the host plant finding or feeding behavior of phytophages, as well as to the many other behavioral processes involving chemical stimuli.

Lack of Appreciation of Interactions Among Ecosystem Components

The chemically mediated conversation existing among and within the various species and trophic levels of crop ecosystems has not been fully appreciated. This is largely because most plant protection research is conducted in highly regimented programs within defined specialties, giving very limited consideration of the interactions with other components of the cropping system. For example, host plant resistance programs are generally designed for extensive breeding and screening, with evaluations made only on the basis of the interactions between the plant and phytophage. Little, if any, consideration is given to the potential effects on the third trophic level. However, we know there are numerous ways in which changes in plant varieties could affect the performance of entomophages. Various chemical and physical cues emitted by the plants are essential to effective location of the host habitat by entomophages (Monteith 1958a, 1958b).

Also, the host diet dramatically affects the body odor and consequently the host-finding behavior of entomophages (Sauls et al. 1979).

Another example can be seen from our sex pheromone studies. Despite the knowledge that pheromones in some instances serve as kairomones, little consideration has been given to the potential impact of the employment of pheromones in pest control programs on beneficial organisms.

Proper incorporation of the knowledge of these interactions among various components of the ecosystem can open many new avenues for our pest management strategies, whereas failure to adequately consider them can result in negative net effects.

Lack of Adequate Appreciation for Individual Variability

We have well recognized the capability of insects to detect and respond to minute amounts of sex pheromones and other behavioral chemicals. However, we have too often visualized individual insects as constants, consequently failing to appreciate the extent to which various factors can greatly influence their behavior. Many conditions, some of them very subtle, can dramatically affect an individual insect's response to chemical cues. These include such factors as previous experience with the chemical cue, hunger, food availability, environmental regimes under which the insect is reared, and proper balance of other interacting cues. An adequate knowledge and incorporation of these influences is essential for effective development of programs employing behavioral chemicals.

Registration Regulations

The complexity of the requirements that must be fulfilled to register (with the U.S. Environmental Protection Agency) naturally occurring behavioral chemicals that present no known threat to our environment is well known. Pest management researchers are also aware that safeguards designed to protect our environment against conventional pesticides are now hampering the development of alternative safer solutions to our pest problems; and the topic receives much discussion. It is apparent that there is no easy solution to this dilemma. However, it is urgent that pest management scientists and environmentalists strive together zealously for an equitable balance between the concern for our environment and the need for more effective methods of controlling pests that compete for our food and fiber and threaten our health.

Role of Semiochemicals and Other Forms of Biological Technology in Integrated Pest Management

As a result of the recognition of our need to protect the integrity of our environment and reduce the indiscriminate use of hazardous chemicals, the concept of integrated pest management (IPM) has emerged and gained strong government and industrial support. Extensive, supervised scouting programs have been implemented, and action thresholds have been established so that application of insecticides will be properly timed and will occur only when necessary.

The IPM concept is based on the recognition that no single approach to pest control offers a universal cure-all, and that our best defense can be provided by a fusion of various tactics into practices based on sound ecological principles. The objective of the concept is to integrate all suitable techniques and information into the most economical and ecologically acceptable system, either to reduce and maintain pest populations below damaging levels or to manipulate them in such a manner as to prevent their causing damage. The system includes non-chemical measures such as cultural practices, crop rotation, use of biological agents, use of resistant varieties, exploitation of weak links in the biology of the pests, and interference with their reproduction and/or communication.

Ultimate development of this system will be a multidisciplinary enterprise involving not only the traditional plant protection fields of entomology, pathology, and weed sciences, but also agronomy, agricultural economics, plant physiology, sociology, and other disciplines (Smith and Adkisson 1979). This concept has served and surely will continue to serve effectively and valuably in optimizing the strategy of pesticide use. Yet, to date, the extent to which alternative biological technology has been implemented as a component of the IPM programs is extremely limited, and the programs continue to be centered more on pesticide management than on pest management. A fuller incorporation of these biologically based ingredients into the IPM programs is essential to development of this concept to its full potential as a long-range acceptable and effective approach to pest control.

POTENTIAL ROLE OF SEMIOCHEMICALS

Effective control programs of the future must reflect an appreciation of ecological principles. In designing such programs, however, we must realize and to some extent accept that in many situations man's activities create an environment so unnatural that the regulating forces of nature cannot operate normally. Because of the increasing demand to produce greater amounts of food and fiber on areas of ever-decreasing land and given the high degree of mechanization that prevails in developed countries, ecosystems with low diversity will continue to dominate our agricultural practices. Consequently, effective pest management will require much more than conservation-oriented programs designed for maximum utilization of natural controls by establishment of economic thresholds and "treating" only when necessary.

The real challenge of the future, therefore, will be the development of techniques for augmenting and manipulating the performance of natural enemies, in other ways enhancing the benefit of natural weapons, or somehow artificially intervening with selective, nondisruptive measures. The development of such technology will come only with extensive and intensive investigations employing the talents of visionary scientists who have a broad appreciation of the complex and subtle interactions among the various components of our agricultural ecosystems, and who are willing to freely cross disciplinary lines for the sake of exploring the effectiveness and soundness of imaginative pest management tactics.

Information presented in this book demonstrates the strategic importance and potential value of semiochemicals. These chemical mediators are vitally involved at key points throughout the intertwining relations amog the various components of the ecosystem. The challenge of deciphering their role and devising effective approaches for their employment will require zeal, innovativeness, persistence, and long-term monetary investments. The result will be precise, specific, ecologically sound, and effective techniques essential for dealing with the pest control problems of the future.

REFERENCES

Klassen, W. 1976. A look forward in pest management. Proc. Tall Timbers Conf. Ecological Animal Control by Habitat Management. 6:173–193.

Knipling, E. F. 1979. The Basic Principles of Insect Population Suppression and Management. U.S. Department of Agriculture, Agricultural Handbook No. 512.

Lodeman, E. G. 1903. The Spraying of Plants. Macmillan, New York.

Metcalf, C. L. 1930. Obituary, Steven Alfred Forbes: May 29, 1844–March 13, 1930. Entomol. News 41:175–178.

Metcalf, R. L. 1968. Methods of estimating effects. P. 17–29. *In* C. O. Chichester (ed.), Researching Pesticides. Academic Press, New York.

Monteith, L. G. 1958a. Influence of host and its food plant on host-finding by *Drino bohemica* Mesn. (Diptera: Tachinidae) and interactions of other factors. Proc. Tenth Int. Congr. Entomol. 2:603–606.

Monteith, L. G. 1958b. Influences of food plant of host on attractiveness of the host to tachinid parasites with notes on preimaginal conditioning. Can. Entomol. 90:478–482.

Muir, W. R. 1978. Pest control—A perspective. P. 3–7. *In* E. H. Smith and D. Pimentel (eds.), Pest Control Strategies. Academic Press, New York.

Newsom, L. D. 1971. The end of an era and future prospects for insect control. Proc. Tall Timbers Conf. Ecological Animal Control by Habitat Management. 2:117–136.

Price, P. W. (ed.) 1975. Evolutionary Strategies of Parasitic Insects and Mites. Plenum Press, New York.

Sauls, C. E., D. A. Nordlund, and W. J. Lewis. 1979. Kairomones and their use for management of entomophagous insects. VIII. Effect of diet on the kairomonal activity of frass from *Heliothis zea* (Boddie) larvae for *Microplitis croceipes* (Cresson). J. Chem. Ecol. 5:363–369.

Smith, R. F. 1975. The origin of integrated control in California—An account of the contributions of C. W. Woodworth. Pan-Pac. Entomol. 50:426–429.

Smith, R. F., and P. L. Adkisson. 1979. Expanding horizons of integrated pest control in crop production. Proc. Opening Session and Plenary Session Symp. IX Int. Congr. Plant Protection. Washington, D.C. 29–30.

Smith, R. F., J. L. Apple, and D. G. Bottrell. 1976. The origins of integrated pest management concepts for agricultural crops. P. 1–16. *In* J. L. Apple and R. F. Smith (eds.), Integrated Pest Management. Plenum Press, New York.

Smith, R. F., and R. van den Bosch. 1967. Integrated control. P. 295–340. *In* W. W. Kilgore and R. L. Doutt (eds.), Pest Control: Biological, Physical, and Selected Chemical Methods. Academic Press, New York.

Stevens, R. B. 1960. Cultural practices in disease control. P. 357–429. *In* J. G. Horsfall and A. E. Dimond (eds.), Plant Pathology: An Advanced Treatise, Vol. 3. Academic Press, New York.

van den Bosch, R., and P. S. Messenger. 1973. Biological Control. Intext Press, New York.

van den Bosch, R. 1979. The Pesticide Conspiracy. Doubleday, Garden City, N.Y.

Weiss, H. B. 1936. The Pioneer Century of American Entomology. New Brunswick, NJ.

Wright, J. S., R. F. Fritz, and J. Haworth. 1972. Changing concepts of vector control in malaria eradication. Annu. Rev. Entomol. 17:75–102.

CHAPTER TWO

SEMIOCHEMICALS: A REVIEW OF THE TERMINOLOGY

DONALD A. NORDLUND

AR–SEA–USDA
Southern Grain Insects Research Laboratory
Tifton, Georgia

With the improvements in chemical technology during the past 20–30 years, chemical ecology has developed very rapidly. As with any rapidly growing field of scientific endeavor, that is, "the attempt to make the chaotic diversity of our sense-experience correspond to a logically uniform system of thought" (Einstein 1940), there have been some growing pains. The development of a suitable system of terminology has been, and will probably continue to be, a problem in chemical ecology.

With this in mind, let us review the classification system that is used in this book and seems to be in general use today. Since we deal here with semiochemicals, the discussion in this chapter is restricted to this group of chemical-releasing stimuli. The system was developed over the years, and many synonymous and overlapping terms have been proposed during its development (Table 2.1). One look at Figure 13.1 will go a long way toward explaining the need for a system of this nature and why its development was so difficult and confusing.

Several points need to be remembered. One is that we are dealing with chemical ecology. Chemical communication is generally accepted as a more restricted area of chemical ecology, unless you use a broad definition of communication, such as that of Shorey (1977a), a definition that according to Burghardt (1970) is so broad that it renders the concept of communication meaningless. There has been some concern expressed because information transferred in some of the interactions that are discussed here does not constitute true communication (Blum 1974, Burghardt 1970). Otte (1974) found that a common and critical difficulty in treatments of communication is the failure to distinguish between

TABLE 2.1. Terms Used to Designate Various Semiochemicals

- Semiochemical (Law and Regnier 1971)
  - Ektohormone (Beth 1932)
  - Ecomone (Florkin 1965)
  - Telergone (Kirschenblatt 1962)
- Pheromone (Karlson and Butenandt 1959, Karlson and Luscher 1959)
  - Homoiohormone (Beth 1932)
  - Homotelergone (Kirschenblatt 1962)
    - Gonophyone (Kirschenblatt 1962)
    - Gamophone (Kirschenblatt 1962)
    - Epagone (Kirschenblatt 1962)
    - Odmicone (Kirschenblatt 1962)
    - Thorybone (Kirschenblatt 1962)
  - Allelochemic (Whittaker 1970a, 1970b)
    - Alloiohormone (Beth 1932)
    - Heterotelergone (Kirschenblatt 1962)
      - Amynone (Kirschenblatt 1962)
      - Prohaptone (Kirschenblatt 1962)
      - Lichneomone (Kirschenblatt 1962)
      - Xenoblaptone (Kirschenblatt 1962)
    - Xeonmone (Chernin 1970)
    - Coactone (Florkin 1965)
      - Exocoactone (Florkin 1965)
      - Endocoactone (Florkin 1965)
- Allomone (Brown 1968)
- Kairomone (Brown et al. 1970)
  - Blaptone (Beth 1932)
- Synomone (Nordlund and Lewis 1976)
  - Allomone-kairomone (Borden 1977)
- Apneumone (Nordlund and Lewis 1976)

evolved functions and incidental effects. This is not the place to discuss what constitutes communication, however; thus the first point that needs to be made is that the interactions described do exist and that the chemicals involved in the interactions may be classified by this system regardless of whether the interactions involve communication. The second point is that the classification system is based on the function or effect of the chemical in each specific interaction (Whittaker and Feeny 1971). The third point is that these functions are not mutually exclusive (Brown et al. 1970), and thus it is possible for a single chemical to function in several types of interaction, as any of a number of types of semiochemicals, or even as a hormone.

SEMIOCHEMICALS

Organisms interact with others of the same and different species. The interactions may be of many different kinds, ranging from predator-prey interactions to mating interactions. Law and Regnier (1971) proposed the term "semiochemicals" (Gk. *semeon,* a mark or signal) for chemicals that mediate interactions between organisms. The semiochemicals are subdivided into two major groups, pheromones and allelochemics, depending on whether the interactions are intraspecific or interspecific, respectively. See Sondheimer and Simeone (1970), Beroza (1970, 1976), van Emden (1973), Shorey and McKelvey (1977), Ritter (1979), for more information on semiochemicals.

Pheromones

The term "pheromone" (Gk. *phereum,* to carry; *horman,* to excite or to stimulate) was originally proposed by Karlson and Butenandt (1959) and Karlson and Luscher (1959). Their definition restricted the use of the term to chemicals from animals. Nordlund and Lewis (1976) broadened the definition to include chemicals from plants and gave it as follows: a substance secreted by an organism to the outside that causes a specific reaction in a receiving organism of the same species. Pheromones may be further classified on the basis of the type of interaction mediated, such as sex pheromone (discussed in Chapter 11), alarm pheromone, and epideictic pheromone (discussed in Chapters 9 and 10).

The existence of pheromones has been known for many years. Charles Butler, for example, in 1609, described how lone bees are attracted and provoked to mass sting by a substance released by a single sting. The first isolation and identification of a pheromone was not reported until 1959 (Butenandt et al. 1959). Since that time, hundreds of pheromones have been identified for organisms ranging from algae to some primates (Starr 1968, Siegel and Cohen 1962, Mueller et al. 1971, Michael et al. 1971, Curtis et al. 1971), and have even been postulated for humans (Michael et al. 1974).

The identification of the first pheromone sparked a burst of activity in chemical ecology, as is evidenced by the voluminous literature. Concern about the use of conventional insecticides was mounting, so entomologists, in particular, began optimistically searching for pheromones and hormones that could be used to control pests. As scientists learned more about the systems, they began to become more aware of just how complex the systems are, and their optimism, in many cases, began to wane.

This complexity in pheromones is due to a number of factors, which are also applicable to other semiochemicals. Perception of a pheromone by an organism, for example, may result in the releaser effect, an immediate behavioral response, such as is common with sex attractant pheromones, or in the primer effect, a complex physiological response that is simply set in motion by the initial perception (Wilson and Bossert 1963). An example of a pheromone with a primer effect would be the "queen substance," which is produced by honeybee (*Apis mellifera*) queens and inhibits ovarian development in workers (C. G. Butler 1954). Another cause of complexity is the fact that a single chemical may have more than one pheromonal function (Morse 1972, Rudinsky, 1973a, 1973b, Renwick and Vité 1969, 1970, and others), depending on the context in which it is perceived, or on the dosage. Also, a pheromone may act as another type of semiochemical in interspecific interactions, or as a hormone. The receiving organism must be physiologically ready to respond and in many cases environmental conditions must be appropriate, also adding to the complexity of the various systems. Then, there is the almost unbelievable number of chemical messages in the environment, as discussed in Chapter 13.

A decade or so ago, most workers investigating pheromones felt that single chemicals were responsible for the complex intraspecific interactions that were being demonstrated. Thus much of the work on pheromones was directed toward identifying these chemicals; behavioral and physiological studies beyond those necessary to develop a bioassay generally took a back seat. In tests on sex pheromones, for example, often a glass rod was smeared with a particular chemical, and if a male attempted to copulate with the rod, we generally assumed that this chemical was the pheromone. As progress in the research effort was made, trapping experiments were initiated, but the results were usually disheartening because the chemical did not elicit a response comparable to the response elicited by live females or even crude extracts. Eventually it was demonstrated that the male responded to several chemicals released by the female. Thus the view that pheromones were single chemicals gave way to the view that pheromones consisted of several chemicals, and workers began searching for the multiple components of pheromones. However, work was still focused on the end result of behavior.

Although the existence of a hierarchy of behavioral or physiological responses to various pheromones has been demonstrated (Shorey 1977b, Shorey et al. 1968, and references therein), most workers have thought in terms of an interaction that was mediated by a multicomponent sex pheromone. This is at least partly because Kalmus (1965) suggested that after the word "substance" in the definition of pheromone, one should add "or mixture of substances," and instead of "a specific action" (reaction), one should say "one or several specific actions" (reactions). This approach, which has carried over to other semiochemicals, has unfortunately tended to concentrate efforts at the end result of interactions, such as arrival of males at a pheromone source, rather than on each of the behavioral or physiological responses that make up the interaction. It might be better to think in terms of interactions mediated by several pheromones, each of which elicits a specific behavioral or physiological reaction, each a part of the total interaction.

If behavior or physiology is to be manipulated, it must be understood. To effectively use pheromones or other semiochemicals, one should, ideally, be aware of each behavioral or physiological response in the sequence of responses so that the weakest link in the sequence could be selected for manipulation. It would appear that considering a pheromone as a substance that releases one or several specific responses in any specific interaction inhibits the study of the specific responses and the individual chemicals that elicit those responses.

Allelochemics

The allelochemics make up the other major group of semiochemicals. The term "allelochemic," proposed by Whittaker (1970a, 1970b) and used to describe chemicals that mediate interspecific interactions, is defined as a chemical that is significant to organisms of a species different from its source for reasons other than food as such. At the present time, four types of allelochemics are recognized: allomones, kairomones, synomones, and apneumones.

Allomones

An allomone is a substance produced or acquired by an organism that, when it contacts an individual or another species in the natural context, evokes in the receiver a behavioral or physiological response that is adaptively favorable to the emitter but not to the receiver (Brown 1968, Nordlund and Lewis 1976). The term was derived from "alloiohormone," which has essentially the same meaning and was proposed by Beth (1932). Numerous types of interactions are mediated by allomones. Venoms, for example, are used in prey capture and defense (Habermann 1972). The neotropical social wasp *Mischocyttarus drewseni* applies a secretion to the stem of its nest that repels foraging ants (Jeanne 1970). Larvae of the family Papilionidae typically possess an eversible cervical gland or osmeterium that produces a defensive secretion. When the caterpillar is disturbed, it protrudes the gland and attempts to apply some of the secretion to the source of the disturbance (Burger et al. 1978 and references therein). Adult *Pycnopsyche scabripennis* produce a defensive secretion that contains *p*-cresol, indole, and skatole in a pair of exocrin glands located on the fifth abdominal sternite (Duffield et al. 1977). Allomones also serve plants as defense mechanisms against herbivores and reduce competition from other plants (Rhoades and Cates 1976, Rice 1974, and references therein).

Defense is not the only role allomones play. Certain predators are able to use allomones to attract or otherwise manipulate prey (Hölldobler 1971, Weaver et al. 1975, also see Chapter 7). Leaving insects for a moment, we find that allomones can be important in some very complex interactions. For example, the infective stage of the fish tapeworm, *Ligula intestinalis,* releases an allomone that causes its intermediate host, the rudd, to leave its school and swim in the surface water, making it easy prey for fish-eating birds and mammals in which the tapeworm's cycle can be completed (Croll 1966).

From the few examples listed above we can see that allomones can be involved in many different types of interactions, some of them extremely complex. The following are a few of the available reviews on allomones: Roth and Eisner, 1962, Weatherston 1967, Eisner, 1970, Weatherston and Percy, 1970, Whittaker 1970a, Went 1970, Dethier 1970, Williams 1970, Whittaker and Feeny 1971, Pavan and Dazzini 1971, Bucherl et al. 1968–1971; Habermann 1972, Happ 1973, and Duffey 1976.

Kairomones

In contrast to the allomones, some chemicals benefit the receiver rather than the emitter in interspecific interactions. The term "kairomone" (Gk. *kairos,* opportunistic) was proposed by Brown et al. (1970) to cover the chemicals that mediate these interactions. A kairomone is a substance produced or acquired by an organism that, when it contacts an individual of another species in the natural context, evokes in the receiver a behavioral or physiological response that is adaptively favorable to the receiver but not to the emitter (Brown et al. 1970, Nordlund and Lewis 1976).

The term "kairomone" has received some criticism but generally appears to have been adopted. Blum (1974) argued that the "so-called" kairomones appear to be pheromones and allomones that have "evolutionarily backfired" and as such "do not represent a class of chemical signals distinct from allomones and pheromones." He used the term "kairomonal effect" (Blum 1977). Wilson (1975) is of the option that the allomone-kairomone dichotomy should be dissolved by employing "allomone" in a broader sense and dropping "kairomone." He reasons that distinguishing what is advantageous to the receiver from what benefits the emitter is "a difficult and occasionally impossible choice to make in practice."

Though in some cases it is difficult to determine advantage, this is generally not the case. Host-parasitoid or predator-prey interactions, for example, are very clear-cut, so in these situations dropping the dichotomy would serve no useful purpose. When a determination of benefit has not been made, the term "allelochemic" will serve nicely, while indicating that we need to learn more. Blum's concern that kairomones are maladaptive to the releaser is justified but seems to miss the other half of the interaction. The ability to respond to a particular chemical may be highly adaptive to the receiver. Thus, although the chemical may originally have evolved as an allomone or pheromone, hence is beneficial to the emitting organism in some situations, it has, in evolutionary time, become adaptively beneficial to the receiver and detrimental to the emitter in other situations. The definitions of these terms have always been couched in relation to the specific interaction that involved them, reflecting the realization that the involvement of a chemical need not be limited to any one interaction. Also, it is important to remember that we are witnesses to an infinitesimally small portion of the cat-and-mouse game that goes on in evolutionary time (see Chapter 13).

Although some chemicals stimulated a limited number of herbivores to feed on a plant, they also function and probably originally evolved to deter a vast multitude of herbivorous species from feeding on that plant (Fraenkel 1959); however, this turn of events does not prevent "kairomone" from serving as meaningful terminology. In interactions in which such a substance serves as a feeding stimulant it is a kairomone, and when it is a feeding deterrent it is an allomone. Even a hormone can be a kairomone, in a different interaction, as in the case of corticosteroids of the rabbit and the rabbit flea, *Spilopsyllus cuniculi* (Rothschild 1965). Kairomones may be chemical cues that are hormones, pheromones, allomones, and so on, to the legitimate receiver that are used by an illegitimate receiver (Otte 1974), or they may be incidental cues such as waste products.

Many beneficial insects use kairomones in their host or prey selection behavior. The larval parasitoid *Microplitis croceipes* is stimulated into intensive searching behavior by kairomones found in the frass of *Heliothis zea, Heliothis virescens,* and *Heliothis subflexa* (Lewis and Jones 1971). Vinson (1968) found that a substance in the mandibular gland secretion of *H. virescens* is used by *Cardiochiles nigriceps* in its host location behavior. *Trichogramma pretiosum* females are stimulated into an intensive search behavior by chemicals found in the scales of *H. zea* moths (Lewis et al. 1975). Vité and Williamson (1970) found that brevicomin, the bark beetle sex pheromone, was used by the predator *Thanasimus dubius* to locate its prey.

In a few cases, it has been shown that prey use kairomones to detect the presence of a predator. For example, in a marine environment, *Dendrastes excentricus, Pecten,* and *Nassa* all exhibit an escape response when exposed to chemicals from a star fish (von Uexküll 1921, Hoffman 1930, and Tinbergen 1951). L-Serine from several mammalian predators, including man, deters the upstream movement of migrating salmon, *Oncorhynchus* (Idler et al. 1961). I could continue with numerous examples, but this subject is covered extensively in Chapters 5–8.

Synomones

The term "synomone" (Gk. *syn,* with or jointly) was proposed by Nordlund and Lewis (1976) for chemicals that mediate mutualistic interactions. The term was defined as a chemical substance produced or acquired by an organism that, when it contacts an individual of another species in the natural context, evokes in the receiver a behavioral or physiological response that is adaptively favorable to both the emitter and the receiver. These chemicals have been regarded as allomones (Brown et al. 1970), as either allomones or kairomones (Whittaker and Feeny 1971), or as allomone-kairomones (Borden 1977).

Synomones often play subtle roles in symbiotic relationships (Henry 1966, Nutmann and Mosse 1963, and others). For example, to survive on a diet of wood, the wood-eating cockroach, *Cryptocercus punctulatus*, requires wood-digesting protozoa in its gut. The hormone ecdysone, which regulates molting in the cockroach, also acts as a synomone that induces the sexual cycle of some of these protozoa, allowing them to reproduce (Cleveland 1959). Synomones also may play important roles in maintaining the species specificity of response to certain pheromones. For example, *Ips paraconfusus* males produce *cis*-verbenol, ipsenol, and ipsdienol as pheromones to attract other individuals to trees (Silverstein et al. 1966), while *Ips pini* males produce *cis*-verbenol and ipsdienol (Vité et al. 1972). *I. pini* males also produce linalool (Young et al. 1973). In laboratory and field tests, ipsenol inhibited the response of *I. pini* to *I. pini* pheromone extracts or male beetles, and linalool at least partially inhibited the response of *I. paraconfusus* to *I. paraconfusus* (Birch and Wood 1975). Thus ipsenol and linalool each act as synomones by allowing these species to avoid competitive situations. I also consider such substances as floral scents that attract pollinators of various types to be synomones.

Apneumones

There are some interesting chemical-mediated interactions in which the chemical or chemicals mediating the interactions between individuals of different species originate from a nonliving source. For example, Thorpe and Jones (1937) reported that the ichneumonid parasitoid *Venturia canescens* is attracted to the odor of oatmeal, its host's food. Also, Laing (1937) demonstrated that the braconid *Alysia manducator* and the chalcid *Mormoniella (Nasonia) vitripennis* are attracted to meat, even though it never contained their dipteran hosts. Laing's results with regard to *M. vitripennis* have been questioned by several authors (Jacobi 1939, Wylie 1958, Edwards 1954), but the concept is still valid. These chemicals cannot be classified as allomones, kairomones, or synomones because a nonliving material cannot derive any biological benefit from them. Nordlund and Lewis (1976) proposed the term "apneumone" (Gk. *a-pneum*, breathless or lifeless) for chemicals that are emitted by a nonliving material and evoke a behavioral or physiological reaction that is adaptively favorable to a receiving organism but detrimental to an organism of another species that may be found in or on the nonliving material. Dealing with chemicals from nonliving materials, in terms of terminology at least, is rather new. There may well be other types of chemically mediated interactions involving nonliving materials and various other organisms, but we have really just begun to consider such phenomena.

At this time, for the sake of simplicity, apneumones are considered a subgroup of allelochemics. If more interactions involving chemicals from nonliving materials are found, it may be convenient to reclassify them in another major group.

Allelochemics in Multiple Species Interactions

We often find interactions involving organisms of more than two species. The terminology for describing the chemical-releasing stimuli involved in these interactions is the same as that for two species' interactions. A good example of an interaction involving more than two species is one involving a plant, a phytophagous insect feeding on the plant, and a parasitoid attacking that insect. Chemicals from the plant that attract the phytophagous species and stimulate it to feed are, as pointed out above, kairomones. Those chemicals released by the phytophagous species that are used by the parasitoid in its host selection behavior are also kairomones. It is also known that many parasitoids utilize chemical cues emitted by the host's food plant to locate the host's habitat (Arthur 1962, Monteith 1958, Ullyett 1953, see also Chapter 4). These chemicals are synomones in this interaction, since the receiving organism benefits because it comes into a habitat likely to contain suitable hosts, and the emitter benefits from the reduction in damage from the phytophagous species. The plant-released chemical, serving as a kairomone for the phytophagous species, may also serve as the synomone utilized by the parasitoid.

Some of these interactions are fairly clear-cut, and the terms to describe the allelochemics involved are easily determined. This is not always the case, however. As mentioned by Wilson (1975), there are some interactions in which the determination of benefit can be particularly difficult. Greany et al. (1977), for example, reported that females of *Biosteres (Opius) longicaudatus,* a parasitoid of tephritid fruit fly larvae, are attracted to various fermentation products of the fungus *Monilinia fructicola,* including acetaldehyde, ethanol, and acetic acid. This fungus inhabits rotting fruit, a likely habitat for the host of the parasitoid. The problem of classifying these chemicals as kairomones or synomones is in determining whether the fungus benefits from the interaction. Since this is not yet known, we can only say that these fermentation products are allelochemics in this interaction. In other cases it is difficult to satisfactorily determine the releaser. For example, Vinson (1975) showed that *Cardiochiles nigriceps,* a larval parasitoid of *H. virescens,* is attracted to damaged tobacco plants and that it is stimulated into an intensive search of the surrounding area if the damage was caused by the feeding of *H. virescens*. Nordlund and Lewis (1976) suggested that we consider kairomones to be chemicals that (1) originate from the plant as a result of the activities of, or in combination with substances from, the phytophagous species, and (2) act as cues to the parasitoid. "Release as the result of the activities of" may be considered to be a special case of "acquired by."

COMPLEMENTARY TERMINOLOGY

When dealing with these chemical-releasing stimuli, it is often of considerable value to be able to indicate the response pattern that the receiving organism exhibits. For example, we might wish to indicate that an allomone repels an organism or that a kairomone attracts an organism. The terms "attractant" and "repellent" have a long history of application to chemical-releasing stimuli, but these two terms do not apply to all possible response patterns.

Dethier et al. (1960) proposed several terms to describe the response pattern of insects to chemical stimuli:

1 *Arrestant* A chemical that causes an organism to aggregate in contact with it, the mechanism of aggregation being kinetic or having a kinetic component. An arrestant may slow the linear progression of the organism by reducing actual speed of locomotion, or by increasing the turning rate.

2 *Attractant* A chemical that causes an organism to make oriented movements towards its source.

3 *Repellent* A chemical that causes an organism to make oriented movements away from its source.

4 *Stimulant* (a) *Locomotor stimulant*—a chemical that causes, by a kinetic mechanism, organisms to disperse from a region more rapidly than if the area did not contain the chemical. (b) *Feeding, mating,* or *ovipositional stimulant*—a chemical that elicits feeding, mating, or oviposition in an organism. Feeding stimulant is synonymous with "phagostimulant," which was coined by Thorsteinson (1953, 1955).

5 *Deterrent* A chemical that inhibits feeding, mating, or oviposition when in a place where an organism would, in its absence, feed, mate, or oviposit.

These workers noted that we can make several generalizations about the use of the foregoing terms: (1) the same compound may have multiple effects on behavior; (2) any given effect may be elicited by chemicals that in other respects act differently; (3) movements involving orientation can be evoked by concentration gradients or by currents carrying a chemical that may not necessarily have to be present as a gradient.

These terms can be used in conjunction with the preceding terms to indicate exactly what behavior pattern is involved in the response. Thus we may have a sex attractant pheromone that attracts male moths from a distance or a mating stimulant pheromone that stimulates copulation; a kairomone may be an attractant, an arrestant, or a stimulant; an epideictic pheromone may be a repellent or a locomotor stimulant.

CONCLUSION

Terminology is a tool that assists in the development of a "logically uniform system of thought." An appropriate system of terminology is particularly helpful as we attempt to communicate the results of our work. It is also true that ". . . our interpretive abilities can be clouded by the nature of our languages" (Duffey 1976). This is evident from the research on pheromones.

Terminology, thus, must be subject to improvement as our knowledge increases, and it must be discarded if we find that it no longer serves a useful purpose or if a better system can be developed. The free market of ideas will ensure that terms will be discarded when necessary.

The terminology presented in this chapter appears to be serving us well in that it allows us now to communicate our ideas rather precisely; it stimulates further study because it forces us to ask more questions about the interaction and about the possibility that other interactions are mediated by the same chemical. The definitions are all listed in Table 2.2. Without some system of terminology, it would be impossible to unify this volume.

TABLE 2.2. Chemical-Releasing Stimuli

Hormone. A chemical agent, produced by tissue or endocrine gland, that controls various physiological processes within an organism.

Semiochemical. A chemical involved in the interaction between organisms.

- *Pheromone.* A substance that is secreted by an organism to the outside and causes a specific reaction in a receiving organism of the same species.
- *Allelochemic.* A substance that is significant to organisms of a species different from its source, for reasons other than food as such.
  - *Allomone.* A substance produced or acquired by an organism that, when it contracts an individual of another species in the natural context, evokes in the receiver a behavioral or physiological reaction that is adaptively favorable to the emitter but not to the receiver.
  - *Kairomone.* A substance produced or acquired by an organism that, when it contacts an individual of another species in the natural context, evokes in the receiver a behavioral or physiological reaction that is adaptively favorable to the receiver but not to the emitter.

TABLE 2.2 *(Continued)*

Synomone. A substance produced or acquired by an organism that, when it contacts an individual of another species in the natural context, evokes in the receiver a behavioral or physiological reaction that is adaptively favorable to both emitter and receiver.

Apneumone. A substance emitted by a nonliving material that evokes a behavioral or physiological reaction that is adaptively favorable to a receiving organism but detrimental to an organism of another species that may be found in or on the nonliving material.

REFERENCES

Arthur, A. P. 1962. Influence of host tree on abundance of *Itoplectis conquisitor* (Say) (Hymenoptera: Ichneumonidae), a polyphagous parasite of the European pine shoot moth, *Rhyacionia buoliana* (Schiff.) (Lepidoptera: Olethreutidae). Can. Entomol. 94:337–347.

Beroza, M. (ed.) 1970. Chemicals Controlling Insect Behavior. Academic Press, New York.

Beroza, M. 1976. Pest Management with Insect Sex Attractants and Other Behavior-Controlling Chemicals. American Chemical Society, Washington, D.C.

Beth, A. 1932. Vernachlassigte Hormone. Naturwissenschaften 20: 177–181.

Birch, M. C., and D. L. Wood. 1975. Mutual inhibition of the attractant pheromone response by two species of *Ips* (Coleoptera: Scolytidae). J. Chem. Ecol. 1:101–113.

Blum, M. S. 1974. Deciphering the communicative rosetta stone. Bull. Entomol. Soc. Am. 20:30–35.

Blum, M. S. 1977. Behavioral responses of Hymenoptera to pheromones and allomones. P. 149–167. *In* H. H. Shorey and J. J. McKelvey, Jr. (eds.), Chemical Control of Insect Behavior. Wiley, New York.

Borden, J. H. 1977. Behavioral responses of Coleoptera to pheromones, allomones, and kairomones. P. 169–198. *In* H. H. Shorey and J. J. McKelvey, Jr. (eds.), Chemical Control of Insect Behavior. Wiley, New York.

Brown, W. L., Jr. 1968. An hypothesis concerning the function of the metapleural glands in ants. Am. Nat. 102:188–191.

Brown, W. L., Jr., T. Eisner, and R. H. Whittaker. 1970. Allomones and kairomones: Transspecific chemical messengers. BioScience 20:21–22.

Bucherl, W., E. Buckley, and V. Deulofeu (eds.) 1968–1971. Venomous Animals and Their Venoms, 3 vols. Academic Press, New York.

Burger, B. V., M. Röth, M. le Roux, H. S. C. Spies. V, Truter, and H. Geertsma. 1978. The chemical nature of the defensive larval secretion of the citrus swallowtail *Papilio demodocus*. J. Insect Physiol. 24:803–805.

Burghardt, G. M. 1970. Defining "communication." P. 5–18. *In* J. W. Johnston, Jr., D. G. Moulton, and A. Turk (eds.), Advances in Chemoreception, Vol. 1, Communication by Chemical Signal. Appleton-Century-Crofts, New York.

Butenandt, A., R. Beckmann, D. Stamm, and E. Hecker, 1959. Über den Sexuallockstoff des Seidenspinnter, *Bombyx mori*. Reindarstellung und Konstitutution. 2. Z. Naturforsch. 146:283–284.

Butler, C. 1609. The Feminine Monarchie. On a Treatis Concerning Bees, and the Due Ordering of Them. Joseph Barnes, Oxford. Cited in E. O. Wilson, 1971. The Insect Societies. P. 235–236. Belknap (Harvard University) Press, Cambridge, MA.

Butler, C.G. 1954. The method and importance of the recognition by a colony of honeybees (*A. mellifera)* of the presence of its queen. Trans. R. Entomol. Soc. London 105:11–29.

Chernin, E. 1970. Interspecific chemical signals. BioScience 20:845.

Cleveland, L. R. 1959. Sex induced with ecdysone. Proc. Natl. Acad. Sci. U.S.A. 45:747–753.

Croll, N. A. 1966. Ecology of Parasites. Harvard University Press, Cambridge, MA.

Curtis, R. F., J. A. Ballantine, E. B. Keverne, R. W. Bonsall, and R. P. Michael. 1971. Identification of primate sexual pheromones and the properties of synthetic attractants. Nature (London) 233:396–398.

Dethier, V. G. 1970. Chemical interactions between plants and insects. P. 83–102. *In* E. Sondheimer and J. B. Simeone (eds.), Chemical Ecology. Academic Press, New York.

Dethier, V. G., L. Barton Browne, and C. N. Smith. 1960. The designation of chemicals in terms of the responses they elicit from insects. J. Econ. Entomol. 53:134–136.

Duffey, S. S. 1976. Arthropod allomones: Chemical effronteries and antagonists. Proc. XV Int. Congr. Entomol. 323–394.

Duffield, R. M., M. S. Blum, J. B. Wallace, H. A. Lloyd, and F. E. Regnier, 1977. Chemistry of the defensive secretion of the caddisfly *Pycnopsyche scabripennis* (Trichoptera: Limnephilidae). J. Chem. Ecol. 3:649–656.

Edwards, R. L. 1954. The host-finding and oviposition behavior of *Mormoniella vitripennis* (Walker), a parasite of muscoid flies. Behavior 7:88–112.

Einstein, A. 1940. Considerations concerning the fundamentals of theoretical physics. Science 91:487–492.

Eisner, T. 1970. Chemical defenses against predation in arthropods. P. 157–218. *In* E. Sondheimer and J. B. Simeone (eds.), Chemical Ecology. Academic Press, New York.

Florkin, M. 1965. Approches moléculaires de l'intégration écologique. Problèmes de terminologie. Bull. Acad. R. Belg. (5th Ser.) 51:239–256.

Fraenkel, G. S. 1959. The raison d'être of secondary plant substances. Science 129:1234–1237.

Greany, P. D., J. H. Tumlinson, D. L. Chambers, and G. M. Boush. 1977. Chemically mediated host finding by *Biosteres (Opius) longicaudatus,* a parasitoid of tephritid fruit fly larvae. J. Chem. Ecol. 3:189–195.

Habermann, E. 1972. Bee and wasp venoms. Science 177:314–322.

Happ, G. M. 1973. Chemical signals between animals. Allomones and pheromones. P. 149–190. *In* G. M. Happ, J. Lobue, and A. S. Gordon (eds.), Humoral Control of Growth and Differentiation. Vol 2, Nonvertebrate Neuroendocrinology and Aging. Academic Press, New York.

Henry, S. M. (ed.) 1966. Symbiosis, Vol. 1. Academic Press, New York.

Hoffmann, H. 1930. Über den Fluchtreflex bie *Nassa*. Z. Vgl. Physiol. 2:662–688.

Hölldobler, B. 1971. Communication between ants and their guests. Sci. Am. 224:86–95.

Idler, D. R., J. R. McBride, R. E. E. Jonas, and N. Tomlinson. 1961. Olfactory perception in migrating salmon. II. Studies on a laboratory bioassay for homestream water and a mammalian repellant. Can. J. Biochem. Physiol. 39:1575–1584.

Jacobi, E. F. 1939. Über Lebensweise, auffinden des Wirtes und Regulierung des individuenzaul von *Mormoniella vitripennis* Walker. Arch. Neerl. Zool. leiden 3:197–282.

Jeanne, R. L. 1970. Chemical defense of brood by a social wasp. Science 168:1465–1466.

Kalmus, H. 1965. Possibilities and constraints of chemical telecommunication. Proc. 2nd Int. Cong. Endocrinol. London 181–192.

Karlson, P., and A. Butenandt. 1959. Pheromones (Ectohormones) in insects. Annu. Rev. Entomol. 4:39–58.

Karlson, P., and M. Luscher. 1959. "Pheromones," a new term for a class of biologically active substances. Nature (London) 183:155–176.

Kirschenblatt, J. 1962. Terminology of some biologically active substances and validity of the term "pheromones." Nature (London) 195:916–917.

Laing, J. 1937. Host-finding by insect parasites. I. Observations on finding of hosts by *Alysia manducator, Mormoniella vitripennis* and *Trichogramma evanescens*. J. Anim. Ecol. 6:298–317.

Law, J. H., and F. E. Regnier. 1971. Pheromones. Annu. Rev. Biochem. 40:533–548.

Lewis, W. J., and R. L. Jones. 1971. Substance that stimulates host-seeking by *Microplitis croceipes* (Hymenoptera: Braconidae), a parasite of *Heliothis* species. Ann. Entomol. Soc. Am. 64:471–473.

Lewis, W. J., R. L. Jones, D. A. Nordlund, and H. R. Gross, Jr. 1975. Kairomones and their use for management of entomophagous insects. II. Mechanisms causing increase in rate of parasitization by *Trichogramma* spp. J. Chem. Ecol. 1:349–360.

Michael, R. P., E. B. Keverne, and R. W. Bonsall. 1971. Pheromones: Isolation of male sex attractants from a female primate. Science 172:964–966.

Michael, R. P., R. W. Bonsall, and P. Warner. 1974. Human vaginal secretions: Volatile fatty acid content. Science 186:1217–1219.

Monteith, L. G. 1958. Influence of host and its food plant on host-finding by *Drino bohemica* Mesn. (Diptera: Tachinidae) and interaction of other factors. Proc. 10th Int. Congr. Entomol. 2:603–606.

Morse, R. A. 1972. Honeybee alarm pheromones: Another function. Ann. Entomol. Soc. Am. 65:1430.

Mueller, D. G., L. Jaenicke, M. Donike, and L. Akintobi. 1971. Sex attractant in brown algae: Chemical structure. Science 171:815–817.

Nordlund, D. A., and W. J. Lewis. 1976. Terminology of chemical releasing stimuli in intraspecific and interspecific interactions. J. Chem. Ecol. 2:211–220.

Nutman, P. G., and B. Mosse (eds.) 1963. Symbiotic Associations. Cambridge University Press, Cambridge.

Otte, D. 1974. Effects and functions in the evolution of signaling systems. Annu. Rev. Ecol. Syst. 5:385–417.

Pavan, M., and M. V. Dazzini. 1971. Toxicological and pharmacology—Arthropods. Chem. Zool. 6:365–409.

Renwick, J. A. A., and J. P. Vité. 1969. Bark beetle attractants: Mechanism of colonization by *Dendroctonus frontalis*. Nature (London) 224:1222–1223.

Renwick, J. A. A., and J. P. Vité. 1970. Systems of chemical communication in *Dendroctonus*. Contrib. Boyce Thompson Inst. 24:283–292.

Rhoades, D. F., and R. G. Cates. 1976. Toward a general theory of plant antiherbivore chemistry. P. 168–213. *In* J. W. Wallace and R. L. Mansell (eds.), Biochemical Interactions Between Plants and Insects. Plenum Press, New York.

Rice, E. L. 1974. Allelopathy. Academic Press, New York.

Ritter, F. J. (ed.) 1979. Chemical Ecology: Odour Communication in Animals. Elsevier/North-Holland Biomedical Press, Amsterdam.

Roth, L. M., and T. Eisner. 1962. Chemical defenses of arthropods. Annu. Rev. Entomol. 7:107–136.

Rothschild, M. 1965. The rabbit flea and hormones. Endeavour 24:162–168.

Rudinsky, J. A. 1973a. Multiple functions of the southern pine beetle pheromone verbenone. Environ. Entomol. 2:511–514.

Rudinsky, J. A. 1973b. Multiple functions of the Douglas fir beetle pheromone 3-methyl-2-cyclohexen-1-one. Environ. Entomol. 2:575–585.

Shorey, H. H. 1977a. Interactions of insects with their chemical environment. P. 1–5. *In* H. H. Shorey and J. J. McKelvey, Jr. (eds.), Chemical Control of Insect Behavior. Wiley, New York.

Shorey, H. H. 1977b. The adaptiveness of pheromone communication. Proc. XV Int. Congr. Entomol. 294–307.

Shorey, H. H., and J. J. McKelvey, Jr. (eds.) 1977. Chemical Control of Insect Behavior. Wiley, New York.

Shorey, H. H., L. K. Gaston, and R. N. Jefferson. 1968. Insect sex pheromones. P. 57–126. *In* R. L. Metcalf (ed.), Advances in Pest Control Research, Vol. 8. Wiley, New York.

Siegel, R. W., and L. W. Cohen. 1962. The intracellular differentiation of cibia. Am. Zool. 2:558.

Silverstein, R. M., J. O. Rodin, and D. L. Wood. 1966. Sex attractants in frass produced by male *Ips confusus* in Ponderosa Pine. Science 154:509–510.

Sondheimer, E., and J. B. Simeone (eds.) 1970. Chemical Ecology. Academic Press, New York.

Starr, R. C. 1968. Cellular differentiation in volvox. Proc. Natl. Acad. Sci. U.S.A. 59:1082–1088.

Thorpe, W. H., and F. G. W. Jones. 1937. Olfactory conditioning in a parasitic insect and its relations to the problems of host selection. Proc. R. Soc. London, Ser. B. 124:56–81.

Thorsteinson, A. J. 1953. The chemotactic responses that determine host specificity in an oligophagous insect *(Plutella maculipennis* (Curt.) Lepidoptera). Can. J. Zool. 31:52–72.

Thorsteinson, A. J. 1955. The experimental study of the chemotactic basis of host specificity in phytophagous insects. Can. Entomol. 87:49–57.

Tinbergen, N. 1951. The Study of Instinct. Oxford University Press, New York.

Uexküll, J. von. 1921. Umwelt und Innenwelt der Tiere. Springer, Berlin.

Ullyett, G. C. 1953. Biomathematic and insect population problems. Mem. Entomol. Soc. S. Afr. 2:1–89.

van Emden, H. V. 1973. Insect/Plant Relationships. Blackwell Scientific Publications, London.

Vinson, S. B. 1968. Source of a substance in *Heliothis virescens* (Lepidoptera: Noctuidae) that elicits a searching response in its habitual parasite *Cardiochiles nigriceps* (Hymenoptera: Braconidae). Ann. Entomol. Soc. Am. 61:8–10.

Vinson, S. B. 1975. Biochemical coevolution between parasitoids and their hosts. P. 14–48. *In* P. Price (ed.), Evolutionary Strategies of Parasitic Insects and Mites. Plenum Press, New York.

Vité, J. P., and D. L. Williamson. 1970. *Thanasimus dubius:* Prey perception. J. Insect Physiol. 16:233–239.

Vité, J. P., A. Bakke, and J. A. A. Renwick. 1972. Pheromones in *Ips* (Coleoptera: Scolytidae): Occurrence and production. Can. Entomol. 104:1967–1975.

Weatherston, J. 1967. The chemistry of arthropod defensive substances. Q. Rev. Chem. Soc. London 21:287–313.

Weatherston, J., and J. E. Percy. 1970. Arthropod defensive secretions. P. 95–144. *In* M. Beroza (ed.), Chemicals Controlling Insect Behavior. Academic Press, New York.

Weaver, E. C., E. T. Clarke, and N. Weaver. 1975. Attractiveness of an assassin bug to stingless bees. J. Kansas Entomol. Soc. 48:17–18.

Went, F. W. 1970. Plants and the chemical environment. P. 71–82. *In* E. Sondheimer and J. B. Simeone (eds.), Chemical Ecology. Academic Press, New York.

Whittaker, R. H. 1970a. The biochemical ecology of higher plants. P. 43–70. *In* E. Sondheimer and J. B. Simeone (eds.), Chemical Ecology. Academic Press, New York.

Whittaker, R. H. 1970b. Communities and Ecosystems. Macmillan, New York.

Whittaker, R. H., and P. P. Feeny. 1971. Allelochemics: Chemical interactions between species. Science 171:757–770.

Williams, C. M. 1970. Hormonal interactions between plants and insects. P. 103–132. *In* E. Sondheimer and J. B. Simeone (eds.), Chemical Ecology. Academic Press, New York.

Wilson, E. O. 1975. Sociobiology: The New Synthesis. Harvard University Press, Cambridge, MA.

Wilson, E. O., and W. H. Bossert, 1963. Chemical communication among animals. Recent Prog. Horm. Res. 19:673–716.

Wylie, H. G. 1958. Factors that affect host finding by *Nasonia vitripennis* (Walk.) (Hymenoptera: Pteromalidae). Can. Entomol. 90:597–608.

Young, J. C., R. G. Brownlee, J. O. Rodin, D. N. Hildebrand, R. M. Silverstein, D. L. Wood, M. C. Birch, and L. E. Brown. 1973. Identification of linalool produced by two species of bark beetles of the genus *Ips*. J. Insect Physiol. 19:1615–1622.

SECTION II

ROLE AND SIGNIFICANCE OF ALLELOCHEMICS

CHAPTER THREE

CHEMICAL MEDIATORS BETWEEN PLANTS AND PHYTOPHAGOUS INSECTS

LOUIS M. SCHOONHOVEN

Department of Animal Physiology
Agricultural University
Wageningen
The Netherlands

Plant tissues contain a large variety of chemicals. Many of them, such as amino acids, carbohydrates, and some flavonoids, are found in all plants, though in varying concentrations. Others, representatives of the so-called secondary plant substances or allelochemics, occur only in certain plant taxa. Impressive compilation works, like Hegnauer's series (1962–1973) and Karrer's (1958) list of known allelochemics, perplex us by the enormous chemical diversity found in the Plant Kingdom. Yet probably more than 90% of the organic constituency of the angiosperms still remains to be discovered (Schultes 1972).

Insects have profited by using plants as a source of food throughout evolutionary time. To the present-day observer, it is striking that most insects are more or less specialized and eat only members of one plant family or genus (oligophagous species). Even most "generalists", for example many acridids, under natural conditions prefer a fairly restricted diet. This dependence of insects on certain plants is thought to be the result of a long coevolution. The distinct food preferences are to a large extent attributable to the insect's ability to identify acceptable host plants amid a multitude of nonhosts on the basis of chemical composition. Our present insights on the role of chemical mediators between plants and phytophagous insects, as expressed by Fraenkel (1969), Dethier (1970), Whittaker and Feeny (1971), Swain (1977), and several others, may be recapitulated in the following theorems:

1 All plants have allomones that protect them from insects (and other organisms).

2 Some insect species tolerate the presence of the allomones of certain plant species.

3 Some insect species exploit these substances and use them as kairomones in the recognition of their host plants.

4 Concomitant with behavioral preferences, physiological adaptations are present, maximizing the insect's nutritional efficiency on specific food plants.

5 Plant-insect relationships continuously evolve, and an apparent status quo exists only at the instant of our observation.

These five postulates together seem to form a plausible theory for explaining insect-plant interactions in terms of a chemical intertwinement, but we must ask whether our present knowledge is sufficient to support these views. Although several observations may be cited from an extensive literature (e.g. Beck and Reese 1976, Feeny 1976, Kogan 1976, Cooper-Driver 1978) that tend to confirm the theory mentioned above, I prefer in this chapter to indicate some gaps in our knowledge and point to facts that are difficult to reconcile with our current concepts. The tendency to neglect negative results (e.g. failure of an insect to show any reaction to a certain plant constituent) or results that are hard to understand in view of existing theories leads to a bias in the published literature. However, it may be argued that negative results are more informative than evidence confirming our hypotheses, and such data certainly merit special attention.

ALLOMONES PROTECT PLANTS

Under the pressure of natural selection, plants have developed a large variety of chemicals that are toxic to herbivores. The best protection for insects against inadvertently consuming plants with possible toxic constituents is provided by the possession of receptors detecting these compounds or other compounds that indicate their presence. That insects are only rarely observed feeding on plants that are toxic to them indicates that in some way the insects recognize such plants as unsuitable. Exceptions to this rule may occur on imported toxic plants or when insects are introduced into a new habitat and have had no time to evolve avoidance reactions. Thus a cicada species was found to feed in large numbers on an imported oleander bush (Janzen 1978), and the bug *Oncopeltus fasciatus* may suck from tobacco leaves (Eggermann and Bongers 1971), although in both cases the death of the insect results.

Indeed most if not all insects appear to have "deterrent" receptors, reacting to a wide spectrum of substances that prevent them from ingesting possible toxic substances. Interestingly, all known deterrent receptors respond to many compounds that are usually absent from the insect's normal host plants, but certainly do not respond to all apparent allelochemics. Table 3.1 shows that among lepidopterous larvae, amygdalin, for instance, is perceived by only some species, whereas salicin stimulates many more species. It could be argued that the deterrent cell responds only to allelochemics occurring in plants found in the insect's biotope. If that were the case, however, for most species studied a sensitivity to amygdalin might be expected, rather than to strychnine, to which they have never been exposed.

It should be realized that deterrents do not inhibit feeding completely. At low or even moderate concentrations they may be acceptable to an insect that is hungry, although in a choice situation the insect would avoid the compound. The polyphagous peach aphid (*Myzus persicae*), for instance, will readily accept and develop well on a diet containing allomones, but it prefers a control diet when given a choice (Schoonhoven and Derksen-Koppers 1976). Probably normal host plants also contain some allelochemics, which are weakly deterrent to insects living on them. In such cases the deterrent effect is overridden by feeding stimulating compounds. Thus hordenine and gramine decrease food uptake in *Melanoplus bivittatus,* though these compounds occur in acceptable food plants (Harley and Thorsteinson 1967).

The notion that allomones are generally toxic does not hold absolutely. On the contrary, several exceptions are known and awaiting explanations. Condensed tannins, present in almost all gymnosperms and woody angiosperms, are generally thought to represent broad-spectrum antinutritional allomones. However, the acridids *Locusta migratoria* and *Zonocerus variegatus,* which normally do not encounter these substances in their food, tolerate 10% quebracho in their diets without any noticeable effects (Bernays 1978), thus casting doubts on the presumed function of this compound. In another case each of 32 compounds known to be allomones in some interactions was added to the diet of the black cutworm (*Agrotis ipsilon*). Although the majority exerted some deleterious effect on the insect's growth or food utilization, six compounds did not show any appreciable influence (Reese 1978). Some nonhost plants may be nutritionally adequate but are normally not accepted because the distasteful allomones are present. Larvae of the cucumber looper (*Anadevidia peponis*) that have been maxillectomized, hence lack their deterrent receptors, feed well and show normal development on certain nonhost plants, like cotton and willow (Ichinosé and Sasaki 1975). Some other anomalies in the concept that toxic compounds are also frequently deterrents were revealed by Harley and Thorsteinson (1967). They showed that hyoscyamine and lobeline were strongly deterrent to a grasshopper (when tested in a choice situation), but when incorporated in an artificial diet, these substances did not significantly reduce rate of weight gain or adult weight.

TABLE 3.1. Electrophysiological Evidence for the Presence of Deterrent-Sensitive Receptors in Some Lepidopterous Larvae[a,b]

	A. orana (1)	*D. pini* (1)	*M. sexta* (1,7)	*C. euphorbiae* (1)	*C. cossus* (1)	*O. brumata* (1)	*L. salicis* (1)	*L. populi* (1)	*V. atalanta* (1)	*V. urticae* (1)	*L. dispar* (1,5,6)	*P. brassicae* (2,7)	*M. brassicae* (3)	*B. mori* (4)	*S. ligustri* (1)	*H. zea* (6)	*E. acrea* (6)	*M. americana* (6)	*D. plexippus* (6)	*C. catalpae* (6)	*P. polyxenes* (6)	*C. ethlius* (6)	*P. rapae* (6)	*I. isabella* (6)
Conessine	+	+			+	o	+	+		o	+	+		+	+									
Strychnine	+		+			+		o		+	+	+	+	+	o									
Phlorizin	+		+			o	o							o	o									
Salicin	+	o	+	+	+			+	+	+	+	o	+	+	o	o	+	o	+	+	+	o	+	+
Caffeine		o	+	+				+				o			o									
Gossypol		o																						
Quinine		o	o								+	+		+	o		+	o	+					
Quebrachitol			o			o																		
Azadirachtin			o								+	+												
Cucurbitacin			o																					
Rutin			o									o		+										
Solanocapsin			o																					
Glucocapparin			o										+			+	o	o	o	+	+			o
Sinigrin											+		+			+	+	o	o	+	+	+		+
Sinalbin			+										+											
Glucotropaeolin			+								o		+		o		o	o	o		o			o
Betulin			o																					
Naringin			o																					
Diosgenin			o																					
Quercitrin											o			+		+	+	o	o		o	+	o	
Quercetin			o										o											
Capsaicin			o																					
Morin			o																					
Amygdalin			o			o	o				o		+	o			+	o	o	+	o	o	o	+
Arbutin			o								+		+		o									

Data from: (1) Schoonhoven, unpublished; (2) Ma, 1969; (3) Wieczorek, 1976; (4) Ishikawa, 1966; (5) Schoonhoven and Jermy, 1977; (6) Dethier and Kuch, 1971; (7) Schoonhoven; 1972.

[a] + = an electrophysiological reaction, o = no discernible reactions.

[b] Most Chemicals were tested at concentrations of 10^{-2}–10^{-3} M or in saturated solutions.

On the other hand, some compounds that did affect growth negatively, for example, the steroids saponin, solanine, and tomatine, were not discriminated against.

It may be concluded with certainty that not all minor compounds found in nonhost plants are in principle noxious, and also that several allelochemics show an incongruence between toxicity and capacity to inhibit feeding.

INSECTS TOLERATE ALLOMONES IN THEIR HOST PLANTS

In spite of the vast array of allomones present in the plant world, "there is scarcely a plant that does not harbor some insect pest" (Frost 1942), and even genera containing well-known, broad-spectrum insecticides like *Chrysanthemum, Derris,* and *Nicotiana* house various insect species (Brues 1946). The impression that some plant taxa are better protected than others, presumably by their chemical constituency, often appears to be erroneous when scrutinized. For instance, the idea that ferns are relatively free from insect attack in comparison to angiosperms, a statement repeatedly found in the literature, probably requires revision (Cooper-Driver 1978).

It follows that phytophagous insects which tolerate in their hosts the presence of substances that are allomones for other species, possess physiological mechanisms to evade the injurious effects of these compounds, and may in fact even utilize them in some way. Only a few cases have been analyzed in sufficient detail to elucidate the nature of such mechanisms. Many chemicals can be broken down by mixed-function oxidases (MFO), which occur mainly in the midgut wall, but also in other organs like fat body. Adding terpenoids (e.g. pinene) or glucosinolates (e.g. sinigrin) to the artificial diet of *Spodoptera* larvae raises MFO levels in their midgut within a short time, promoting the degradation of these potentially harmful substances (Brattsten et al. 1977). One would expect polyphagous insects, which have to deal with a wider variety of toxic chemicals, to be equipped with a better degradation system than specialized feeders. Krieger et al. (1971) have compared the MFO levels of larval midguts in 35 lepidopterous species. MFO activity in polyphagous species indeed appear on the average to be 30 times higher than in monophagous species, whereas oligophagous species show intermediate values.

Undoubtedly, enzymatic detoxification systems play a major role in insects, enabling them to neutralize many highly toxic substances. In addition, some other physiological solutions, such as storage in special organs or selective excretion, safeguard some other species against being poisoned by particular toxins in their host. For an excellent review on our present knowledge of biochemical defense systems, one may consult Brattsten (1979).

KAIROMONES USED TO IDENTIFY HOSTS

Some insects have not only developed resistance against the noxious effects of specific chemicals in their food plants, but also have taken advantage of the presence of these substances and use them as kairomones to recognize their hosts. Over evolutionary time, these compounds probably came to offer the plant protection from some phytophages, but some insect species evolved the ability to use these compounds to stimulate their food intake or oviposition. It should be added that these compounds, though phagostimulants (kairomones) for some insects, at the same time remain phagodeterrent and/or are toxic (allomones) to other species. In a number of cases the insect really depends on such specific phagostimulants. The classical example since Verschaffelt (1911) is the presence of glucosinolates in cruciferous plants. Several insects, belonging to different orders, depend on the presence of sinigrin or related compounds to recognize their cruciferous hosts (e.g. Nielsen et al. 1979, Thorsteinson 1953, Wensler 1962). At the same time sinigrin, when added by perfusion to parsley, a food plant of *Papilio polyxenes,* seriously affects the growth of this insect (Erickson and Feeny 1974). It seemed that at least for cruciferous-feeding animals, the wonder of the "botanical instinct" had become intelligible: the insect needs only a glucosinolate receptor to discriminate between host plants and nonhosts. Although such receptors have been found (Schoonhoven 1967, Ma and Schoonhoven 1973), evidence has accumulated that other chemicals also play a role in host recognition and acceptance. A neutral substrate to which sinigrin has been added is eaten to some degree, but the very common carbohydrate sucrose evokes a much stronger reaction (Thorsteinson 1953, Ma 1972, Blom 1978), and a combination of sucrose and sinigrin is still more acceptable. The concept that specific allelochemics alone determine acceptance (or rejection) had to be reviewed, and the function of common nutrients had to be included in theories on host recognition. Such a role of nutritive compounds was underestimated for some time because of the belief that all plants are nutritionally more or less identical and form an adequate food for all phytophagous insects. The view that "in general, one can regard certain organs of all plants (especially young leaves and stems, mature fruits and seeds), as being potentially nutritious to every insect . . ." (Swain 1976) is seriously doubted now. This problem deserves the special attention of insect nutritionists.

Living plant tissues vary considerably in quantitative ratios of nutritive compounds according to plant species, locality, age, plant organ, and site of the plant. One problem faced by herbivores is the low proportion of protein in most plant tissues. This may be a particularly important problem for fast-growing organisms, like insects (McNeill and Southwood 1978). In view of the foregoing, it is not surprising that electrophysiology has revealed the almost general presence of sucrose and/or glucose receptors, enabling insects to obtain quantitative information on the levels of these essential nutrients in different plants or in specific sites of the same plant. Moreover, many insects have specific amino acid receptors (Table 3.2), or in some other cases, amino acids act on or modify the activity of other receptors. Thus a sucrose receptor in *Entomoscelis americana* is also stimulated by some amino acids (Mitchell and Gregory 1979). In the case of insects specialized on Cruciferae, not only the presence of glucosinolates, but also the quality and quantity of several nutritive components determine the acceptability of a particular plant. Although one attempt has been made to analyze in detail the total sensory input required for optimal feeding intensity (Blom 1978), the complexity of the natural stimulus situation thus far has prevented a complete understanding of host recognition, even in the species studied most extensively: *Pieris brassicae*.

The glucosinolates success story has stimulated the search for specific phagostimulants in several other plant families. Some remarkable relationships have been elucidated, such as the dependence of the beetle *Chrysolina brunsvicensis* on hypericin, which typically occurs in its food plant *Hypericum* (Rees 1969). Other examples are given by Kogan (1976) and Städler (1976), but in none of them is such a wide variety of insects dependent on a chemical specific for a certain plant taxon. The closest other case may be found in the relationship between a number of insect species restricted to Rosaceae. The larvae of seven lepidopterous species that feed only or at least mainly on rosaceous plants have specific receptors for sorbitol, a sugar alcohol typically occurring in large quantities in this family, whereas such receptors are not known from other insects (Table 3.3). The relative paucity of examples discovered up till now of specific kairomones that serve as feeding (and/or oviposition) stimulants may be due to the limited research capacity devoted to this subject, but it could also very well be that clear and simple relationships between insects and plants based on one or a few chemicals are much rarer than we tend to think. It should be realized that most groups of minor plant compounds have fairly wide distributions, and only a few have a distribution sufficiently restricted to give them a striking role in the chemical characterization of a certain plant taxon (Swain 1972). This implies that when an insect cannot dispose of a chemical flag that is sufficiently specific, it requires more subtle and at the same time more complex information of a plant's chemical composition to distinguish hosts from nonhosts. Our present knowledge of the properties of insect chemoreceptors, obtained through the electrophysiological approach, allow the conclusion that even for the species having very small numbers of receptors, like lepidopterous larvae, highly detailed images of the plant's chemical makeup may be obtained (Dethier 1973, Schoonhoven 1969).

TABLE 3.2. Presence of Specific Sorbitol Receptors in Various Lepidopterous Larvae

Species	Feeding Range[a]	Food Plant	Sorbitol Receptors[b]
Estigmene acrea (Arctiidae)	P	Herbaceous plants	○
Malacosoma americana (Lasiocampidae)	O	Mainly Rosaceae	+
Calpodes ethlius (Hesperiidae)	M	*Canna* sp.	○
Lymantria dispar (Liparidae)	P	Rosaceous and other trees	+
Episema caeruleocephala (Plusiinae)	O	Rosaceae	+
Yponomeuta evonymellus (Yponomeutidae)	M	*Prunus padus* (Ros.)	+
Yponomeuta malinellus	M	*Malus* sp. (Ros.)	+
Yponomeuta padellus	O	Rosaceae	+
Yponomeuta mahalebellus	M	*Prunus mahaleb* (Ros.)	+
Yponomeuta cagnagellus	M	*Euonymus europea* (Celastraceae)	○
Yponomeuta rorellus	M	*Salix* sp.	○
Yponomeuta irrorellus	M	*Euonymus europea*	○
Yponomeuta vigintipunctatus	M	*Sedum telephium*	○
Yponomeuta plumbellus	M	*Euonymus europea*	○

Data from Dethier and Kuch (1971), van Drongelen (1979), Schoonhoven (1972), Schoonhoven et al. (1977).

[a] P = polyphagous, O = oligophagous, M = monophagous.

[b] ○ = no discernible reaction, + = reaction occurs.

TABLE 3.3 Amino Acid Receptors in Various Insect Larvae[a]

	P. brassicae (1)	*A. orana* (2)	*H. zea* (3)	*E. acrea* (3)	*M. americana* (3)	*D. plexippus* (3)	*M. brassicae* (4)	*P. polyxenes* (3)	*L. dispar* (3)	*C. ethlius* (3)	*P. rapae* (3)	*L. decemlineata* (5)	*E. americana* (6)
γ-Amino butyric acid	++											++	○
Proline	+++		○	++	++	+++		○	○	++	++	+	+
Alanine	+		○	+++	+++	++		○	—		++	++	++
Serine	++		○	+++	○	+++		++	—		+++	+	+
Glycine	+		○	++	—	○		—	○			○	+
Cysteine	+++			++	○			—	○	++			
Cystine			++	○	+	+		—	○	+	○		
Tryptophan		+	○	+	○	○		+	+		+		○
Tyrosine			+	+	○	○		○	—				
Arginine			○	++	+	—		—	—	++	++		+
Methionine	+++	+	++	+	○	++	+	—	—	++	++		+
Leucine	++	+++	++	++	○	○		—	—			○	○

TABLE 3.3. *(Continued)*

	P. brassicae (1)	*A. orana* (2)	*H. zea* (3)	*E. acrea* (3)	*M. americana* (3)	*D. plexippus* (3)	*M. brassicae* (4)	P. polyxenes (3)	*L. dispar* (3)	*C. ethlius* (3)	*P. rapae* (3)	*L. decemlineata* (5)	*E. americana* (6)
Isoleucine		+++	○	○	○	○		−	+		++	○	○
Aspartic acid			○	−	+	++		−	○	++	+		+
Glutamic acid	+		○	++	++	−		−	○	++			+
Histidine		+	○	○	○			+	○		○		○
Valine		++	−	+	++	++		++	○		++	○	
Phenylalanine		○	+	++	○	○	+	○	○		+	○	○
Threonine		○	○	+	○	++		++	+			○	○
Asparagine													+

Data from: (1) Schoonhoven, 1967; (2) Schoonhoven, unpublished; (3) Dethier and Kuch, 1971; (4) Wieczorek, 1976; (5) Mitchell and Schoonhoven, 1974; (6) Mitchell and Gregory, 1979.

[a] + + + = strong reaction, + +, medium Reaction, +, mild Reaction, ○ = reaction, — = inhibition as compared to control.

ALLELOCHEMICS AND FOOD UTILIZATION

Allelochemics not only guide insects in their food selection behavior, but because of their toxicity, they affect the insect's physiology once ingested. How much energy is used in neutralization processes in comparison to total energy intake, and how much is the insect willing to spend on detoxification activities before it loses interest in a certain plant species because the costs are too high? Despite current interest in the implications of plant chemical defense mechanisms for insects in terms of energy budgets (e.g. Feeny 1976), only a small body of experimental data is available.

In several instances insects have been made to accept nonhost plants after their maxillae, which carry deterrent receptors, were removed. In the case of the tobacco hornworm (*Manduca sexta*), it appeared that insects thus treated have a reduced efficiency of food utilization on some nonhost plants as compared to host plants (Waldbauer 1964). This may be because the insect's digestive system is suboptimally adapted to the strange food. Alternatively, the toxic compounds may reduce the efficiency of the digestive processes directly, or the insect may have to spend extra energy on the elimination of the unfamiliar compounds. Direct effects of allelochemics on food utilization can be measured by adding them to an artificial diet.

In an extensive study involving the polyphagous black cutworm, it appeared that many allelochemics negatively affect various nutritional indices (Beck and Reese 1976, Reese 1978). Likewise, in an oligophagous insect it could be shown that food utilization is influenced by allelochemics that either are new to the insect or are known to occur in their host plant. The food conversion rate in the tobacco hornworm is lowered considerably when canavanine, a structural analogue of arginine in many leguminous plants, is added to the insect's diet (Dahlman 1977). Supplementation of the diet with the alkaloid nicotine or atropine results in less spectacular reductions in food utilization (Schoonhoven and Meerman 1978). It thus appears that generally the effects of host plant specific allelochemics are relatively small, in contrast to substances from nonhost plants. It is unclear whether small reductions of digestive efficiency are important in an insect-plant relationship. It seems reasonable to assume that reductions of a few percent in the efficiency of food conversion are of considerable relevance to the insect's energy budget. On the other hand, it is not known whether energy is always the limiting factor in the growth and fitness of insects in nature (Feeny 1976). Since under slightly different experimental conditions metabolic indices may vary considerably, perhaps small losses in metabolic cost due to host-borne allelochemics may be neglected. Likewise it is thought that storage of certain host specific allelochemics for protection—as, for instance, *Danaus plexippus* does with cardenolides from its food—does not demand an appreciable amount of energy (Dixon et al. 1978). It is concluded, on the basis of rather scarce experimental evidence, that the metabolic cost to eliminate host-specific allelochemics is low, whereas it is high, up to fatal levels, for compounds from nonhosts.

EVOLUTION

Present insect-plant associations can hardly be understood if not considered in a historical perspective. Unfortunately any attempt to reconstruct evolutionary pathways along which such relationships have developed is hampered by the fact that physiological mechanisms leave few traces after the organisms die, and we depend mainly on indirect evidence and comparative studies on presently living systems.

Food selection behavior is based on two physiological processes: details in the sensory message sent to the central nervous system, and the neural translation of this message into behavioral activity. Changes in both steps may be involved when a species changes its feeding behavior. Since central processing mechanisms are not yet readily accessible to direct experimental investigation, little information of possible changes taking place here in relation to food selection is available. On the other hand, our rapidly growing knowledge of sensory coding in various phytophagous insects already helps to explain some evolutionary changes in feeding behavior. Two approaches to the study of modifications of the sensory code (concomitant with changes in behavior) are possible: (1) a comparative study of closely related species may yield insight into the significance of changes in the sensory system on feeding preferences, and (2) a study of the receptor systems of different strains of one species may give an impression about the evolutionary plasticity of receptor characteristics and an understanding of the relation between sensory code and behavior. Both approaches seem to offer very promising ways of deepening our insights into insect-plant relationships.

When we compare the sensory characteristics of various phytophagous insects, the most striking feature is that no two species have identical receptor systems (Dethier and Kuch 1971, Schoonhoven 1973). Apparently each species has evolved a (physiologically) unique chemosensory setup tuned to recognizing its specific range of host plants in its particular biotope. Thus even related species like *Pieris brassicae* and *P. rapae*, both feeding on Cruciferae, show differences. *P. rapae*, for example, has receptors that are sensitive to salicin and inositol (Dethier and Kuch 1971), whereas *P. brassicae* is insensitive to these substances (Ma 1969).

An interesting group of insects for the study of evolutionary processes consists of eight closely related *Yponomeuta* species, all of which have narrow, and often different food preferences. This group is considered to be in the phase of differentiation, some members already being taxonomically more widely separated than are others (Wiebes 1976). All species tested have different receptors, though some overlap occurs (Schoonhoven et al. 1977, van Drongelen 1979). Three *Yponomeuta* species live on *Euonymus* (Celastraceae). These species as well as some other species, have a receptor cell that is sensitive to dulcitol, which typically occurs in Celastraceae. Presumably this plant family is the ancestral host plant of most *Yponomeuta* species (Wiebes 1976), and the fact that some species still have dulcitol sensitivity, though living on dulcitol-free plants, probably points to a physiological ''relic'' (dulcitol stimulates feeding in these species) (van Drongelen 1978). Several *Yponomeuta* species are found on certain rosaceous plant species. They all have sorbitol-sensitive receptors (sorbitol is a stereomer of dulcitol), in contrast to all remaining *Yponomeuta* species specialized on nonrosaceous plants (van Drongelen 1979). Since the dulcitol sensitivity is probably located in a cell other than the one possessing sorbitol sensitivity (W. van Drongelen, personal communication), the species that moved from celastraceous to rosaceous host plants presumably have developed a new sorbitol receptor without discarding the dulcitol receptor in some cases. In other species dulcitol sensitivity has been lost. It is interesting to note in this context that *Malacosoma americana,* oligophagous on Rosaceae, also has a sorbitol receptor, but is insensitive to dulcitol (Dethier and Kuch 1971).

The evolution of the sensory code may also be studied within one species. Especially, strains of the same insect, differing in feeding preferences, would present excellent material for the study of receptor evolution. Several insect species show geographic differences in host preferences (Hsiao 1978). The beetle *Syneta betulae,* for instance, lives on birch in Western Europe, but in Norway it feeds on pine, and birch is actually avoided (Jolivet 1954). It would be interesting to analyze the basis of such variations. Van Drongelen (1979) compared the contact chemoreceptory system of two races of *Yponomeuta padellus,* one race feeding on *Crataegus* sp., the other on *Prunus spinosa.* No differences between the two forms were found, but subtle quantitative differences between both host races may have escaped the investigator's attention.

In an impressive study of the glycoside receptor in larvae of *Mamestra brassicae,* Wieczorek (1976) observed quantitative differences between two genetically different strains (unfortunately no biological details of these strains are given). Although the glycoside-sensitive cell in both strains is stimulated by the same compounds, its relative sensitivity varies. Strain I is more sensitive for some stimulants, whereas other compounds act more strongly on strain II. This result may be interpreted by assuming that this receptor has two types of receptor site and that the relative numbers of each differ for the two strains. It could be conjectured that an evolutionary change of a chemoreceptor proceeds via a gradual change of relative frequencies of certain receptor sites. The *Mamestra* case exemplifies, in this view, a transient phase, and it would be interesting to study food selection of the two strains in some detail to see whether the receptor differences are expressed in feeding preferences. Differences between the two strains of *P. rapae,* one of them preferentially feeding on kale and the other on mustard, have been correlated with preferences for different concentrations of allyl isothiocyanate (Hovanitz 1969). The strain showing a predilection for kale prefers mustard oil concentrations that are 100 times lower than what the other strain would choose. Crossing of the two strains gives an F_1 that shows a selection exactly intermediate between the two parental strains, indicating a simple genetic basis for these differences. An analysis of the sensory system involved might elucidate the mechanism underlying the variations in food preferences. Behavioral differences were also used to investigate changes in olfactory receptors of two geographically isolated populations of the oriental fruit fly (*Dacus dorsalis*). It was concluded that perceptible evolution of the receptor protein has occurred during a period of 50 years (Metcalf et al. 1979).

Of course, as mentioned before, changes in food selection behavior are not necessarily based on changes in the sensory system. Alterations in the central processing of peripheral information could also be involved, and some cases are known of such modifications within the brain. Silkworm (*Bombyx mori*) larvae are notorious food specialists, but one strain is known to be nonselective. Since the sensory system of this strain does not differ from the normal type, it is concluded that a mutation has in some way affected central processing (Ishikawa et al. 1963). In *Drosophila melanogaster* three mutants are known that tolerate high salt levels in their feeding solutions. Differences between the reactions to salt stimuli applied to the insect's tarsae and to its labellar receptors, allow a conclusion that the altered process is within the central nervous system (Falk and Atidia 1975).

Thus it appears that several systems are presently available that may reveal certain aspects of the evolution of an insect's host selection behavior. Moreover, small steps in such evolutionary trends may manifest themselves in relatively short periods, facilitating their analysis. The transfer of *P. brassicae* from cruciferous hosts to *Tropaeolum,* as shown by the *cheiranthi* strain on the Canary Islands, may have taken at the most a few hundred years (Gardiner 1979), and the differences in oviposition behavior observed between Canadian and Australian populations of *P. rapae* also probably developed within a few centuries (Ives 1978). Under laboratory conditions, comparable changes may be obtained within even much shorter periods.

CONCLUSION

Phytophagous insects for which the diet is restricted face the difficult problem of selecting their food plants from among many unsuitable species. The major plant constituents are chemicals of general occurrence, though their relative proportions may vary in different plant species. In addition, many chemicals more or less specific for certain plant taxa are present in variable concentrations. The insect uses representatives of both classes of chemicals to identify its host. Among the minor plant constituents, some may act as a positive signal (kairomones) to the insect, and others are negative (allomones). The limited capacity of the chemosensory system has forced the insect to concentrate on important cues only. These are different for different species, depending on the chemical composition of the preferred food and the array of unacceptable plants with which the particular insect may be confronted. For an optimal discrimination, each insect species has evolved a unique sensory apparatus, designed to distinguish hosts from nonhosts with reasonable reliability. The interspecific differences in the receptor system are strikingly manifested in the stimulus spectrum of amino acid receptors of various species (Table 3.3), and also the responses to deterrents differ from species to species (Table 3.1). This extensive differentiation requires great flexibility of the physiological characteristics of the chemoreceptors. Evidence presented above seems to show that such flexibility indeed exists.

Moreover, a second mechanism aids the insect in preventing mistakes. The process of food selection involves a chain of reactions, and each step requires a specific stimulus situation. Orientation toward a food source, the induction by "incitants" or "biting factors" to take a bite, the swallowing of the food (stimulated by "swallowing factors"), followed by a continuous feeding activity, all represent phases that take place only in the presence of the appropriate chemical stimuli. This system is similar to that shown for entomophagous insects in other chapters of this book.

When explaining food selection behavior, the metaphor of a "key-lock" system is often used, in which the key stands for a complex sensory pattern. Only when this pattern sufficiently corresponds to some innate standard, is a certain behavioral response triggered (Schoonhoven 1977). Since host selection consists of a series of steps, a set of keys is required, each one unlocking only one reaction. Lack of detail in one key will be compensated by details of another sensory activity pattern. Each phase requires a complex stimulus situation. Orientation, for example, is sometimes elicited by a complex mixture of plant volatiles (Visser and Avé 1978), and several authors (e.g. Städler 1976) have stressed the complexity of the adequate stimulus by referring to the Gestalt theory. In view of the series character of plant selection behavior, the dimension of time should be included in the Gestalt concept. The idea of a simple key then has to be changed to one of a complex multidimensional key and an equally intricate corresponding lock system.

When we realize that chemical stimuli may stimulate different phases in feeding behavior, some seeming anomalies in our knowledge of receptor characteristics as related to feeding activity may become comprehensible. The larvae of *Spodoptera exempta* and of *M. brassicae* both have receptors that are strongly stimulated by inositol. In *S. exempta* addition of this compound to a diet does not increase its consumption, whereas in *M. brassicae* inositol acts as a powerful phagostimulant (Ma 1976, 1977, Blom 1978). This difference could be due to the different functions of the inositol receptors in the two species. In *S. exempta* for instance, it might serve only as a biting factor or swallowing factor, whereas in *M. brassicae* this substance might be involved only or mainly in continuous feeding activity. Amino acids are also strong stimulants for specific receptors in a number of insects, but they hardly affect food intake directly. When expressed per action potential, the amount of diet eaten by *P. brassica* because of stimulation by proline is negligible compared to the effect of sucrose (Blom 1978). Again, this may be due to a different function of both receptor types in food selection behavior. Choosing a certain plant is different from eating that plant. An alternative explanation may be found by assuming differences in the central processing of the sensory code. Action potentials in the "sucrose line" may receive more attention than those in the "amino acid line." However, although there is no reason to believe that action potentials in different lines all lead to quantitatively exactly the same behavioral responses, it seems illogical to suppose that information from, for example, the inositol receptor in *S. exempta* conveys some message that is not used by the insect at some time.

One last unresolved matter with regard to the role of allelochemics in host plant selection concerns the question whether host plant recognition is primarily realized via the perception of host-specific kairomones or whether a host is essentially characterized by the absence of repellent and deterrent allomones. The idea that specific compounds are largely or wholly responsible for host recognition has received much attention (e.g. Verschaffelt 1911, Fraenkel 1959), and it is appealing because of its simplicity. Jermy (1965, 1966), on the other hand, has stressed the importance of feeding inhibitors in host selection, and Chapman (1974) has presented evidence for the general occurrence of such compounds. The failure, despite intensive research, to trace host-specific stimulants in the food of some oligophagous insects, such as the silkworm and the Colorado potato beetle (*Leptinotarsa decemlineata*) (Ritter 1967), seems to support Jermy's view. Nevertheless, it is hard to imagine that such specialists as the insects mentioned select their food merely by refraining from eating nonhost plants. If we were to design a food selection system, it would be logical in the case of a specialist to select for a receptor system that is specialized in recognizing typical food plant allelochemics. However, there may be historical reasons for the apparent tendency of nature, in a number of cases, to follow a different line. A long period of coevolution would not a priori lead to the most logical system, as judged from a momentary viewpoint. The finding that deterrent receptors are almost universally met in phytophagous insects, even in species that are known to recognize their host plants by specific substances, such as the cruciferous-inhabitating species (Ma 1969, Nielsen et al. 1979), points to the important function of allomones in most if not all plants. Not only in food recognition, but also in the selection of oviposition substrates, repellents and deterrents may play a more important role than has been hitherto realized (Jermy and Szentési 1978). Therefore, it seems that deterrent allomones play a crucial role in insect-plant relationships, whereas additionally insects obtain a fairly complex sensory impression of host plant composition, using chemical cues that vary in different insect species.

REFERENCES

Beck, S. D., and J. C. Reese. 1976. Insect-plant interactions: Nutrition and metabolism. Recent Adv. Phytochem. 10:41–92.

Bernays, E. A. 1978. Tannins: An alternative viewpoint. Entomol. Exp. Appl. 24:244–253.

Blom, F. 1978. Sensory activity and food intake: A study of input-output relationships in two phytophagous insects. Neth. J. Zool. 28:277–340.

Brattsten, L. B. 1979. Biochemical defense mechanisms in herbivores against plant allelochemics. P. 190–270 *In* G. A. Rosenthal and D. H. Janzen (eds.), Herbivores, Their Interaction with Secondary Plant Metabolites. Academic Press, New York.

Brattsten, L. B., C. F. Wilkinson, and T. Eisner. 1977. Herbivore-plant interactions. Mixed function oxidases and secondary plant substances. Science 196:1349–1352.

Brues, C. T. 1946. Insect Dietary. Harvard University Press, Cambridge, MA.

Chapman, R. F. 1974. The chemical inhibition of feeding by phytophagous insects: A review. Bull Entomol. Res. 64:339–363.

Cooper-Driver, G. A. 1978. Insect-fern associations. Entomol. Exp. Appl. 24:310–316.

Dahlman, D. L. 1977. Effect of L-canavanine on the consumption and utilization of artificial diet by the tobacco hornworm, *Manduca sexta*. Entomol. Exp. Appl. 22:123–131.

Dethier, V. G. 1970. Chemical interactions between plants and insects. P. 83–102. *In* E. Sondheimer and T. B. Simeone (eds.), Chemical Ecology. Academic Press, New York.

Dethier, V. G. 1973. Electrophysiological studies of gustation in lepidopterous larvae. II. Taste spectra in relation to food-plant discrimination. J. Comp. Physiol. 82:103–134.

Dethier, V. G., and J. H. Kuch. 1971. Electrophysiological studies of gustation in lepidopterous larvae. I. Comparative sensitivity to sugars, amino acids and glycosides. Z. Vgl. Physiol. 72:343–363.

Dixon, C. A., J. M. Erikson, D. N. Kellet, and M. Rothschild, 1978. Some adaptations between *Danaus plexippus* and its food plant, with notes on *Danaus chrysippus* and *Euploea core* (Insects: Lepidoptera). J. Zool. 185:436–467.

Eggermann, W., and J. Bongers. 1971. Wasser- und Nahrungsaufnahme an Pflanzen unter besonderer Berücksichtigung der Wirtsspezifität von *Oncopeltus fasciatus* Dallas. Oecologia 6:303–317.

Erickson, J. M., and P. Feeny. 1974. Sinigrin: A chemical barrier to the black swallowtail butterfly, *Papilio polyxenes*. Ecology 55:103–111.

Falk, R., and J. Atidia. 1975. Mutation affecting taste perception in *Drosophila melanogaster*. Nature (London) 254:325–326.

Feeny, P. 1976. Plant apparency and chemical defense. Recent Adv. Phytochem. 10:1–40.

Fraenkel, G. S. 1959. The raison d'être of secondary plant substances. Science 129:1466–1470.

Fraenkel, G. S. 1969. Evaluation of our thoughts on secondary plant substances. Entomol. Exp. Appl. 12:473–486.

Frost, S. W. 1942. General Entomology. McGraw-Hill, New York.

Gardiner, B. O. C. 1979. A review of variation in *Pieris brassicae* (L.) (Lep., Pieridae). Proc. Br. Entomol. Nat. Hist. Soc. 12:24–46.

Harley, K. L. S., and A. J. Thorsteinson. 1967. The influence of plant chemicals on the feeding behavior, development and survival of the two-striped grasshopper, *Melanoplus bivittatus* (Say), Acrididae: Orthoptera. Can. J. Zool. 45:305–319.

Hegnauer, R. 1962–1973. Chemotaxonomie der Pflanzen, Vols. I–VI. Birkhäuser, Basel.

Hovanitz, W. 1969, Inherited and/or conditioned changes in host-plant preference in *Pieris*. Entomol. Exp. Appl. 12:729–735.

Hsiao, T. H. 1978. Host plant adaptations among geographic populations of the Colorado potato beetle. Entomol. Exp. Appl. 24:437–447.

Ichinosé, T., and N. Sasaki. 1975. An experimental analysis and integrated evaluation of various factors involved in the host plant specificity of the cucumber looper, *Anadevidia peponis* (Fabricius) (Lepidoptera: Noctuidae). Appl. Entomol. Zool. 10:284–297.

Ishikawa, S., Y. Tazima, and T. Hirao. 1963. Responses of the chemoreceptors of maxillary sensory hairs in a "non-preference" mutant of the silkworm. J. Sericult. Sci. Jpn. 32:125–129.

Ives, P. M. 1978. How discriminating are cabbage butterflies? Aust. J. Ecol. 3:261–276.

Janzen, D. H. 1978. Cicada (*Diceroprocta apache* (Davis)) mortality by feeding on *Nerium oleander*. Pan-Pac. Entomol. 5:69–70.

Jermy, T. 1965. On the nature of the oligophagy in *Leptinotarsa decemlineata* Say (Coleoptera: Chrysomelidae) Acta Zool. Acad. Sci. Hung. 7:119–132.

Jermy, T. 1966. Feeding inhibitors and food preference in chewing phytophagous insects. Entomol. Exp. Appl. 9:1–12.

Jermy, T., and A. Szentési. 1978. The role of inhibitory stimuli in the choice of oviposition site by phytophagous insects. Entomol. Exp. Appl. 24:458–471.

Jolivet, P. 1954. Phytophagie et selection trophique. Vol. Jubilé, Inst. R. Sci. Nat. Belg. 2:1101–1134.

Karrer, W. 1958. Konstitution und Vorkommen der organischen Pflanzenstoffe. Birkhäuser, Basel.

Kogan, M. 1976. The role of chemical factors in insect/plant relationships. Proc. XV Int. Congr. Entomol. 211–227.

Krieger, R. I., P. P. Feeny, and C. F. Wilkinson, 1971. Detoxication enzymes in the guts of caterpillars: An evolutionary answer to plant defenses? Science 172:579–581.

Ma, W. C. 1969. Some properties of gustation in the larva of *Pieris brassicae* Entomol. Exp. Appl. 12:584–590.

Ma, W. C. 1972. Dynamics of feeding responses in *Pieris brassicae* L. as a function of chemosensory input: A behavioral, ultrastructural and electrophysiological study. Med. Landbouwhogesch., Wageningen. 72–11, 1–162.

Ma, W. C. 1976. Mouth parts and receptors involved in feeding behavior and sugar perception in the African armyworm, *Spodoptera exempta* (Lepidoptera, Noctuidae). Symp. Biol. Hung. 16:139–151.

Ma, W. C. 1977. Electrophysiological evidence for chemosensitivity to adenosine, adenine and sugars in *Spodoptera exempta* and related species. Experientia 33:356–357.

Ma, W. C., and L. M. Schoonhoven. 1973. Tarsal contact chemosensory hairs of the large white butterfly *Pieris brassicae* and their role in oviposition behavior. Entomol. Exp. Appl. 16:343–357.

McNeill, S., and T. R. E. Southwood. 1978. The role of nitrogen in the development of insect plant relationships. P. 77–98. *In* J. B. Harborne (ed.), Biochemical Aspects of Plant and Animal Coevolution. Academic Press, New York.

Metcalf, R. L., E. R. Metcalf, W. C. Mitchell, and L. W. Y. Lee, 1979. Evolution of olfactory receptor in oriental fruit fly *Dacus dorsalis*. Proc. Natl. Acad. Sci. U.S.A. 76:1561–1565.

Mitchell, B. K., and P. Gregory. 1979. Physiology of the maxillary sugar sensitive cell in the red turnip beetle, *Entomoscelis americana*. J. Comp. Physiol. A, 132:167–178.

Nielsen, J. K., L. Dalgaard, L. M. Larsen, and H. Srensen. 1979. Host plant selection of the horse-radish flea beetle *Phyllotreta armoraciae* (Coleoptera: Chrysomelidae): Feeding responses to glucosinolates from several crucifers. Entomol. Exp. Appl. 25:227–239.

Rees, C. J. C. 1969. Chemoreceptor specificity associated with choice of feeding site by the beetle, *Chrysolina brunsvicensis* on its food plant *Hypericum hirsutum*. Entomol. Exp. Appl. 12:565–583.

Reese, J. C. 1978. Chronic effects of plant allelochemics on insect nutritional physiology. Entomol. Exp. Appl. 24:625–631.

Ritter, F. J. 1967. Feeding stimulants for the Colorado beetle. Meded. Rijksfac. Landbouwwet. Gent. 32:291–305.

Schoonhoven, L. M. 1967. Chemoreception of mustard oil glucosides in larvae of *Pieris brassicae*. Proc. K. Ned. Akad. Wet. Amsterdam C, 70:556–568.

Schoonhoven, L. M. 1969. Gustation and food-plant selection in some lepidopterous larvae. Entomol. Exp. Appl. 12:555–564.

Schoonhoven, L. M. 1973. Plant recognition by lepidopterous larvae. Symp. R. Entomol. Soc. London 6:87–99.

Schoonhoven, L. M. 1977. On the individuality of insect feeding behavior. Pro. K. Ned. Akad. Wet. Amsterdam C, 80:341–350.

Schoonhoven, L. M., and I. Derksen-Koppers. 1976. Effects of some allelochemics on food uptake and survival of a polyphagous aphid, *Myzus persicae*. Entomol. Exp. Appl. 19:52–56.

Schoonhoven, L. M., N. M. Tramper, and W. van Drongelen. 1977. Functional diversity in gustatory receptors in some closely related *Yponomeuta* species (Lep.). Neth. J. Zool. 27:287–291.

Schoonhoven, L. M., and T. Jermy. 1977. A behavioral and electrophysiological analysis of insect feeding deterrents. P. 133–146. *In* N. R. McFarlane (ed.), Crop Protection Agents, Their Biological Evaluation. Academic Press, New York.

Schoonhoven, L. M., and J. Meerman. 1978. Metabolic cost of changes in diet and neutralization of allelochemics. Entomol. Exp. Appl. 24:689–693.

Schultes, R. E. 1972. The future of plants as sources of new biodynamic compounds. P. 103–124. *In* T. Swain (ed.), Plants in the Development of Modern Medicine. Harvard University Press, Cambridge, MA.

Städler, E. 1976. Sensory aspects of insect plant interactions. Proc. XV Int. Congr. Entomol. 228–248.

Swain, T. 1972. The significance of comparative phytochemistry in medical botany. P. 125–160. *In* T. Swain (ed.), Plants in the Development of Modern Medicine. Harvard University Press, Cambridge, MA.

Swain, T. 1976. The effect of plant secondary products on insect plant co-evolution. Proc. XV Int. Congr. Entomol. 249–256.

Swain, T. 1977. Secondary compounds as protective agents. Annu. Rev. Plant Physiol. 28:479–501.

Thorsteinson, A. J. 1953. The chemotactic responses that determine host specificity in an oligophagous insect (*Plutella maculipennis*) (Cuer.) (Lepidoptera). Can. J. Zool. 31:52–72.

van Drongelen, W. 1978. The significance of contact chemoreceptor sensitivity in the larval stage of different *Yponomeuta* species. Entomol. Exp. Appl. 24:343–347.

van Drongelen, W. 1979. Contact chemoreception of host plant specific chemicals in the larvae of various *Yponomeuta* species (Lepidoptera) J. Comp. Physiol. A, 134:265–280.

Verschaffelt, E. 1911. The cause determining the selection of food in some herbivorous insects. Proc. Acad. Sci. Amsterdam 13:536–542.

Visser, J. H., and D. Avé. 1978. General green leaf volatiles in the olfactory orientation of the Colorado beetle, *Leptinotarsa decemlineata*. Entomol. Exp. Appl. 24:738–749.

Waldbauer, G. P. 1964. The consumption, digestion, and utilization of solaneceous and non-solanaceous plants by larvae of the tobacco hornworm *(Protoparce sexta* (Johan.) (Lepidoptera: Sphingidae). Entomol. Exp. Appl. 7:253–269.

Wensler, R. 1962. Mode of host selection by an aphid. Nature (London) 195:830–831.

Whittaker, R. H., and P. P. Feeny, 1971. Allelochemics. Chemical interactions between species. Science 171:757–760.

Wiebes, J. T. 1976. The speciation process in the small ermine moths. Neth. J. Zool. 26:440.

Wieczorek, H. 1976. The glycoside receptor of the larvae of *Mamestra brassicae* L. (Lepidoptera: Noctuidae). J. Comp. Physiol. A 106:153–176.

CHAPTER FOUR

HABITAT LOCATION

S. BRADLEIGH VINSON

Department of Entomology
Texas A&M University
College Station, Texas

Upon emergence, a female parasitoid (see Vinson 1982 for definition and discussion of parasitoidism) may be located far from potential hosts. To ensure the reproduction of the species, the female must have evolved a successful means by which potential hosts are located. The successful location of hosts by parasitoids depends on a number of complex and interacting factors. The processes necessary for successful parasitism were first divided into two steps by Salt (1935) and Laing (1937) and were later added to by Flanders (1953). Doutt (1964) combined these suggestions into four steps: host habitat location, host location, host acceptance, and of host suitability. Vinson (1975) added host regulation as a fifth step. Similar steps are necessary for prey finding by predacious insects and plant finding by herbivorous insects.

These various steps, important to the reproductive success of parasitoids, have been reviewed in recent years. They include host selection, which consists of host habitat location, host location, and host acceptance (Vinson 1976, 1977b, Lewis et al. 1976, Saladin 1980); determination of host suitability (Salt 1970, Vinson 1977a, Vinson and Iwantsch 1980a); and host regulation (Stoltz and Vinson 1979, Vinson and Iwantsch 1980b). Although successful parasitoidism is divided into steps, these divisions are primarily for our convenience in thought and communication. In nature there is a great deal of overlap, and certain steps may be absent, modified, or subdivided.

In this chapter I discuss host habitat location with regard to its importance to the reproductive success of the parasitoid. Additional discussions are directed toward habitat location by predacious and phytophagous species. Habitat location is a problem shared by most species of insects involved in the location of food or oviposition sites. Even though habitat location is the initial step in the location of food or oviposition sites, there has been relatively little research as to the stimuli involved.

IMPORTANCE OF HOST HABITAT LOCATION

Host habitat location is achieved by the adult stage in almost all species of phytophagous and entomophagous insects. Although host location and acceptance by parasitoids is usually performed by the female, in some species this task may be performed by the immatures. For example, *Prosena siberita* larviposits in the soil, and these first instar larvae tunnel through the soil in search of their host, scarabid grubs, into which they burrow (Clausen et al. 1927). The tachinid *Lixophaga diatraeae* larviposits in response to host frass, and the larvae must seek and enter the host (Roth 1976).

Other than catastrophic disturbances such as floods, storms, high winds, or habitat destruction, two major factors are important in placing a female at a point removed from the host habitat—a phenomenon that necessitates a habitat location ability. These are the emergence and the dispersal of females.

Emergence

Parasitoids usually destroy the host prior to the completion of their development. In fact, host destruction is one of the key differences between parasitoids and parasites (Vinson 1982). Furthermore, upon emergence of female parasitoids, the host population from which they developed may have completed their development, hence are no longer present or are present in an unacceptable life stage. Insect populations are usually clumped in time and/or space (Southwood 1966), and the distribution of these clumps changes from season to season and from year to year. Thus even parasitoids that are synchronized with their host population upon emergence may be in synchrony with the susceptible host stage in time but not space. Such a phenomenon has necessitated the evolution of efficient mechanisms for the location of suitable host habitats. The same is true for phytophagous species, particularly the species feeding or ovipositing on herbaceous annuals.

Dispersal

The second factor that results in a female parasitoid's location at a point distant from a host population is dispersal. Several factors effect the dispersal of females.

Preoviposition Period

In some species the newly emerged female is not reproductively ready to search for hosts and oviposit in them. Instead, her energies are directed toward the location of a suitable environment for food, shelter, and, in some cases, a mate.

The former situation is particularly characteristic of synovigenic species (Flanders 1950). During the preoviposition period the female may not respond to host-associated cues or to stimuli from the host's habitat. For example, *Opius fletcheri* females are not attracted to the host habitat until 3 days postemergence (Nishida 1956). This results in the undirected movement of the parasitoid with regard to potential host habitats and tends to remove the female from the habitat from which she emerged.

In some species dispersal is ensured by the behavioral response of the female to a host's habitat upon emergence. For example, *Pimpla ruficollis* attacks the larvae of the European pine shoot moth (*Rhyaciona buoliana*), but upon emergence the female parasitoid is repelled by the odor of the pine oil in an olfactometer (Thorpe and Caudle 1938). This reaction tends to disperse the female from the host's habitat. In 3 or 4 weeks the ovaries of *P. ruficollis* mature, and the female is attracted to pine oil (Thorpe and Caudle 1938). A similar response was observed for the tachinid *Eucarcelia rutilla* (Zwolfer and Kraus 1957). There are, of course, some species that upon emergence are capable of attacking and ovipositing in a host and are thus attracted to the host habitat within hours of emergence (Taylor 1974).

Host Marking

Another phenomenon that may lead to dispersal is host marking. Host marking or host discrimination is common in parasitoids, particularly the Hymenoptera, as well as in phytophagous species (van Lenteren 1976, van Lenteren and Bakker 1978, van Lenteren et al. 1978, Chapters 9 and 10). The term "host discrimination" has been applied to the ability of a parasitoid to avoid attacking or accepting a potential host that has been parasitized, as well as to reduce the searching of previously searched environments. Salt (1935) was the first to report that a parasitoid left a "factor" that inhibited further attack; this substance was later called spoor factor (Flanders 1951).

The spoor factor has been referred to by a number of terms, including dispersal pheromone, marking pheromone, and epideictic pheromone (Price 1970, Corbet 1971, Guillot and Vinson 1972, Vinson and Guillot 1972, Chapters 9 and 10).

Host marking is important to the evolutionary strategies of parasitoids in several ways. These include the reduction in competition between conspecifics; reduction in egg wastage, time, and energy; and dispersal. All these strategies are interrelated, and a brief discussion of them is appropriate.

Because the food reserves provided by the host are limited, it is usual for only one or a limited number of eggs deposited in or on a host to survive to maturity. Although gregarious species of parasitoids occur, multiple infection is rarer than randomly expected (King and Rafai 1970). Even in these situations only one generation is capable of being produced per host, with rare exceptions (Vinson

1982). Many species are solitary, and only one adult emerges from a host. Even if super- or multiple parasitism occurs in or on the same host, the supernumeraries are destroyed by physical combat or physiological suppression (Fisher 1963, 1971, Vinson 1972a, Vinson and Iwantsch 1980b). Thus host marking has been reported to function in reducing competition for possession of the host by conspecifics (Fiske 1910, Smith 1916, Rabb and Bradley 1970, King and Rafai 1970, Wylie 1971). In view of the role of host marking in reducing competition, the occurrence of supernumerary ovipositions by solitary species is in itself surprising, and the reasons for its occurrence should be more carefully examined.

To understand the importance of host marking to dispersal, the behavior of the female after oviposition must be examined in relation to host distribution. Following oviposition many parasitoids exhibit intense searching of the surrounding area (Edwards 1954, Chabora 1967, Gerling and Schwartz 1974).

The importance of intense searching after oviposition becomes evident when the distribution of hosts is considered. Host densities may exist in the following types of distributions: high host density over a wide area where hosts are plentiful, host density clumped in a distribution where host density is high in a local area, and a low host density that is evenly distributed (Southwood 1966). In the latter case the intensive search of the area around a host after oviposition would be unproductive. In the other cases the intense searching of an area around an encountered host would be expected to lead to the discovery of additional hosts.

Such strategies have been documented in many other organisms, particularly birds (Smith 1974) and insect predators (Chandler 1969, Evans 1976, Marks 1977, Rowlands and Chapin 1978), and have been referred to as success-motivated searching (Croze 1970). Organisms such as predators or scavengers that practice success-motivated searching remove the located food item, thus eliminating a possible future food encounter.

If food supplies are slowly renewed, it is advantageous to avoid inspecting or searching a place that has recently been harvested. This strategy has been documented for birds (Gill and Wolf 1977), and insects (Marks 1977, Carroll and Jenzen 1973). There is an added dimension to the role of success-motivated searching by parasitoids because the host remains in the environment. Thus success-motivated searching increases the frequency of encounter with an already accepted and attacked host, particularly as more hosts become parasitized. In addition, the host may be capable of movement, and if movement occurs, it reduces the importance of substrate marking in favor of host marking. The deposition of eggs in an already parasitized host wastes eggs, time, and energy. Thus the second function of host marking is to avoid such negative activities (Price 1970, van Lenteren 1976, Waage 1978).

It is also counterproductive for a female to remain in an area when the frequency of encounter of parasitized hosts is increased: it would be advantageous for a female to leave such a habitat and locate a potentially more productive one.

Thus the third function of host marking is to stimulate dispersal and search for new potential host habitats. Because search for food or oviposition sites requires an expenditure of energy and time, one would expect selection to favor behaviors that increase searching efficiency (Schoener 1971, Heinrich 1976, Pyke et al. 1977). Such behavior as host marking, and the insect's response to marked hosts, maximizes the location and utilization of productive habitats.

Host Density

The lack of hosts in a habitat or the presence of unsuitable hosts may reduce the time spent in the habitat. Although a few hosts may be present in the habitat, they may represent an insufficient number of contacts to retain a female. A female in an unproductive habitat would expend a certain amount of energy in host searching, but at some point it would become advantageous to leave rather than to continue to search the habitat in the chance that a host would be encountered. A female in a habitat devoid of hosts would be expected to invest little energy and to leave rapidly. If, however, the presence of hosts were detected (e.g. by kairomones), the investment of additional time and energy might be warranted. In fact, the retention of parasitoids increases when these chemicals are applied to plant foliage (Lewis et al. 1975a, 1975b). However, if hosts are not encountered, a female may disperse even in the presence of kairomones.

Environmental Factors

Environmental changes may lead to the dispersal of females. Although there are few data, it is reasonable to assume that a female located in a productive habitat could be forced to leave by extreme conditions of temperature or other environmental changes. For example, Juillet (1964) reported that wind may reduce searching in some species. Adverse light intensity has been reported to reduce searching and lead to dispersal or failure to orient to certain habitats (Miller 1959, Vinson 1975).

Biological Factors

A female may leave a productive habitat and disperse if additional mature eggs are lacking; she may leave in search of food, or she may be chased away by the presence of certain predators or competitors.

SELECTION AND USE OF FOOD OR OVIPOSITIONAL RESOURCES

In all cases the successful insect must be coincident in time and space with its food or ovipositional site. Proper temperature, humidity, wind, light conditions, and time of day must also prevail. If these conditions are met and the insect is in "proper physiological condition" (a catchall for factors that collectively stimulate the host-searching response or appetite), the female will search for, locate, and accept a suitable food or ovipositional site. In addition, the food or ovipositional site maybe altered in a way that serves to increase its suitability. The sequence of behavioral and physiological responses of the insect engaged in the selection and utilization of its food or ovipositional resources has been described for several groups of insects (Jacobson 1966; Schoonhoven 1972, Kogan 1975, Vinson 1975, 1976, Hagen et al. 1976b). Although the division of the process of locating and utilizing these resources is arbitrary, as stated earlier, these divisions serve to categorize the various behaviors.

The strategies utilized by insects for the exploitation of food or ovipositional resources may appear different at first glance, but they have a great deal of similarity (Figure 4.1). A phytophagous insect involved in the selection and efficient use of an oviposition site is similar to the parasitoid involved in a similar task. The females of both proceed through steps Ia to IIIa, at which point oviposition occurs; then the female proceeds to step VII (Figure 4.1). In some species, the female may proceed through only one or two steps prior to oviposition and leave her progeny to complete the steps. For example, the gypsy moth *Lymantria dispar* (a phytophagous species), is similar to the tachinid *Lixophaga diatraeae* in that both deposit their progeny some distance from the progeny's food. In both cases the progeny must seek the host and proceed through the remainder of the food selection chain beginning at IIa (Figure 4.1).

Predacious, phytophagous, and scavenger species seeking food appear to follow a similar scheme, proceeding through steps Ia to IIIa of the food selection chain and then to VII (Figure 4.1). Although suitability (step IVa, Figure 4.1) is important to predacious, phytophagous, and scavenger species, host regulation may be less important than it is for species involved in oviposition site selection. However, when prey are considered on a populational level, host regulation by predators may be involved (Elton 1936, Cameron 1956), and the same may be true for the food plants of phytophagous species.

Host regulation often appears to be involved in successful oviposition for both phytophagous and parasitoid species. As pointed out by Miles (1968), many insects, when ovipositing or feeding, deposit in or on plants secretions that may have injurious or trivial effects. As boll weevil (*Anthonomus grandis*) larvae molt, they appear to release chemicals that cause the boll in which they were developing to abscise from the plant (Coakley et al. 1969). Other examples of host regulation by ovipositing phytophagous insects include the alteration of *Pinus radiata* by the woodwasp *Sirex noctilio,* which appears to be necessary for the development of the woodwasp progeny (Madden 1977). Coulson (1979) points out that bark beetles of the genus *Dendroctonus* must colonize the host tree in sufficient numbers to overcome the tree's resistance for the successful development of the immature stages. This results in the death of the tree, or some

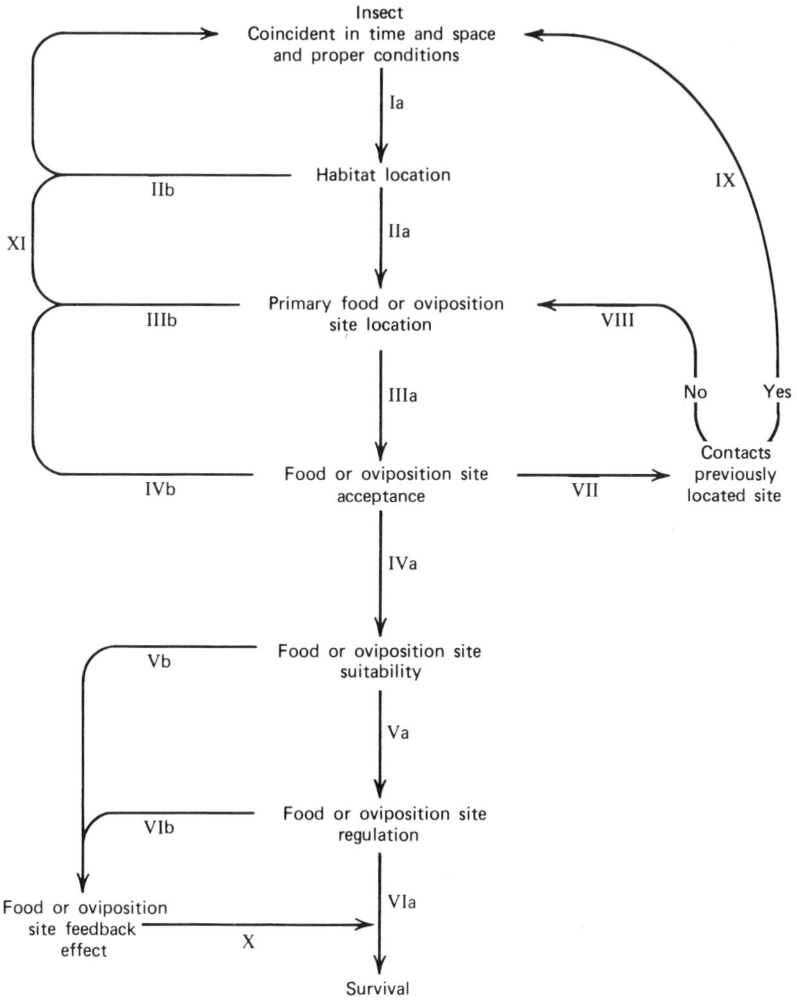

FIGURE 4.1. Diagrammatic representation of the food and oviposition selection and utilization chain.

portion of the tree. The initiation and development of galls by secretions from the ovipositing female sawfly and her progeny is an example of host regulation (Hovanitz 1959). Insect parasitoids also regulate their host. For example, *Microplitis croceipes* injects a virus into the host that elevates the trehalose level of the host hemolymph (Dahlman and Vinson 1975), and *Campoletis sonorensis* reduces the growth of its host, *Heliothis virescens* (Vinson 1972b). Other examples are provided by Vinson and Iwantsch (1980b).

If the insect has been successful in oviposition or location and acceptance of food, it may be stimulated to search for additional food resources or ovipositional sites; that is, success motivated searching (step VII, Figure 4.1). If the insect contacts a previously located food or ovipositional site indicated by the presence of an epideictic pheromone, the insect may disperse and seek out a new habitat. It should also be noted that at any point in the food or oviposition selection chain the female may fail to receive the appropriate cues or stimuli and disperse (step XI, Figure 4.1).

HOST HABITAT LOCATION

Effective orientation of an insect from a distance of a meter or more would likely involve volatile chemicals, sound, light (sight), or other forms of electromagnetic radiation. Although there has been relatively little work toward elucidating the factors important in habitat location, the potential source of such cues consists of the host's food or shelter, the host, non-host-associated organisms, or interaction between these factors (Figure 4.2). Of these factors, plants may have a major role in providing cues to habitats, since they have a dominant role in providing unique habitats and are primary producers. Food or ovipositional site finding by phytophagous and saprophagous insects probably involves electromagnetic (visible) and chemical cues, whereas movement and vibrational information such as sound may be an added dimension for predators and parasitoids.

However, chemicals appear to play the major role in the orientation of many insects. Host habitat location of distant plants or other sources of habitat cues via responses to odor can be divided into long-range and short-range factors (Kennedy 1977). In comparison to "short-range" orientation, which usually occurs

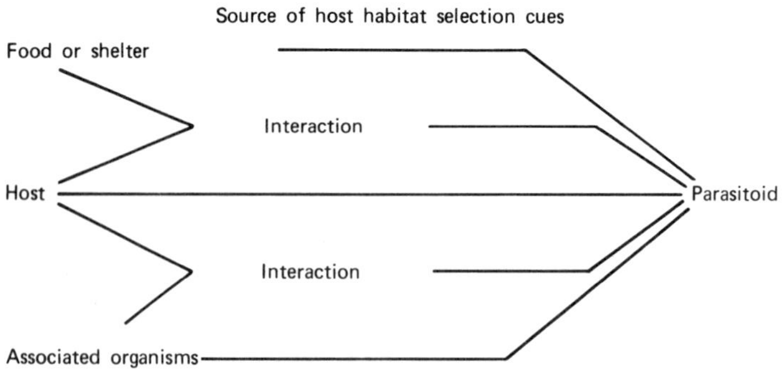

FIGURE 4.2. Diagrammatic representation of the sources of possible host habitat tion cues.

within a few centimeters of the source of the chemical cue (Kennedy 1977), long-range orientation to habitat cues has been generally neglected as a field of study (Schoonhoven 1968, Dethier 1970, Kennedy 1977). Although relatively few in-depth studies involving the long-range orientation to plants by phytophagous insects (Schoonhoven 1968), parasitoids (Vinson 1976), or predators (Hagen et al. 1976b) have been reported, there is substantial support for this phenomenon and it is probably widespread (Jermy 1976).

Habitat Location by Phytophagous Insects

Both color and form have a role in the orientation of phytophagous insects. For example, several species of aphids are attracted to a yellowish to whitish surface (Muller 1958, Kennedy et al. 1961, Moericke 1962), and color with high light remission attracts numerous other phytophagous species (Ilse 1937, Horber 1955, Callahan 1957, Moericke 1962, 1969, Prokopy 1968, 1972, Prokopy and Boller 1971, Prokopy and Haniotakis 1975, Verschoor−van der Poel and van Lenteren 1978, Röttger 1979). Other species also orient by vision, such as the acridids (Williams 1954, Mulkern 1969), some caterpillars (Hundertmark 1937), and beetles (Hierholzer 1950), all of which orient to vertical lines. Shape may also influence orientation (Prokopy 1972, 1975, Prokopy and Bush 1973).

Although form of the perception of various portions of the electromagnetic spectrum may influence habitat location by phytophagous insects, the literature is less extensive (Markl 1974, Stadler 1977) than the literature on the role of chemicals (Hedin et al. 1977, Kogan 1977, Stadler 1977). Various plant extracts are attractive to adult phytophagous insects for feeding and oviposition and to the immature, as well (Moore 1928, Kennedy 1965, Green et al. 1980, Thorsteinson 1960, Yamamoto and Fraenkel 1960, Dethier 1966, Schoonhoven 1968). Plant compounds not only attract insects but may incite biting, stimulate swallowing, and lead to continued feeding (Chapter 3). As Hamamura and Naito (1961) and Hamamura et al. (1962) reported, the silkworm (*Bombyx mori*) was attracted to 3-hexen-1-al, linalol, citral, terpinyl acetate, and linalyl acetate. They also found that 3-hexen-1-ol attracted young larvae, whereas 2-hexen-1-al attracted older larvae. Isoquercitrin and morin were found to induce biting, and cellulose induced swallowing (Hamamura et al. 1962). Nayar and Fraenkel (1962) reported that *n*-aliphatic alcohols, together with a mixture of sterols, stimulated larval biting; continuous feeding occurred only when sugars were present.

Although many plant extracts "attract" insects, many of these studies have been conducted in situations that attract the insect over distances of a few centimeters at most. This is particularly true of larval feeding behavior (Grier 1963, Sutherland 1972), although Steiner (1939) showed that newly hatched codling moths (*Laspeyresia pomonella*) could reach and enter fruit on a tree 3 m from where they had hatched.

As stated by Beroza and Jacobson (1963), the literature abounds with references to the attraction of insects from an undefined distance to such substances as sliced cucumbers, crshed bananas, fermented sugar solution, corn steep liquor, vitamin preparations, bacterial cultures, and decaying organic matter. Although such reports are fairly common, little has been accomplished to identify the attractive agents.

Plant odors are usually complex mixtures. Schoonhoven (1968) noted that many odorous factors, often in specific combinations, are typical for a single plant species or higher taxon. An attempt to analyze the relationship between an insect and the odor of a plant could involve the reaction of the insect to the combined odor of all the various plant volatiles. Kamm and Fronk (1964) obtained 95 chemicals either as liquid or solubilized from alfalfa. Of these, 38 were attractive to the alfalfa seed chalcid (*Bruchophagus roddi*), nine were repellent, and the remainder had no apparent effect. Because a combination of a number of attractants and repellents may be involved, the isolation and identification of odorous plant components that attract insects may be extremely difficult. Also, the context of the bioassay used will influence the type of chemical stimulus isolated.

The orientation to a habitat by long-range factors may occur in two ways. The first is odor-conditioned positive anemotaxis, which was demonstrated by Haskell et al. (1962). The authors found that desert locusts *(Schistocerca gregaria)* changed from downwind to upwind walking when held in a wind tunnel and exposed to grass odor in an air stream. Similar positive anemotaxis toward host plant odor has been reported for walking potatoe beetles *(Leptinotarsa decemlineata)* (de Wilde et al. 1969), and flying *Drosophila* (Kellogg et al. 1962).

The second case involves klino-, tropo-, or telotaxis, which is in contrast to the speculation of Thorsteinson (1960) that the effect of host plant odors was to arrest the insect after its arrival at the plant. There is increasing evidence that plants produce chemicals that "attract" insects (chemotaxis). For example, Kullenberg (1961) found that some orchid flowers produce a scent similar to the sex pheromone of certain euglosine bees. Various flowers in the genera *Arum, Arisarum,* and *Arisaema* attracts certain Coleoptera and Diptera with odors resembling decaying plant and animal matter (Yeo 1972). Although these may be considered rather "special" cases, it is becoming increasingly clear that many plants produce odors that play an important role in the orientation of insects. De Wilde (1947) suggested that while in flight, the cabbage rootfly *(Hylemya brassicae)* responds to either olfactory or visual stimuli specific to its host plant. Five volatile compounds obtained from cotton plants, α-pinene, β-bisabolol, caryophyllene, limonene, and caryophyllene oxide, have been found to be individually and collectively attractive to the boll weevil (Hedin et al. 1973, McKibben et al. 1977).

Keller et al. (1963) and Neff and Vanderzant (1963) reported the presence of attractants to the boll weevil in cotton squares and seedlings. The odor of apples was reported to attract apple maggot flies *(Rhagoletis pomonella)* (Prokopy et al. 1973). Red pine bark was found attractive to pine beetles *(Cryphalus fulvus)* because it contains benzoic acid (Yasunaga et al. 1963). Douglas fir oleoresin and several terpene hydrocarbons attracted several species of bark and timber beetles associated with Douglas fir (Rudinsky 1966). Munakata et al. (1959) isolated an aliphatic, unsaturated ketone from rice plants that was attractive to the rice stem borer *(Chilo suppressalis)*.

In the Scolytidae a number of host tree components have been isolated and identified that either independently or in conjunction with pheromones attract the Scolytidae to their host tree (Borden 1974).

Host Habitat Location by Parasitoids

As with phytophagous insects, little is known about habitat location by parasitoids. Much of what has been discussed concerning the orientation of phytophagous insects to plants applies to parasitoids. There are several reports of parasitoids being attracted to a habitat, regardless of the presence or absence of hosts. This provides circumstantial evidence that the host plant plays a role in orientation to the habitat. For example, *Cardiochiles nigriceps* was observed searching host-free tobacco plants removed to the field (Vinson 1975), and *O. fletcheri* was found to be attracted to the habitat of its host whether hosts were present or not (Nishida 1956).

In other cases parasitoids search and attack hosts on some species of plants but not others. Taylor (1932) reported that *Heliothis armigera,* which feeds on a variety of plants in South Africa, was attacked by *Microbracon brevicornis* only when it fed on *Antirrhinum*. Another example is provided by *C. nigriceps* a solitary endoparasitoid of *H. virescens*. *C. nigriceps* has been recovered from hosts collected from tobacco *(Nicotiana tabacum),* spider flower *(Cleome spinosa),* ground cherry (*Physalis* spp.), toadflac *(Linaria canadensis),* deergrass *(Rhexia mariana),* and cotton *(Gossypium hirsutum)* (Snow et al. 1966, Neunzig 1969, Vinson, personal observation). Shepard and Sterling (1972) reported *C. nigriceps* to be the most abundant parasitoid of *H. virescens* on cotton in Texas, and Neunzig (1969) found *C. nigriceps* to be the most abundant parasitoid on cotton and tobacco in North Carolina. However, *C. nigriceps* has not been observed in fields of peanuts *(Arachis hypogaea),* where *H. virescens* is sometimes found (J. W. Smith, personal communication). Zwolfer and Kraus (1957) reported that the fir budworm *(Choristoneura muranana),* which occurs on fir, was attacked by the ichneumonid *Ephialtes (Apechthis) rufata* when it was rolled up in oak leaves (*E. rufata* usually attacks only the oak tortricid). *Itoplectis conquistor* was found to attack hosts on Scots pine but not the same hosts on red pine (Arthur 1962). Salt (1958) reported that *Apanteles glomeratus* attacks *Pieris brassicae* on cabbage but not on sea rocket or capers.

Stary (1966) observed that parasitoids attacking *Aphis fabae* on the spindle tree in a forest community were different from the parasitoids attacking the same aphid on beets located in the Steppe community of Czechoslovakia. However, whether this difference should be attributed to the difference in host plant or to differences in the species of parasitoids making up the two communities is difficult to decide. General habitat preference is an important factor: not only does it determine which species of parasitoids will attack a given host in a particular habitat, but the habitat preference may influence the type of response that a particular host habitat cue may elicit from a parasitoid. The importance of habitat preference is exemplified by the work of Flanders (1937) and Simmonds (1954). Flanders (1937) noted that *Trichogramma evanescens* preferred a field habitat, whereas *T. embryophagum* was found most frequently in arboreal habitats and *T. semblidis* preferred a marsh habitat. Simmonds (1954) found that *Spalangia drosophilae,* a parasitoid of *Drosophila,* was attracted to dampness and soil level within a grass habitat. *C. nigriceps* readily searches tobacco plants in open-sunlight locations, but tobacco plants removed to forested or heavily shaded locations are not often searched by this parasitoid (Vinson 1975). *Chelonus insularis* (= *texanus*) tends to orient to the lower bushes on which its hosts occur, even though it may be released in the higher shrubs and trees (Ullyett 1953). Weseloh (1972) found that different species of parasitoids that attack the gypsy moth specialize in different regions of the host tree. Although a few examples have been cited, environmental conditions within a habitat may profoundly affect which species of host on a given species of plant will be attacked by a particular species of parasitoid.

The presence of nutrients, particularly the secretion of nectar by plants, may be particularly important in attracting entomophagous insects to a habitat. The presence of floral nectar has been implicated in attracting and retaining parasitoids in a habitat. The parasitoidism of the tarnished plant bug *(Lygus pallipes)* was lower on weedy species of *Oenothera, Daucus, Solidago,* and *Arnaranthus* than *Erigeron* (Streams et al. 1968). Shahjahan (1974) later attributed the difference in parasitoidism observed by Streams et al. (1968) to the presence of nectar in *Erigeron* flowers that attracted and retained the female parasitoids.

The presence of the nectar or pollen sources need not be the principal feature of the plant of the herbivore (host) but may be associated with and add to the attractiveness of the habitat (Simmons et al. 1975, van Emden 1963, Wolcott 1942). Not only may the presence of associate plants in the habitat attract a parasitoid, but such presence may increase the parasitoid's longitivity and fecundity (Leius 1963, 1967, Sundby 1967, Syme 1975, 1977), resulting in increased parasitoidism.

In some cases the associated plants may benefit from the presence of a particular plant species in the habitat. For example, aphids on sugar beets adjacent to collard plants were more frequently parasitized by *Diaeretiella rapae* than aphids on sugar beets when collards were not nearby (Reed et al. 1970). The collards appear to attract the parasitoids, which simply search the nearby sugar beets as well.

Certain associated plants may have an effect opposite to that described above; that is, they may produce masking or repellent chemicals that interfere with host habitat location. Monteith (1960) speculated that the reduced parasitoidism of the larch sawfly *(Pristiphora erichsonii)* by the tachinids *Bessa harveyi* and *Drino bohemica,* when the host sawfly occurred on its host plant *Larix laricina* in association with other trees and understory plants, was due to repellent or masking agents produced by the associated plants. Parasitoidism of *P. erichsonii* was much greater on pure stands of *L. laricina*.

It is difficult to determine whether such differences in parasitoidism are due to the absence of parasitoid attractants, the presence of repellents, or the presence or absence of attractants and repellents associated with other plants in the community. Only through the identification of the stimuli involved in habitat location will these interrelationships be unraveled.

However, Salt (1935) and Zwolfer and Kraus (1957) concluded that the plant does play an important role in the host selection process, most likely because the parasitoid cues on plant volatiles as a means of habitat location. The contention that the evolution of the parasitoid habit in Hymenoptera may stem from a plant parasitic habit (Malyshev 1968) provides support and insight into the possible role of the plant in the host selection process.

The importance of the plant is further supported by observations that there is less tendency for parasitoids to select phylogenetically related hosts than to favor a range of hosts related by a particular habitat or plant (Askew and Shaw 1978). Cushman (1926) found that members of a species of *Sympiesis* that parasitize leaf-mining Lepidoptera will also attack leaf-mining Diptera, Coleoptera, and Hymenoptera. *Scambus (= Epiurus) pterophori* parasitizes Lepidoptera and Coleoptera that occur in plant stems but is known to also parasitize sawfly larvae that enter the stem to pupate (Cushman 1926). The boll weevil parasitoid *Bracon mellitor* will attack various insects found in cotton squares (Cross and Chesnut 1971). Gahan (1933), after studying host selection by *A. glomeratus,* suggested that this parasitoid will attack almost any insect suitably located in the type of plant the parasitoid frequents. Picard and Rabaud (1914) reported that many parasitic Hymenoptera attack larvae of insect species of different families and even different orders, provided they all feed on the same species of food plant. They suggested that the food plant has a profound influence on the host selection process. For example, *I. conquisitor,* which attacks hosts on Scots pine but not red pine, is attracted to the odor of Scots pine but not red pine in an olfactometer (Arthur 1962). Camors and Payne (1971) showed that volatile host tree terpenes, primarily α-pinene, which were released following bark beetle attack, attracted *Heydenia unica,* a parasitoid of the bark beetle. In olfactometric studies, the aphid parasitoid *D. rapae* was attracted to collard leaves, which serve as the food source for their aphid hosts (Reed et al. 1970). Furthermore, Monteith (1955, 1956) reported that in an olfactometer the tachinid *D. bohemica* was attracted to odors from white spruce *(Picea glauca),* the food plant of its host. Monteith (1958) also found that there was a high degree of interaction between the odor of several different food trees and several different species of host in the ability to attract female *D. bohemica. Venturia (= Nemeritis) canescens* was attracted to the odor of oatmeal in which its host *Ephestia* was cultured even if the oatmeal never contained hosts (Thorpe and Jones 1937).

Despite their apparent importance, very little serious effort has been devoted to the isolation and identification of the chemicals from plants that influence the host selection process. Such studies are needed to provide a firmer foundation to an understanding of the parasitoid-host relationship; in addition, they may yield new tools for the manipulation of beneficial insects in plant protection programs. If plants produce factors that are important or necessary for the host selection chain, selective plant breeding could have a profound effect on the host selection process. One of the reasons for trying to increase the host plant's resistance is the desire to breed strains of plants that are less subjected to the pest damage through reduced attractiveness, increased tolerance to the pest, or antibiosis. If such studies yielded plants that reduced the pest population by 50% but at the same time reduced parasitism by 80% through selective breeding, resulting in the loss of a plant attractant, the net effect could be an increase in the pest population via reduction of the parasitoid population. An appreciation of the role of the plant in the host selection process could lead to the opposite situation, that is, breeding plants to increase the factors that attract the parasitoids thus increasing parasitoidism (and predation by predators) and reducing pest damage. Furthermore, the identification of chemicals that attract parasitoids could provide information to plant breeders that would result in the selective breeding of more attractive plants, and the identification and synthesis of these chemicals could offer additional means of beneficial insect manipulation, that is, by application to plants.

A hypothetical example can be developed using the work of Arthur (1962). If the attractant for *I. conquisitor* from Scots pine were isolated and identified, the information could be used to selectively breed a strain of red pine containing the chemical. Alternatively, the chemical could be applied to the trees, resulting in the parasitism of hosts by *I. conquisitor* on red pine where pests have thus far escaped attack by their parasitoid.

Although the host's food plant is important to many parasitoid species in providing cues to the habitat of the host, other factors may play a role in other parasitoid-host relationships. Some entomophagous insects cue on organisms associated with their hosts. The fungus *Monolinia fructicola* produces acetaldehyde in rotting peaches, which attracts the habitual parasitoid *Biosteres*

(Opius) longicaudatus to fruit that may contain tephritid fruit fly larvae (Greany et al. 1977). The parasitoid is attracted to the rotting fruit irrespective of the presence or absence of host larvae.

Many of the parasitoids that attack saprophagous insects appear to be attracted to the habitat by odors associated with carrion. *Alysia manducator* females were found by Laing (1937) to be attracted to *Calliphora* larvae that had been removed from meat but not cleaned off, thus retaining the odor of the carrion; they were also attracted to fresh meat and 2-day meat, even though hosts were not present. These results suggest that odors, regardless of the state of the meat spoilage, are responsible for attracting *A. manducator* to a host habitat. The situation with *Mormoniella (Nasonia) vitripennis* is less clear. Laing (1937), using a two-armed choice chamber with static air, found that *M. vitripennis* was attracted to meat in proportion to its state of spoilage regardless of the presence or absence of hosts. Jacobi (1939), in contrast to Laing (1937), reported that *Mormoniella* was attracted only to meat that had been infested with maggots. Wylie (1958) found that females of *Mormoniella* were equally attracted to both arms in an olfactometer when given a choice between fresh or spoiled meat and clean air. Wylie (1958) also reported that unless dessicated, female *M. vitripennis* chose dry over humid air in an olfactometer. The results suggest that female *Mormoniella* responded by anemotaxis. Wylie's observations were in support of Edwards (1954), who reported that females of *Mormoniella* were positively anemotaxic in an olfactometer with clean air. Edwards (1954) supported Jacobi (1939) rather than Laing (1937) when he reported that *Mormoniella* females became strongly klinokinetic instead of anemotactic when contacting an odor plume from carrion that had contained hosts, but they showed no preference between odor plumes of clean air over spoiled meat. The difference in procedure may explain the differences in results: Laing used static air and steak as an attractant, whereas Edwards and Wylie used the odor of liver in an airstream.

In addition, Jacobi (1939) found that larvae removed and cleaned of carrion and allowed to pupate in clean paper were not attractive to *Mormoniella,* nor were larvae that developed on milk-soaked cotton pads (Edwards 1954). The consensus of the results (see discussions of Saladin 1982) suggested that females of *Mormoniella* are attracted to the odor produced as a result of the interaction of the host maggots and carrion. Other examples demonstrating the importance of the host and plant interactions in orienting parasitoids to a host habitat have been investigated.

Camors and Payne (1973) reported that there was a change in the complex of parasitoids and predators attracted to the tree as bark beetle brood development occurred and as the host tree's physiological state changed in response to the development of bark beetles and their associated organisms. Monteith (1955) reported that the tachinid *D. bohemica* was preferentially attracted to host plants fed on by the host. The aphid parasitoid *D. rapae* was attracted to aphids removed from collard leaves 15 minutes earlier but not to aphids that had been removed from leaves for 24 hours (Reed et al. 1970). Attraction to aphids recently removed from collard leaves may be due to contaminated plant odors rather than to odors produced by an interaction between the aphid and its host plant.

The orientation of a female parasitoid to a habitat by host-induced alterations of plant or food odor, or by plant-induced alterations of host odor may be common and may have an adaptive advantage. Orientation to odors produced by the interactions of a host and its food, rather than odors released by the host food, would ensure that only habitats containing potential hosts would be searched.

Although some parasitoids may orient to a host habitat by odors released by the food of the host, by associated organisms, or by odors produced by an interaction between the host and its food, this may or may not be the situation with all species. If fact Esmaili and Wilde (1971) could not show any involvement of the host plant in orienting or attracting *Aphelinus asychis*, a parasitoid of the greenbug *(Schizaphis graminum)* to the host habitat or host. Ullyett (1953) reported that large numbers of *Pimpla bicolor* would be attracted and cover one's arms and hands if the cocoons of its host *Euproctis terminalis* were opened in a South African pine forest, even though the parasitoids were not in evidence prior to exposure of the host. Although the distinction between host habitat and host location is obscured and may not exist, some parasitoids are attracted or orient from some distance to volatile compounds produced by the host itself. Most of the reports to date indicate that sex pheromones of the host are often involved; however, this may reflect the great amounts of interest in insect sex pheromones rather than having any particular coevolutionary significance. Sternlicht (1973) reported two parasitoids that were responsive to the pheromones of the California red scale *(Aonidiella aurantii)*. Camors and Payne (1973) reported that aggregation pheromones of the southern pine beetle *(Dendroctonus frontalis)* attracted several species of parasitoids. Kennedy (1979) reported that several components of multilure, the aggregation pheromone for the European elm bark beetle *(Scolytus multistriatus)* were attractive to several parasitoids of the bark beetle. Other examples of parasitoids that are attracted to host pheromones have been reported (Rice 1969, Mitchell and Mau 1970, Vité and Williamson 1970), and additions to this list may be expected as more pheromones are identified and field tested.

Prey Habitat Location by Predatory Insects

The information available on various factors involved in the prey selection chain by predators is sparse. In fact, there is little in the literature to suggest that a selection chain analogous to that indicated for parasitoids or phytophagous insects exists. However, predators face problems similar to those of parasitoids in the location of prey habitats and acceptable and suitable prey within the habitat. I would be surprised if a prey selection chain similar to that proposed for parasitoids and phytophagous insects did not exist. In fact, Hagen et al. (1976b) divided the sequence of events leading adult predators to prey into the first four events in Figure 4.1, although they suggest that predacious larvae or nymphs respond to less complex environmental stimuli than do adults in prey finding.

Much of the emphasis of predatory behavior has been directed to sound or sight in ambush or search and attack. For example, sight and movement have been implicated for dragonflies (Corbet et al. 1960, Pritchard 1965), and sound or vibration has been implicated for *Notonecta* (Murphey and Mendenhall 1973) and *Gerris* (Murphey 1971).

Wilbert (1974) concluded that odor over a short distance attracted the aphid predator *Aphidoletes aphidimyza* to its prey. Hölldobler (1969) found that a staphylinid beetle was attracted to the odor of *Myrmica* or *Formica* depending on the beetle's age. However, both these examples, like that of Lewis et al. (1977) for *Chrysopa*, appear to be short-range orient factors; thus they are more in the area of prey location than of prey habitat location.

It is the adult predator capable of moving long distances that must locate the prey's habitat. Hagen et al. (1976b) suggested that this may be particularly true for predators of prey that infest herbaceous annuals, since the predators must disperse or migrate to find suitable habitat. Again, vision may play an important role. Usinger (1963) suggested that dytiscids and notonectids are attracted to aquatic habitats via their positive response to shiny surfaces. *Chrysopa* and syrphid adults are attracted to certain colors of flowers (Ickert 1968, Schneider 1969).

The presence of extrafloral nectaries greatly influences the presence of predators in a habitat (Gilbert 1975, Smiley 1978), particularly ants (Bentley 1977). In fact, Tilman (1978) reported that the extrafloral nectaries of black cherry secrete the majority of secretion when the major herbivore *Malacosoma americana* is available as prey for ants. Whether the presence of nectaries attracts predators is unknown. The presence of nectaries does appear to retain and maintain predator populations. Like phytophagous and parasitoid species, predators are influenced by odors.

Food sprays consisting of artificial honeydew have been known to orient and attract *Chrysopa* adults from downwind (Hagen et al. 1971) and to attract adult *Hippodamia* sp., *Coccinella transversoguttata, Scymnus postpinctus,* and *Geocoris pallens* (Saad and Bishop 1976). A tryptophan product in honeydew was speculated to be the attractive factor to *Chrysopa,* since acidified tryptophan elicited positive anemotaxis in this species (van Emden 1966, Hagen et al. 1976a). Flint et al. (1979) reported that caryophyllene isolated from cotton plants attracted adult *Chrysopa*. These authors also reported that caryophyllene oxide was attractive to the malachiid *Collops vittatus*. The aggregation pheromones of bark beetles and volatile host tree terpenes attract adult bark beetle predators such as trogositids, dolichopodids, clerids, and possibly histerids (Wood et al. 1968, Vité and Williamson 1970, Camors and Payne 1973, Borden 1977). The adult coccinellid *Anatis ocellata* is attracted to the aphid habitat by odors from pine needles rather than the aphids on the pine (Kesten 1969).

Thus, like parasitoid and phytophagous insects, the predators, particularly the hunting species, appear to orient to suitable habitats for prey by odors and electromagnetic stimuli as well as by vibrational stimuli. Such information could be extremely important in the manipulation of predators, as well as parasitoids, in effective pest management strategies.

CONCLUSION

A female insect may emerge at a site removed from potential food or oviposition sites, or she may disperse from an exhausted or a potential site; either behavior necessitates the evolution of an efficient food or ovipositional site location ability. Natural selection minimizes the time and energy spent searching, while maximizing the use of the food or ovipositional resource. Although dispersal from a potential site would appear to be nonproductive, it actually increases the utilization of resources. Females may disperse as a result of the absence of hosts, insufficient host numbers, lack of mature eggs, adverse environmental conditions, frequent encounter with marked hosts, hunger, or encounters with predators or other dangerous situations.

The similarities of the food and oviposition selection and utilization chains among parasitoids, predators, and phytophagous species were discussed. Phytophagous and saprophagous species may locate suitable habitats using chemical and electromagnetic cues; vibrational cues may give an added dimension to the behavior of predators and parasitoids. The role of chemicals in host habitat selection was emphasized for phytophagous, parasitoid, and predatory insects.

ACKNOWLEDGMENTS

Approved as TA 15823 by the Director of the Texas Agricultural Experiment Station. I thank J. R. Ables for his comments and K. Edson for her editorial work.

REFERENCES

Arthur, A.P. 1962. Influence of host tree on abundance of *Itoplectis conquisitor* (Say), a polyphagous parasite of the European pine shoot moth *Rhyacionia buoliana* (Schiff). Can. Entomol. 94:337–347.

Askew, R. R., and M. R. Shaw. 1978. Account of Chalcidoidea parasitizing leaf-mining insects of deciduous trees in Britain. Biol. J. Linn. 6:289–335.

Bentley, B. L. 1977. Extrafloral nectaries and protection by pugnacious bodyguards. Annu. Rev. Ecol. Syst. 8:407–427.

Beroza, M., and M. J. Jacobson. 1963. Chemical insect attractants. World Rev Pest Control 2:36–48.

Borden, J. H. 1974. Aggregation pheromones in the Scolytidae. P. 135–140. *In* M. C. Birch (ed.), Pheromones. American Elsevier, New York.

Borden, J. H. 1977. Behavioral responses of Coleoptera to pheromones, allomones, and kairomones. P. 169–200. *In* H. H. Shorey and J. J. McKelvey, Jr. (eds.), Chemical Control of Insect Behavior: Theory and Application. Wiley, New York.

Callahan, P. S. 1957. Ovipositional response of the imago of the corn earworm, *Heliothis zea* (Boddie), the various wavelengths of light. Ann. Entomol. Soc. Am. 50:444–452.

Cameron, T. W. M. 1956. Parasites and Parasitism. Methuen, London.

Camors, F. B., Jr., and T. L. Payne. 1971. Response in *Heydenia unica* to *Dendroctonus frontalis* pheromones and a host-tree terpene. Ann. Entomol. Soc. Am. 65:31–33.

Camors, F. B., Jr., and T. L. Payne. 1973. Sequence of arrival of entomophagous insects to trees infected with the southern pine beetle. Environ. Entomol. 2:267–270.

Carroll, C. R., and D. H. Jenzen. 1973. Ecology of foraging ants. Annu. Rev. Ecol. Syst. 4:231–257.

Chabora, P. C. 1967. Hereditary behavior variation in oviposition patterns in the parasite, *Nasonia vitripennis*. Can. Entomol. 99:763–765.

Chandler, A. E. F. 1969. Locomotory behavior of first instar larvae of aphidophagous Syrphidae (Diptera) after contact with aphids. Anim. Behav. 17:673–678.

Clausen, C. P., J. L. King, and C. Teranishi. 1927. The parasites of *Papilia japonica* in Japan and Chosen (Korea) and their introduction into the United States. U.S. Department of Agriculture Bulletin No. 1429: 55 pp.

Coakley, J. M., F. G. Maxwell, and J. N. Jenkins. 1969. Influence of feeding, oviposition, and egg and larval development of the boll weevil on abscission of cotton squares. J. Econ. Entomol. 62:244–245.

Corbet, S. A. 1971. Mandibulary gland secretion of larvae of the flour moth, *Anagasta kuehniella*, contains an epideictic pheromone and elicits oviposition movements in a hymenopteran parasite. Nature (London) 232:481–484.

Corbet, P. S., C. Longfield, and N. W. Moore. 1960. Dragonflies. Collins Press, London.

Coulson, R. N. 1979. Population dynamics of bark beetles. Annu. Rev. Entomol. 24:417–447.

Cross, W. H., and T. L. Chesnut. 1971. Arthropod parasites of the boll weevil, *Anthonomus grandis*. I. An annotated list. Ann. Entomol. Soc. Am. 64:516–527.

Croze, H. 1970. Searching Image in Carrion Crows. Parez, Berlin.

Cushman, R. A. 1926. Location of individual hosts versus systematic relation of host-species as a determining factor in parasitic attack. Proc. Entomol. Soc. Washington 28:5–6.

Dahlman, D. L., and S. B. Vinson. 1975. Trehalose and glucose levels in the hemolymph of *Heliothis virescens* parasitized by *Microplitis croceipes* or *Cardiochiles nigriceps*. Comp. Biochem. Physiol. 52B:465–468.

Dethier, V. G. 1966. Feeding behavior. P. 46–50. *In* P. T. Haskel (ed.), Insect Behavior, Symposium No. 3. Royal Entomolgical Society, London.

Dethier, V. G. 1970. Chemical interactions between plants and insects. P. 83–102. *In* E. Sondheimer and J. B. Simeone (eds.), Chemical Ecology. Academic Press, New York.

Doutt, R. L. 1964. Biological characteristics of entomophagous adults. P. 145–167. *In* P. DeBach (ed.), Biological Control of Insect Pests and Weeds. Reinhold, New York.

Edwards, R. L. 1954. The host-finding and oviposition behavior of *Mormoniella vitripennis* (Walker), a parasite of muscoid flies Behavior 7:88–112.

Elton, C. S. 1936. Animal Ecology. Macmillan, New York.

Esmaili, M., and G. Wilde. 1971. Behavior of the parasite *Aphelinus asychis* in relation to the greenbug and certain hosts. Environ. Entomol. 1:266–268.

Evans, H. F. 1976. The searching behavior of *Anthocoris confusus* (Reuter) in relation to prey density and plant surface topography. Ecol. Entomol. 1:163–169.

Fiske, W. F. 1910. Superparasitism: An important factor in the natural control of insects. J. Econ. Entomol. 3:88–97.

Fisher, R. C. 1963. Oxygen requirements and the physiological suppression of supernumerary insect parasitoids. J. Exp. Biol. 40:531–540.

Fisher, R. C. 1971. Aspects of the physiology of endoparasitic Hymenoptera. Biol. Rev. 46:243–278.

Flanders, S. E. 1937. Habitat selection by *Trichogramma*. Ann. Entomol. Soc. Am. 30:208–210.

Flanders, S. E. 1950. Regulation of ovulation and egg disposal in the arasitic Hymenoptera. Can. Entomol. 82:134–140.

Flanders, S. E. 1951. Mass culture of California red scale and its golden chalcid parasites. Hilgardia 21:1–42.

Flanders, S. E. 1953. Variations in susceptibility of citrus-infesting coccids to parasitization. J. Econ. Entomol. 46:266–269.

Flint, H. M., S. S. Salter, and S. Walters. 1979. Caryophyllene: An attractant for the green lacewing *Chrysopa carnea* Stephens. Environ. Entomol. 8:1123–1125.

Gahan, A. B. 1933. The serphoid and chalcidoid parasites of the hessian fly. U.S. Department of Agriculture, Miscellaneous Publication No. 174:1–147.

Geier, P. 1963. The life history of codling moth *Cydia pomonella* (L.) in the Australian Capital Territory. Aust. J. Zool. 11:323–367.

Gilbert, L. E. 1975. Ecological consequences of a coevolved mutualism between butterflies and plants. P. 210–240. *In* L. E. Gilbert and P. H. Raven (eds.), Coevolution of Animals and Plants. University of Texas Press, Austin.

Gill, F. B., and L. L. Wolf. 1977. Nonrandom foraging by sunbirds in a patchy environment. Ecology 58:1284–1296.

Gerling, D., and A. Schwartz. 1974. Host selection by *Telenomus remus,* a parasite of *Spodoptera littoralis* eggs. Entomol. Exp. Appl. 17:391–396.

Greany, P. D., J. H. Tumlinson, D. L. Chambers, and G. M. Bousch. 1977. Chemically mediated host finding by *Biosteres (Opius) longicaudatus,* a parasitoid of tephritid fruit fly larvae. J. Chem. Ecol. 3:189–195.

Green, N., M. Beroza, and S. A. Hall. 1960. Recent developments in chemical attractants for insects. Adv. Pest Control Res. 3:129–179.

Guillot, F. S., and S. B. Vinson. 1972. Sources of substances which elicit a behavioral response from the insect parasitoid *Campoletis perdistinctus*. Nature (London) 235:169–170.

Hagen, K. S., E. F. Sawall, Jr., and R. L. Tassan. 1971. The use of food sprays to increase effectiveness of entomophagous insects. Proc. Tall Timbers Conf. Ecological Animal Control by Habitat Management. 2:59–81.

Hagen, K. S., P. Greany, E. F. Sawall, Jr., and R. L. Tassan. 1976a. Trytophan in artificial honeydews as a source of an attractant for adult *Chrysopa carnea*. Environ. Entomol. 5:458–468.

Hagen, K. S., S. Bombasch, and J. A. McMurty. 1976b. The biology and impact of predators. P. 93–142. *In* C. B. Huffaker and P. S. Messenger (eds.), Theory and Practice of Biological Control. Academic Press, New York.

Hamamura, Y., and K.-I. Naito. 1961. Food selection by silkworm larvae, *Bombyx mori*. Citral, linalyl acetate, linalol and terpinyl acetate as attractants of larvae. Nature (London) 190:879–880.

Hamamura, Y., K. Hayashiya, K.-I. Naito, K. Matsura, and J. Nishida. 1962. Food selection by silkworm larvae. Nature (London) 194:754–755.

Haskell, P. T., M. W. J. Paskin, and J. E. Moorehouse. 1962. Laboratory observations on factors affecting the movements of hoppers of the desert locust. J. Insect Physiol. 8:53–78.

Hedin, P. A., A. C. Thompson, and R. C. Gueldner. 1973. The boll weevil-cotton plant complex. P. 291–351. Toxicological and Environmental Chemistry Reviews, Vol. 1. Gordon and Breach, Science Pub., N.Y.

Hedin, P. A., J. N. Jenkins, and F. G. Maxwell. 1977. Behavioral and developmental factors affecting host plant resistance to insects. P. 231–275. *In* P. A. Hedin (ed.), Host Plant Resistance to Pests, ACS. American Chemical Society Symposium Series No. 62. Washington, D.C.

Heinrich, B. 1976. Bumblebee foraging and the economics of sociality. Am. Sci. 64:384–395.

Hierholzer, O. 1950. Ein Beitrag zur Frage der Orientierung von *Ips curvidens* Germ. Z. Tierpsychol. 7:588–620.

Hölldobler, B. 1969. Host finding of odor in the myrmecophilous beetle *Atemeles pubicollis* Bris. (Coleoptera: Staphylinidae). Science 166:757–781.

Horber, E. 1955. Oviposition preference of *Meromyza americana* Fitch for different small grain varieties under greenhouse conditions. J. Econ. Entomol. 48:426–430.

Hovanitz, W. 1959. Insects and plant galls. Sci. Am. 201:151–162.

Hundertmark, A. 1937. Das Formen unterscheidungsvermögen der Eiraupe der Nonne *(Lymantria monacha* L.*)* Z. Vgl. Physiol. 24:563–592.

Ickert, G. 1968. Beitrage zur Biologie einkeimischer Chrysopiden. Entomol. Abh. 36:132–192.

Ilse, D. 1937. New observations on responses to colours in egg laying butterflies. Nature (London) 140:544–545.

Jacobi, E. F. 1939. Über Lebensweise, auffinden des Wirtes und Regulierung des Individuenzahl von *Mormoniella vitripennis* Walker. Arch. Neer. Zool. Leiden. 3:197–282.

Jacobson, M. 1966. Chemical insect attractants and repellents. Annu. Rev. Entomol. 11:403–422.

Jermy, T. (Ed.) 1976. The Host-Plant in Relation to Insect Behavior and Reproduction. Akademiai Kiado, Budapest.

Juillet, J. A. 1964. Influences of weather on flight activity of parasitic Hymenoptera. Can. J. Zool. 42:1133–1141.

Kamm, J. A. and W. D. Fronk. 1964. Olfactory response of the alfalfa-seed chalcid, *Bruchophagus roddi* Gus., to chemicals in alfalfa. University of Wyoming Agriculture Experiment Station Bulletin No. 413. 36 pp.

Keller, J. C., F. G. Maxwell, J. N. Jenkins, and T. B. Davich, 1963, A boll weevil attractant from cotton. J. Econ. Entomol. 56:110–111.

Kellogg, F. E., D. E. Frizel, and R. H. Wright. 1962. The olfactory guidance of flying insects. IV. *Drosophila*. Can. Entomol. 94:884–888.

Kennedy, B. H. 1979. The effect of multilure on parasites of the European elm bark beetle, *Scolytus multistriatus*. Bull. Entomol. Soc. Am. 25:116–118.

Kennedy, J. S. 1965. Mechanisms of host plant selection. Ann. Appl. Biol. 56:317–322.

Kennedy, J. S. 1977. Olfactory responses to distant plants and other odors. P. 67–91. *In* H. H. Shorey and J. J. McKelvey, Jr. (eds.), Chemical Control of Insect Behavior: Theory and Application. Wiley, New York.

Kennedy, J. S., C. O. Booth, and W. J. S. Kertshaw. 1961. Host finding by aphids in the field. III. Visual attraction. Ann. Appl. Biol. 49:1–24.

Kesten, V. 1969. Zur Morphologie und Biologie von *Anatis ocellata* (L.). (Coleoptera: Coccinellidae). Z. Angew. Entomol. 63:412–445.

King, P. E., and J. Rafai. 1970. Host-discrimination in a gregarious parasitoid, *Nasonia vitripennia* (Walker). J. Exp. Biol. 43:245–254.

Kogan, M. 1975. Plant resistance in pest management. P. 103–146. *In* R. L. Metcalf and W. H. Luckman (Eds.), Introduction to Insect Pest Management. Wiley, New York.

Kogan, M. 1977. The role of chemical factors in insect/plant relationships. Proc. Int. Congr. Entomol. Washington. 1976:211–227.

Kullenberg, B. 1961. Studies in *Ophrys* pollination. Zool. Bidr. Upps. 34:1–340.

Laing, J. 1937. Host-finding by insect parasites. I. Observations the finding of host by *Alysia manducator, Mormoniella vitripennis,* and *Trichogramma evanescens*. J. Anim. Ecol. 6:298–317.

Leius, K. 1963. Effects of pollens on fecundity and longevity of adult *Scambus buotianae* (Htg.) (Hymenoptera: Ichneumonidae). Can. Entomol. 95:202–207.

Leius, K. 1967. Influence of wild flowers on parasitism of tent caterpillar and codling moth. Can. Entomol. 99:444–446.

Lewis, W. J., and J. W. Snow. 1971. Fecundity, sex ratios, and egg distribution by *Microplitis croceipes,* a parasite of *Heliothis*. J. Econ. Entomol. 64:68.

Lewis, W. J., R. L. Jones, D. A. Nordlund, and A. N. Sparks. 1975a. Kairomones and their use for management of entomophagous insects. I. Evaluation for increasing rate of parasitization by *Trichogramma* spp. in the field. J. Chem. Ecol. 1:343–347.

Lewis, W. J., R. L. Jones, D. A. Nordlund, and H. R. Gross, Jr. 1975b. Kairomones and their use for management of entomophagous insects. II. Mechanisms causing increase in rate of parasitization by *Trichogramma* spp. J. Chem. Ecol. 1:349–360.

Lewis, W. J., R. L. Jones, H. R. Gross, Jr., and D. A. Nordlund. 1976. The role of kairomones and other behavioral chemicals in host finding by parasitic insects. Behav. Biol. 16:267–289.

Lewis, W. J., D. A. Nordlund, H. R. Gross, Jr., R. L. Jones, and S. L. Jones. 1977. Kairomones and their use for management of entomophagous insects. V. Moth scales as a stimulus for predation of *Heliothis zea* (Boddie) eggs by *Chrysopa carnea* Stephens larvae. J. Chem. Ecol. 3:483–487.

Madden, J. J. 1977. Physiological reactions of *Pinus radiata* to attack by woodwasp, *Sirex noctilio* F. Bull. Entomol. Res. 67:405–426.

Malyshev, S. I. 1968. Genesis of the Hymenoptera and the Phases of Their Evaluation. O. W. Richards and B. Uvarov (eds.). Translated (from Russian) by the National Lending Library for Science and Technology. Methuen, London.

Markl, H. 1974. Insect Behavior: Functions and mechanisms. P. 3–148. *In* M. Rockstein (ed.), The Physiology of Insects, Vol. 3. Academic Press, New York.

Marks, R. J. 1977. Laboratory studies of plant searching behavior by *Coccinella septempunctata* L. larvae. Bull. Entomol. Res. 67:235–241.

McKibben, G. H., E. B. Mitchell, W. P. Scott, and P. A. Hedin. 1977. Boll weevils are attracted to volatile oils from cotton plants. Environ. Entomol. 6:804–806.

Miller, C. A. 1959. The interaction of the spruce budworm, *Choristoneura fumiferana*, and the parasite *Apenteles fumiferanae* Vier. Can. Entomol. 91:457–477.

Miles, P. W. 1968. Insect secretions in plants. Annu. Rev. Entomol. 6:137–164.

Mitchell, W. C., and F. R. L. Mau. 1970. Response of the female southern green stink bug and its parasite, *Trichopoda pennipes*, to male stink bug pheromones. J. Econ. Entomol. 64:856–859.

Moericke, V. 1962. Über die optische Orientierung von Blattlausen. Z. Angew. Entomol. 50:70–74.

Moericke, V. 1969. Host plant specific color behavior by *Hyalopterus pruni* (Aphididae). Entomol. Exp. Appl. 12:524–534.

Monteith, L. G. 1955. Host preferences of *Drino bohemica* Mesn. with particular reference to olfactory responses. Can. Entomol. 87:509–530.

Monteith, L. G. 1956. Influence of host movement on selection of host by *Drino bohemica* Mesn. as determined in an olfactometer. Can. Entomol. 88:583–586.

Monteith, L. G. 1958. Influence of host and its food plant on host-finding by *Drino bohemica* Mesn. and interaction of other factors. Proc. 10th Int. Congr. Entomol. 2 (1956):603–606.

Monteith, L. G. 1960. Influence of plants other than the food plants of their host on host-finding by tachinid parasites. Can. Entomol. 92:641–652.

Moore, R. H. 1928. Odorous constituents of the corn plant in their relation to the European corn borer. Proc. Oklahoma Acad. Sci. 8:16–18.

Mulkern, G. B. 1969. Behavioral influences on food selection in grasshoppers. Entomol. Exp. Appl. 12:509–523.

Muller, H. J. 1958. The behavior of *Aphis fabae* in selecting its host plants, especially different varieties of *Vicia fabae*. Entomol. Exp. Appl. 1:66–72.

Munakata, K., T. Saito, S. Ogawa, and S. Ishii. 1959. Oryzanone, an attractant of the rice stem borer. Bull. Agr. Chem. Soc. Jpn. 23:64–65.

Murphey, R. K. 1971. Sensory aspects of the control of orientation to prey by the water strider *Gerris semigis*. Z. Vgl. Physiol. 72:168–185.

Murphey, R. K., and B. Mendenhall. 1973. Localization of receptors controlling orientation to prey by the back swimmer *Notonecta undulata*. J. Comp. Physiol. 84:19–30.

Nayar, J. K., and G. Fraenkel. 1962. The chemical basis of host plant selection in the silkworm, *bombyx mori* (L.). J. Insect Physiol. 8:505–525.

Neff, D. L., and E. S. Vanderzant. 1963. Methods of evaluating the chemotropic response of boll weevils to extracts of the cotton plant and various other substances. J. Econ. Entomol. 56:761–766.

Neunzig, H. H. 1969. The biology of the tobacco and the corn earworm in North Carolina. North Carolina Agriculture Experiment Station Technical Bulletin No. 196.

Nishida, T. 1956. An experimental study of the ovipositional behavior of *Opius fletcheri* Silvestri, a parasite of the melon fly. Proc. Hawaii Entomol. Soc. 16:126–134.

Picard, F., and E. Rabaud. 1914. Sur le parasitisme externe des Braconides. Bull. Entomol. Soc. Fr. 1914:266–269.

Price, P. W. 1970. Trail odors: Recognition by insects parasitic on cocoons. Science 179:546–547.

Pritchard, G. 1965. Prey capture in dragonfly larvae (Odonata: Anisoptera). Can. J. Zool. 43:271–290.

Prokopy, R. J. 1968. Visual responses on apple maggot flies, *Rhagoletis pomonella* (Diptera: Tephritidae): Orchard studies. Entomol. Exp. Appl. 11:403–422.

Prokopy, R. J. 1972. Responses of apple maggot flies to rectangles of different colors and shades. Environ. Entomol. 1:720–726.

Prokopy, R. J. 1975. Apple maggot control by sticky red spheres. J. Econ. Entomol. 68:197–198.

Prokopy, R. J., and E. F. Boller. 1971. Stimuli eliciting oviposition of European cherry fruit flies *Rhagoletis cerasi* (Diptera: Tephritidae), into inanimate objects. Entmol. Exp. Appl. 14:1–14.

Prokopy, R. J., and G. L. Bush. 1973. Ovipositional responses to different sizes of artificial fruit flies of *Rhagoletis pomonella* species group. Ann. Entomol. Soc. Am. 66:927–930.

Prokopy, R. J., and G. Haniotakis. 1975. Responses of wild and laboratry-cultured *Dacus oleae* to host plant color. Ann. Entomol. Soc. Am. 68:73–77.

Prokopy, R. J., V. Moericke, and G. L. Bush. 1973. Attraction of apple maggot flies to odor of apples. Environ. Entomol. 2:743–750.

Pyke, G., H. R. Pulliam, and E. L. Charnov. 1977. Optimal foraging: A selective review of theory and tests. Q. Rev. Biol. 52:137–154.

Rabb, R. L., and J. R. Bradley. 1970. Marking host eggs by *Teleomus sphingis*. Ann. Entomol. Soc. Am. 63:1053–1056.

Reed, D. P., P. P. Feeny, and R. B. Root. 1970. Habitat selection by the aphid parasite *Diaeretiella rapae* and hyperparasite, *Charips brassicae*. Can. Entomol. 102:1567–1578.

Rice, R. E. 1969. Response of some predators and parasites of *Ips confusus* (LeC.) to olfactory attractants. Contrib. Boyce Thompson Inst. 23:189–194.

Roth, J. P. 1976. Host habitat location and larviposition stimulation response of the tachinid, *Lixophaga diatraeae* (Townsend). Dissertation, Mississippi State University, Mississippi State.

Röttger, V. V. 1979. Untersuchungen zur Wirtswahl der Ruenfliege *Pegomya batae* Curt. II Optische Orientierung zur Wirtspflanze. Z. Angeu. Entomol. 88:97–107.

Rowlands, M. L. J., and J. W. Chapin. 1978. Prey searching behavior in adults of *Hippodamia convergens*. J. Georgia Entomol. Soc. 13:309–315.

Rudinsky, J. A. 1966. Scolytid beetles associated with Douglas-fir: Response to terpenes. Science 152:218–219.

Saad, A. A. B., and G. W. Bishop. 1976. Attraction of insects to potato plants through use of artificial honeydews and aphid juice. Entomophaga. 21:49–57.

Saladin, K. 1982. Host-finding by parasite animals. *In* M. Rechcigl, Jr. (ed.), Handbook of Nutrition and Food. CRC Press, West Palm Beach, FL., (in press).

Salt, G. 1935. Experimental studies in insect parasitism. III. Host selection. Proc. R. Soc. London, Ser. B. 117:413–435.

Salt, G. 1958. Parasite behavior and the control of insect pests. Endeavor 65:145–148.

Salt, G. 1970. The Cellular Defense Reactions of Insects. Cambridge University Press, Cambridge.

Schneider, F. 1969. Bionomics and physiology of aphidophagous Syrphidae. Annu. Rev. Entomol. 14:103–124.

Schoener, T. W. 1971. Theory of feeding strategies. Annu. Rev. Ecol. Syst. 2:369–404.

Schoonhoven, L. M. 1968. Chemistry basis of host plant selection. Annu. Rev. Entomol. 13:115–136.

Schoonhoven, L. M. 1972. Some aspects of host selection and feeding in phytophagous insects. P. 557–566. *In* J. C. Rodriguez (ed.), Insect and Mite Nutrition: Significance and Implications in Ecology and Pest Management. North-Holland, Amsterdam.

Shahjahan, M. 1974. *Erigeron* flowers as a food and attractive odor for *Peristenus pseudopallipes*, a braconid parasitoid of the tarnished plant bug. Environ. Entomol. 3:69–72.

Shepard, M., and W. Sterling. 1972. Incidence of parasitism of *Heliothis* spp. in some cotton fields of Texas. Ann. Entomol. Soc. Am. 65:759–760.

Simmonds, F. S. 1954. Host finding and selection by *Spalangia drosophilae* Ashm. Bull. Entomol. Res. 45:527–537.

Simmons, G. A., D. E. Leonard, and C. W. Chen. 1975. Influence of tree species density and composition of parasitism of the spruce budworm, *Choristoneura fumiferana* (Clem.) Environ. Entomol. 4:832–836.

Smiley, J. 1978. Plant chemistry and the evolution of host specificity evidence from *Heliconius* and *Passiflora*. Science 201:745–747.

Smith, H. S. 1916. An attempt to redefine the host relationships exhibited by entomophagous insects. J. Econ. Entomol. 9:477–486.

Smith, J. N. 1974. The food searching behavior of two European thrashers. II. The adaptiveness of the search patterns. Behavior 47:1–61.

Snow, J. W., J. J. Hamm, and J. R. Brazzel. 1966. *Geranium carolinianum* as an early host for *Heliothis zea* and *H. virescens* in the Southeastern United States with notes on associated parasites. Ann. Entomol. Soc. Am. 59:506–509.

Southwood, T. R. E. 1966. Ecological Methods. *Methuen, London.*

Stadler, E. 1977. Sensory aspects of insect plant interaction. Proc. Int. Cong. Entomol. Washington. 1976:228–248.

Stary, P. 1966. Aphid Parasites of Czechoslovakia. A review of the Czechoslovak Aphidiidae (*Hymenoptera*). Junk, The Hague.

Steiner, L. F. 1939. Distances traveled by newly hatched codling moth larvae. J. Econ. Entomol. 32:470.

Sternlicht, M. 1973. Parasitic wasps attracted by the sex pheromone of the coccid host. Entomophaga 18:339–342.

Stoltz, D. B., and S. B. Vinson. 1979. Viruses and parasitism in insects. Adv. Virus Res. 24:125–171.

Streams, F. A., M. Shahiahan, and H. G. LeMesuries. 1968. Influence of plants on the parasitization of the tarnished plant bug. J. Econ. Entomol. 61:996–999.

Sundby, R. A. 1967. Influence of food on the fecundity of *Chrysopa carnea* Stephens (Neuroptera: Chrypopidae). Entomophaga 12:475–479.

Sutherland, O. R. W. 1972. The attraction of newly hatched codling moth (*Laspeyresia pomonella*) larvae to apple. Entomol. Exp. Appl. 15:481–487.

Syme, P. D. 1975. The effects of flowers on the longevity and fecundity of two native parasites of the European pine shoot moth in Ontario. Environ. Entomol. 4:337–346.

Syme, P. D. 1977. Observations on the longevity and fecundity of *Orgilus obscurator* (Hymenoptera: Braconidae) and the effects of certain foods on longevity. Can. Entomol. 109:995–1000.

Taylor, J. S. 1932. Report on Cotton Insect and Disease Investigation. II. Notes on the American Bollworm (*Heliothis obsoleta* F.) on Cotton and Its Parasite (*Microbracon brevicornis* Wesm.). Sci. Bull. Rep. Agric. For. Union S. Afr. 113.

Taylor, R. L. 1974. Role of learning in insect parasitism. Ecol. Manag. 44:89–104.

Thorpe, W. H., and Jones, F. G. W. 1937. Olfactory conditioning in a parasitic insect and its relation to the problem of host selection. Proc. R. Entomol. Soc. Ser. B 124:56–81.

Thorpe, W. H., and H. B. Caudle. 1938. A study of the olfactory responses of insect parasites to the food plant of their host. Parasitology 30:523–528.

Thorsteinson, A. J. 1960. Host selection in phytophagous insects. Annu. Rev. Entomol. 5:193–218.

Tilman, D. 1978. Cherries, ants, and tent caterpillars: Timing of nectar production in relation to susceptibility of caterpillars to ant predation. Ecology 59:686–692.

Ullyet, G. C. 1953. Biomathematics and insect population problems. A critical review. Mem. Entomol. Soc. S. Afr. 2:1–89.

Usinger, R. (ed.) 1963. Aquatic Insects of California. University of California Press, Berkeley.

van Emden, H. F. 1963. Observations on the effect of flowers on the activity of parasitic Hymenoptera. Entomol. Mon. Mag. 98:265–270.

van Emden, H. F. 1966. The effectiveness of aphidophagous insects in reducing aphid populations. P. 227–325. *In* I. Hodek (ed.), Ecology of Aphidophagous Insects. Academia, Prague.

van Lenteren, J. C. 1976. The development of host discrimination and the prevention of superparasitism in the parasite *Pseudeucoila bochei* Weld. Neth. J. Zool. 26:1–83.

van Lenteren, J. C., and K. Bakker. 1978. Behavioral aspects of the functional responses of a parasite (*Pseudeucoila bochei* Weld) to its host (*Drosophila melanogaster*). Neth. J. Zool. 28:213–233.

van Lenteren, J. C., K. Bakker, and J. J. M. van Alphen. 1978. How to analyse host discrimination? Ecol. Entomol. 3:71–75.

Verschoor–van der Poel, P. J. G., and J. C. van Lenteren. 1978. Host-plant selection by the greenhouse whitefly *Trialeurodes vaporariorum* (Westwood). Meded. Fac. Landbouwwet. Rijksuniv. Gent. 43:387–396.

Vinson, S. B. 1972a. Competition and host discrimination between two species of tobacco budworm parasitoids. Ann. Entomol. Soc. Am. 65:229–236.

Vinson, S. B. 1972b. Effect of the parasitoid, *Campoletis sonorensis* on the growth of its host, *Heliothis virescens*. J. Insect Physiol. 18:1509–1514.

Vinson, S. B. 1975. Biochemical coevolution between parasitoids and their hosts. P. 14–48. *In* P. Price (ed.), Evolutionary Strategies of Parasitic Insects and Mites. Plenum Press, New York.

Vinson, S. B. 1976. Host selection by insect parasitoids. Annu. Rev. Entomol. 21:109–133.

Vinson, S. B. 1977a. Insect host responses against parasitoids and the parasitoid's resistance with emphasis on the Lepidoptera-Hymenoptera association. Comp. Pathobiol. 3:103–125.

Vinson, S. B. 1977b. Behavioral chemicals in the augmentation of natural enemies. P. 237–279. *In* R. L. Ridgway and S. B. Vinson (eds.), Biological Control by Augmentation of Natural Enemies. Plenum Press, New York.

Vinson, S. B. 1982. Parasitoidism. *In* M. Rechcigl, Jr. (ed.), Handbook of Nutrition and Food. CRC Press, West Palm Beach, FL., (in press).

Vinson, S. B., and F. S. Guillot. 1972. Host marking: Source of a substance that results in host discrimination in insect parasitoids. Entomophaga 17:241–245.

Vinson, S. B., and G. F. Iwantsch. 1980a. Host suitability for insect parasitoids. Annu. Rev. Entomol. 25:397–419.

Vinson, S. B., and G. F. Iwantsch. 1980b. Host regulation by insect parasitoids. Q. Rev. Biol. 55:143–165.

Vité, J. P., and D. L. Williamson. 1970. *Thanasimus dubius:* Prey perception. J. Insect Physiol. 16:233–237.

Waage, J. K. 1978. Arrestment responses of a parasitoid, *Nemeritis canescens,* to a contact chemical produced by its host, *Plodia interpunctella*. Physiol. Entomol. 3:135–146.

Weseloh, R. M. 1972. Spatial distribution of the Gypsy moth and some of its parasitoids within a forest environment. Entomophaga 17:339–351.

Wilbert, H. 1974. Die Wahrnehmung von Beute durch die Eilarven von *Aphidoletes aphidimyza* (Cecidomyiidae). Entomophaga 19:173–181.

de Wilde, J. 1947. De Koolvlieg en zinjbestrijding. Meded. Tuinbouw Voordr. Denst. 45:1–70.

de Wilde, J., K. Hille Ris Lambers-Suverkropp, and A. van Tol. 1969. Responses to air flow and airborne plant odor in the Colorado beetle. Neth. J. Plant Pathol. 75:53–57.

Williams, L. H. 1954. The feeding habits and food preferences of Acrididae and the factors that determine them. Trans. R. Entomol. Soc. London. 105:423–454.

Wolcott, G. N. 1942. The requirements of parasites for more than hosts. Science 96:317–318.

Wood, D. L., L. E. Brown, W. D. Bedard, P. E. Tilden, R. M. Silverstein, and J. O. Rodin. 1968. Response of *Ips confusus* to synthetic sex pheromones in nature. Science 59:1373–1374.

Wylie, H. G. 1958. Factors that affect host finding by *Nasonia vitripennis* (Walk.). Can. Entomol. 90:597–608.

Wylie, H. G. 1971. Oviposition restraint of *Muscidifurax zaraptor* on parasitized housefly pupae. Can. Entomol. 103:1537–1544.

Yamamoto, R. T., and G. S. Fraenkel. 1960. The specificity of the tobacco hornworm *Protoparce sexta* (Johan.), to solanaceous plants. Ann. Entomol. Soc. Am. 53:503–507.

Yasunaga, K., Y. Oshima, and Y. Kinoshita. 1963. Studies on attractants and synergism of benzoic acid derivatives, higher fatty acids and their esters, and terpenoids. J. Agric. Chem. Soc. Jpn. 37:642–644.

Yeo, P. F. 1972. Floral allurements for pollinating insects. P. 51–57. *In* H. F. van Emden (ed.), Insect/Plant Relationships. Blackwell Scientific Publications, Oxford.

Zwolfer, H., and M. Kraus. 1957. Biocoenotic studies on the parasites of two fir- and two oak-tortricids. Entomophaga 2:173–196.

CHAPTER FIVE

HOST LOCATION BY PARASITOIDS

RONALD M. WESELOH

Connecticut Agricultural Experiment Station
New Haven, Connecticut

After a parasitoid has found a suitable habitat (Chapter 4) and before it accepts or rejects a host (Chapter 6), that host must be located. This is so critically important for parasitoids that many have evolved mechanisms that enable them to detect and orient to hosts from a distance. This ability, called host location or finding, is defined as the perception and orientation by parasitoids to their hosts, from a distance, by responses to stimuli produced or induced by the host or its products. The important components of this definition are as follows: the stimuli must be associated with the presence of the host or with its secretions or excretions, and the parasitoid must perceive such stimuli at a distance from the host.

Like most definitions, this one is not perfect. Whether a given mechanism can be categorized under "host location" may depend on one's point of view. For instance, the tachinid parasitoid *Cyzenis albicans* lays its microtype eggs just as readily on artificially damaged leaves as on leaves partially chewed by its host, the winter moth (*Operophtera brunata*). The oviposition response is due to sugars in sap released along the chewed or cut edges of the leaves (Hassel 1968). This behavior does not really fit the definition given above because the sap flows may or may not be induced by the host. In fact, the chewing damage of just about any organism will cause such flows. However, in any area where the winter moth is abundant, the host can be effectively located because there the sap flows often are caused by its chewing activities.

Consider also the braconid *Apanteles melanoscelus,* a parasitoid of the gypsy moth (*Lymantria dispar*). It intensely examines a substrate covered with gypsy moth silk because a water-soluble kairomone is present on the silk (Figure 5.1). This aids the parasitoid in locating hosts (Weseloh 1977). Responses are much weaker to the silk of other lepidopterous species. One such species, the white-marked tussock moth (*Hemerocampa leucostigma*), may even be successfully

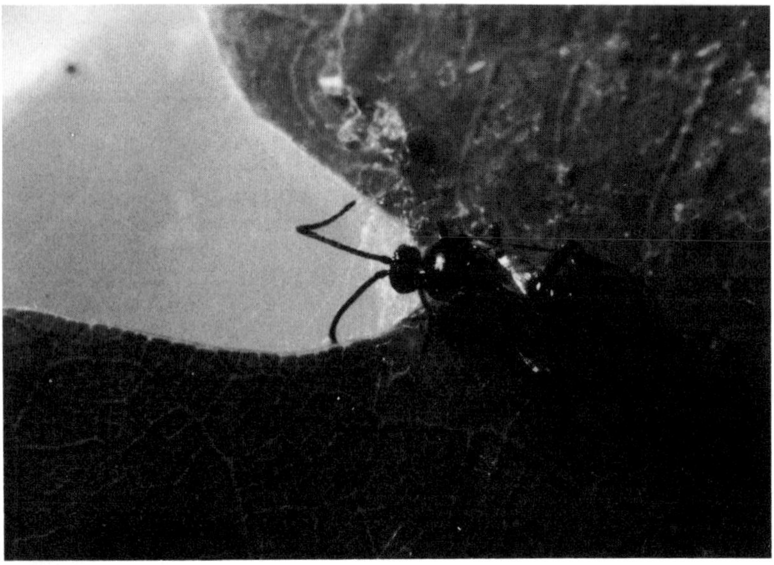

FIGURE 5.1. An *Apanteles melanoscelus* female examining gypsy moth silk on an oak leaf.

parasitized by *A. melanoscelus* under appropriate laboratory or field conditions (Schaffner 1934, Weseloh 1976). Because of the specificity of *A. melanoscelus* for the gypsy moth silk kairomone, hosts may be effectively accepted or rejected before they have even been contacted.

These examples are not given to illustrate the difficulties of defining the process of host location, but to suggest that some aspects of this phenomenon may be justifiably considered in other contexts. My intent is to discuss the mechanisms that are involved in host location, regardless of whether they could also be appropriately considered under other categories.

It should also be kept in mind that host finding is a complex process, with its own hierarchy of behaviors. In a single species of parasitoid it can involve long-distance orientation via chemicals, sound, or light; short-range intensive searching of host traces; and direct detection of hosts through integumentary chemicals or physical characteristics. A good example is *Ventura* (*Nemeritis*) *canescens*. This ichneumonid is attracted from a distance by the odor of its host (i.e. *Ephestia* flour moths) (Thorpe and Jones 1937, Williams 1951). It intensely examines small areas of flour contaminated with host mandibular gland secretions (Corbet 1971, 1973; Mudd and Corbet 1973), and probes the flour with its ovipositor until a host is contacted. These different behaviors may be mediated by different chemical or physical stimuli in different ways for each parasitoid. Furthermore, the same stimulus may have different functions depending on the context. A number of parasitoids have been shown to respond specifically to the sex pheromone of their host. The mandibular gland secretions that

Ventura examines so intensely are used by its host, *Ephestia,* as a dispersal (epideictic) pheromone (Chapter 10). The function assigned to the chemical then depends really on what particular part of the interaction is emphasized.

This presentation is primarily a survey of parasitoid host location processes that involve semiochemicals. However, the discussion also touches on processes involving physical stimuli, the influence of conditioning and learning on parasitoid responses, and the orientation responses of parasitoids due to host semiochemicals.

HOST LOCATION INVOLVING PHYSICAL STIMULI

Although this book is primarily concerned with semiochemicals, a brief discussion on nonchemical means of host location is appropriate for completeness.

Relatively few definitive studies involving physical stimuli have been done. Substrate vibrations or sound have occasionally been evoked to explain how parasitoids find concealed hosts (Deleon 1935, Ryan and Rudinsky 1962, van den Assem and Keunen 1958). Host-produced sound is used by *Euphasiopteryx ochracea* (Tachinidae) to find crickets (Cade 1975, Mangold 1978), and by *Colcondamyia auditrix* (Sacrophagidae) to find cicadas (Soper et al. 1976). In both cases parasitoids were attracted by tape recordings of male host songs, and male hosts were parasitized more often than females.

Vision, as a component of host location, has occasionally been demonstrated, especially in conjunction with host movement (Herrebout 1969, Monteith 1956, Walker 1961, Richerson and DeLoach 1972).

Possible use of infrared radiation in host finding has been reported by Richerson and Borden (1972). They found that *Coeloides brunneri,* a parasitoid of the bark beetle *Dendroctonus pseudotsugae,* attempts to oviposit at points on the bark of logs that have been heated about 2°C higher than the surroundings by resistance wires buried in the bark. Evidently this parasitoid can find its host by means of host activities that the same investigators found to result in local bark temperature increases of 1°C. However, whether the heat is actually perceived because of radiation, convection, or conduction was not determined.

Other physical mechanisms of host location that could be postulated would involve magnetic or electric effects, or perhaps a form of sonar, but as far as I know these have not been demonstrated.

HOST LOCATION INVOLVING CHEMICAL STIMULI

The mechanisms whereby parasitoids use kairomones to find hosts can most conveniently be divided into two categories: long-range and close-range chemoreception. Long-range chemoreception can be equated with olfaction, which is the detection of chemicals in air. It is simply the sense of smell.

Close-range chemoreception involves the perception of chemicals only after direct physical contact with them in solid or liquid form. It functions in host location if the chemicals perceived are in a host or host-induced product removed from but still associated with the host.

Both long- and close-range chemoreception are important in host location. They are arbitrary terms, since even rather nonvolatile chemicals might vaporize enough to be detectable via olfaction over very short distances. This would be very difficult to demonstrate experimentally, however, and so realistically they are best separated by the methods used to study them. Thus we say that long-range chemoreception is involved if it can be shown that an organism responds behaviorally to an obviously airborne chemical. The behavior most often observed is attraction, and olfactometers in the laboratory and baited traps in the field are usually used to demonstrate this phenomenon. Close-range chemoreception is implicated if behavioral changes in an organism are observed when it is seen to contact a substrate on which a chemical has been placed. Typically the behavior observed can be described as "examination behavior." It usually involves substrate sweeping with the antennae, increased turning movements, and a change in locomotor rate by the parasitoid. The behavioral changes are often quite evident and distinct, making the study of close-range chemoreception in parasitoids relatively easy.

The functions of long- and close-range chemoreception are generally distinct as well. Long-range chemoreception enables a parasitoid to localize a host in a large area by attracting it to the host's near vicinity. Short-range chemoreception is important in intensive searches over small areas that eventually result in host contact.

Long-Range Chemoreception

Survey of Examples

A number of investigators have studied parasitoids that locate hosts by means of airborne chemicals (Table 5.1). Some of the evidence is anecdotal. Ullyett (1953) for instance, reported that if one breaks open a pupa of the brown-tailed moth (*Euproctis terminalia*) in a forest, within minutes a swarm of *Pimpla bicolor* females appear, evidently attracted by host odor. More conclusive evidence is presented by Murr (1930) and Jacobi (1939), who each observed that parasitoids in arenas saturated with host odors behaved differently from those in arenas without such odors.

Better evidence in the form of olfactometer data has been gathered by several workers. Probably more such work has been done with *V. canescens* (Thorpe and Jones 1937, Williams 1951) and *Drino bohemica* (Monteith 1958) than with any other parasitoid. In each case the insects were observed to preferentially walk up the olfactometer arm having host odor.

Parasitoid	Family	Experimental Evidence for Olfaction	Host	Reference
Pimpla bicolor	Ichneumonidae	Anecdotal	*Euproctis terminalia*	Ullyett (1953)
Pimpla instigator	Ichneumonidae	Attack of concealed host	*Pieris brassicae*	Carton (1971, 1974)
Spilocryptus extrematis	Ichneumonidae	Ancedotal	*Hyalaphora cecropia*	Marsh (1937)
Therion circumflexum	Ichneumonidae	Ancedotal	Various caterpillars	Slobodchikoff (1973)
Habrobracon juglandis	Braconidae	Observational	*Ephestia kuehniella*	Murr (1930)
Masonia vitripennis	Pteromalidae	Observational	Various fly puparia	Jacobi (1939)
Microbracon gelechiae	Braconidae	Olfactometer	*Gnorimoschema operculella*	Narayanan and Rao (1955)
Bathyplectes curculionis	Ichneumonidae	Olfactometer	*Hypera postica*	McKenney and Pass (1977)
Lysiphlebus testaceipes	Braconidae	Olfactometer	*Schizaphis graminum*	Starks and Schuster (1974)
Drino bohemica	Tachinidae	Olfactometer	*Diprion hercyniae*	Monteith (1955, 1958)
Ventura (Nemeritis) canescens	Ichneumonidae	Olfactometer	*Ephestia kuehniella*	Thorpe and Jones (1937), Williams (1951)
Trichopoda pennipes	Tachinidae	Field trap	*Nezara viridula*	Mitchell and Mau (1971)
Aphytis melinus, A. coheni	Aphelinidae	Field trap, Olfactometer	*Aonidiella aurantii*	Sternlight (1973)
Tomicobia tibialis	Pteromalidae	Field trap	*Ips confusus, I. pini*	Rice (1968, 1969), Lanier et al. (1972), Bedard (1965)
Cheiropachus colon, Cerocephala rufa	Pteromalidae	Field trap	*Scolytus multistriatus*	Kennedy (1979)
Entedon leucogramma	Eulophidae	Field trap	*Scolytus multistriatus*	Kennedy (1979)
Spathius benefactor, Dendrosoter protuberans	Braconidae	Field trap	*Scolytus multistriatus*	Kennedy (1979)

It is interesting to note that in some field trap studies parasitoids were attracted to host sex pheromones (Mitchell and Mau 1971, Sternlight 1973) or to host aggregation pheromones (Rice 1968, 1969, Lanier et al. 1972, Bedard 1965, Kennedy 1979). The original objective of these studies was to assess intraspecific host responses; the parasitoid responses were discovered incidentally. This tantalizingly suggests that more such instances would be known if pheromone traps were routinely examined for specific parasitoids.

Conditioning

Few detailed behavioral investigations have been done on the response of parasitoids to host odor. Usually a study demonstrates only that olfaction occurs. However, some workers have shown that olfaction can be influenced (i.e. conditioned) by the parasitoid's previous experiences. After being reared for 7 years in the laboratory on *Corcyra cephalonica, Microbracon gelechiae* responded to the odor of this insect over the odor of its natural host, *Gnorimoschema operculella* (Narayanan and Rao 1955). Thorpe and Jones (1937) related that *V. canescens* responded strongly to the odor of its preferred host, *Ephestia,* if reared from it, but not so strongly if reared from another host, *Meliphora.* Also, Arthur (1971) found that *V. canescens* was able to associate the odor of geranol with the presence of hosts when the parasitoid was exposed to geranol and hosts simultaneously. Thus, as in other aspects of the host selection process, olfactory responses of parasitoids are variable and may depend somewhat on prior conditions.

Orientation Behavior

The manner in which olfaction enables parasitoids to actually find their hosts is not known. It is possible that anemotaxis and/or chemotaxis as outlined by Shorey (1976) is operative. For instance, Marsh (1937) stated that *Spilocryptus extrematis* approached its host, the cecropia moth (*Platysamis cecropia*), from downwind. On the other hand, Murr (1930) noted that *Habrobracon juglandis,* when in an arena permeated with host odor, made numerous turning movements (klinokinesis). This behavior would be appropriate if the odor were perceived over only a short range, since a host would be likely to be near when the odor was detected. Long-range chemoreception may also interact with other stimulus modalities. When host odor was present in both arms of an olfactometer, *D. bohemica* went up the arm in which a small feather was fluttering in the airstream. This response occurred only if the host odor was present, suggesting that odor perception stimulated the parasitoid to orient visually (Monteith 1956). It also demonstrates the hierarchy of behavioral responses that can occur during the process of host selection.

Close-Range Chemoreception

Investigators have found many more parasitoids that locate their hosts through close-range chemicals in host-related products than users of long-range ones, probably because the former are much easier to document. All one needs do is show that the parasitoid changes behavior when it comes in contact with the chemical of interest (which is usually obtained by extracting a host product with solvent) as compared to a control. The behavioral change is typically a very distinct and intense examination of the chemically impregnated substrate.

A variety of host-related products have been shown to contain such chemicals (Table 5.2).

Host Traces Near Eggs

Adult host traces left by ovipositing females near eggs have been found to influence a number of egg or egg-larval parasitoids. Although in most cases only the existence of such chemicals has been demonstrated, some studies have been more intensive. Spradbery (1970a) and Madden (1968), for instance, showed that the probing response of *Ibalia* spp. into holes made in wood by ovipositing *Sirex* females is due to chemicals in a symbiotic fungus introduced by *Sirex*, which *Ibalia* uses as a cue to find host eggs or young larvae. The examination and probing behavior of the egg-larval parasitoid *Chelonus texanus* is due to a water-soluble, heat-stable substance deposited by *Heliothis virescens* females when they oviposit (Vinson 1975). Much work has been done on *Trichogramma* spp., first by Laing (1937), then by Jones et al. (1973), Lewis et al. (1971, 1972, 1975), and Nordlund et al. (1977). Briefly, these investigators showed that scales from female moths that rub off onto the substrate during oviposition contain a lipid-soluble kairomone that leads to greater parasitization by *Trichogramma* when it is present.

Host Feces

Host fecal matter is also commonly used by parasitoids as an aid in host location (Table 5.2). Again, most listed studies have demonstrated only that active chemicals are present, and they are not specifically noted here. The three tachinids listed in Table 5.2 are of interest because they deposit living maggots directly on feces. Thus the larvae rather than the adult parasitoids actually locate the hosts, but without the adult's ability to deposit larvae nearby, the latter would never be able to survive. The kairomone in host feces to which one of these tachinids (*Archytas marmoratus*) responds has been partially characterized as a large molecular weight chemical (Nettles and Burks 1975).

Investigations on *Orgilus lepidus* by Hendry et al. (1973), *Microplitis*

TABLE 5.2. Parasitoids that Locate Hosts by Close-Range Chemoreception

Parasitoid	Family	Host	References
		ADULT HOST TRACES NEAR EGGS	
Asolcus grandis	Scelionidae	*Aelia germani*	Lairachi and Voegele (1975)
Ooencyrtus fecundus	Encyrtidae	*Aelia germani*	Lairachi and Voegele (1975)
Ibalia spp.	Ichneumonidae	*Sirex*	Spradbery (1970a), Madden (1968)
Chelonus texanus	Braconidae	*Heliothis virescens*	Vinson (1975)
Trissolcus viktorovi	Scelionidae	*Eurydema* spp.	Buleza (1973)
Trissolcus spp., *Telenomus chloropus*	Scelionidae	*Eurygaster integriceps*	Viktorov et al. (1975)
Trichogramma evanescens	Trichogrammidae	*Sitotroga* spp., *Heliothis zea*	Laing (1937), Jones et al. (1973), Lewis et al. (1971, 1972)
Trichogramma achaeae	Trichogrammidae	*Heliothis zea*	Lewis et al. (1975)
Trichogramma pretiosum	Trichogrammidae	*Heliothis zea*, *Trichoplusia ni*	Lewis et al. (1975), Nordlund et al. (1977)
		HOST FECES	
Lydella grisescens	Tachinidae	*Ostrinia nubilalis*	Hsiao et al. (1966)
Lixophaga diatraeae	Tachinidae	*Diatraea saccharalis*	Roth et al. (1978)
Archytas marmoratus	Tachinidae	*Heliothis virescens*	Nettles and Burks (1975)
Spilocrytus extrematis	Ichneumonidae	*Hyalophora cercropia*	March (1937)
Campoplex haywardi	Ichneumonidae	*Phthorimaea operculella*	Leong and Oatman (1968)
Bracon mellitor	Braconidae	*Anthonomus grandis*	Henson et al. (1977)
Orgilus lepidus	Braconidae	*Phthorimaea operculella*	Hendry et al. (1973)

Apanteles sesamilae	Braconidae	*Busseola fiusca*	Ullyett (1935)
Apanteles chilonis, *A. flavipes*	Braconidae	*Chilo suppressalis*	Kajita and Drake (1969)
Apanteles dignus	Braconidae	*Keiferia lycopersicella*	Cardona and Oatman (1971)
Microplitis croceipes	Braconidae	*Heliothis* spp.	Lewis (1970), Lewis and Jones (1971), Jones et al. (1971)
Microterys flavus	Encyrtidae	*Coccus hesperidum*	Vinson et al. (1978)
Trichomalus perfectus	Pteromalidae	*Ceuthorrhynchus assimilis*	Dmoch and Rutkowska-Ostrowski (1978)
Rhyssa persuasoria	Ichneumonidae	*Sirex* spp.	Spradbery (1970b), Madden (1968)
Megarhyssa spp.	Ichneumonidae	*Sirex* spp.	Madden (1968)
HOST MANDIBULAR AND LABIAL GLAND SECRETIONS			
Cardiochiles nigriceps	Braconidae	*Heliothis* spp.	Vinson and Lewis (1965), Vinson (1968), Vinson et al. (1975)
Ventura (Nemeritis) canescens	Ichneumonidae	*Ephestia kuehniella*	Corbet (1971, 1973), Mudd and Corbet (1973)
Campoletis sonorensis	Ichneumonidae	*Heliothis* spp.	Wilson et al. (1974), Schmidt (1974)
Apanteles melanoscelus	Braconidae	*Lymantria dispar*	Weseloh (1976, 1977)
Pseudorhyssa sternata	Ichneumonidae	*Sirex* spp.	Spradbery (1968)
Temelucha interruptor	Ichneumonidae	*Rhyacionia buoliana*	Arthur et al. (1964)
HOST–MARKING PHEROMONE			
Opius lectus	Braconidae	*Rhagoletis pomonella*	Prokopy and Webster (1978)
HOST TRAIL			
Habrobracon juglandis	Braconidae	*Ephestia kuehniella*	Murr (1930)

croceipes by Lewis (1970), Lewis and Jones (1971), and Jones et al. (1971), and *Microterys flavus* by Vinson et al. (1978) are noteworthy because active compounds in host feces (or in the last case, host honeydew) have been chemically identified.

The work by Madden (1968) and Spradbery (1970b) deserves special mention. For years people have wondered how the large *Rhyssa* and *Megarhyssa* ichneumonid parasitoids of *Sirex* are able to locate host larvae through 2–3 inches of wood. Both Madden and Spradbery showed that the symbiotic fungus that develops on host feces (*Amylostereum* sp.) causes examination and drilling behavior by the parasitoids, as do water, methanol, ethanol, and acetone extracts of the fungus.

Host Mandibular and Labial Gland Secretions

Fewer examples of parasitoids responding to host mandibular and labial gland secretions are known than for the host products considered so far, but a greater proportion have received careful study. *Cardiochiles nigriceps* examines with its antennae areas contaminated by chemicals from the mandibular gland salivary secretions of its host, *Heliothis* (Vinson and Lewis 1965, Vinson 1968, Vinson et al. 1975). The mandibular gland secretions of *Ephestia kuehniella* contain an epideictic (dispersal) pheromone (Chapter 10) that is used by *V. canescens* for host location (Mayer 1934; Williams 1951; Corbet 1971, 1973; Mudd and Corbet 1973). At least two parasitoids respond to host silk produced in the labial gland. In the case of *Campoletis sonorensis* (Wilson et al. 1974, Schmidt 1974) the response to silk is evidently due to contamination from other host products. The example is included here because contaminated webbing may function in host location. For *A. melanoscelus* the evidence is strong that a water-soluble kairomone is present in the labial gland itself (Weseloh 1976, 1977).

Parasitoid Traces

Cleptoparasitoids preferentially attack hosts that have already been parasitized by another species. The original parasitoid is destroyed by the immature cleptoparasitoid, which then devours the original host. The cleptoparasitoids listed in Table 5.2 respond to chemicals left by the original parasitoid in the host vicinity, thereby enabling them to preferentially locate already parasitized hosts (Spradbery 1968, Arthur et al. 1964).

Other Host Products

Other host products known to aid in host location by parasitoids include the host marking pheromone of *Rhagoletis pomonella*. Prokopy and Webster (1978) showed conclusively that *Opius lectus* searches fruits that have been marked by female *Rhagoletis,* and therefore are likely to contain host eggs. Finally, Murr (1930) reported that *H. juglandis* can follow a trail previously laid over an arena by an *Ephestia* larva.

Conditioning

Some evidence suggests that as with olfaction, chemotactile responses may be variable. This was shown convincingly for *Bracon mellitor,* a parasitoid of the boll weevil (*Anthonomus grandis*). In the field, *B. mellitor* probes into cotton squares in response to esters of cholesterol and fatty acids in host frass (Hensen et al. 1977). If reared on laboratory hosts fed an artificial diet, the parasitoid responds to the methyl ester of *p*-hydroxybenzoic acid, a diet supplement not found in natural host foods (Vinson et al. 1976). This response was due to associative learning. Female parasitoids learned to associate hosts with the diet supplement if they had previously been exposed to this chemical along with hosts (Vinson et al. 1977). Such variability may occur in other parasitoids, and this suggests that care is needed when interpreting host selection studies.

This example also shows that parasitoids may respond to components of the diet of their host. The phenomenon occurs also for *M. croceipes,* which responds preferentially to frass of *Heliothis zea* reared on plants as compared to hosts reared on artificial diet (Sauls et al. 1979). However, Hsiao et al. (1966) found no evidence of a similar plant effect for the tachinid *Lydella grisescens.*

Orientation Behavior

The behavioral changes that occur when a parasitoid contacts an active host product have seldom been described in detail. Waage (1978) has made a useful start in this direction by interpreting the often vague reported descriptions of parasitoid chemotactile behavior in terms of orientation behavior. He found that when a parasitoid touches a chemical trace, it usually either slows its walking speed or stops (orthokinesis). As it examines the substrate, it also increases its rate of turning, which may either be in a random direction (klinokinesis) or oriented with respect to the geometry of the chemical "patch" (klinotaxis). Waage explored this further in a study on *V. canescens,* essentially confirming his general conclusions and showing that at patch edges parasitoids tend to make directed 180° turns that serve to orient them back onto the chemical spot. This behavior functions to retain the parasitoid on a chemical patch and also ensures that the patch will be explored thoroughly. Any host on the area would almost certainly be located.

Other evidence suggests that at least some parasitoids are induced to thoroughly search areas near the chemical patch as well as the patch itself. Lewis et al. (1975) found that *Trichogramma pretiosum* parasitized host eggs near areas treated with kairomone to a greater extent than they did eggs removed completely from the vicinity of the kairomone. Intermittent contact with the chemical was enough to increase the parasitoid's searching behavior. A similar interpretation can be given to the work by Weseloh (1977), where host larvae placed near kairomone-treated substrates were attacked by *A. melanoscelus* with the same frequency as were hosts on treated substrates. Also, Gross et al. (1975), showed that when *M. croceipes* was exposed to host feces before being released, it subsequently searched for hosts more intensely. That this kind of behavior is adaptive seems obvious. A host product is a cue that a host is present or at least likely to be nearby. Anything that causes a parasitoid to thoroughly search areas contaminated with host-related products and their surroundings will increase the probability of contacting a host, and so of parasitizing it.

These behaviors should result in parasitoids spending more time in areas where the kairomone is abundant than in other areas. It obviously takes more time to examine a group of patches thoroughly than it does to walk across the same area. This means that areas with more kairomone will have a greater density of parasitoids, which may lead to greater parasitism of hosts. This is exactly what happened when kairomones of *Trichogramma* spp. or *M. croceipes* were artificially distributed in fields (Lewis et al. 1976). The percentage of parasitism of *Heliothis* spp. was increased as compared to check areas. This property of kairomones has obvious practical importance, and it is discussed elsewhere in this volume (Chapter 8).

CONCLUSION

The investigation of host location mechanisms used by parasitoids has barely begun. That this is so can be readily appreciated by an examination of Tables 5.1 and 5.2. Ichneumonids and braconids comprise more than 60% of the parasitoids listed, probably because their relatively large size and their importance as control agents have made them the most extensively studied subjects. Certainly it is not implied that other parasitoids lack host location mechanisms. Also, examples of close-range chemoreception are twice as numerous as examples of long-range chemoreception, a state of affairs one would expect in an immature field where easier studies are done first. Even so, the simple expedient of examining pheromone traps for species-specific parasitoids could lead to the reporting of many more examples of host location by olfaction. Most investigations have documented only that responses to chemical stimuli from the host occur. Any in-depth studies have usually concentrated on characterizing the kairomone, which is a narrow, though worthwhile goal.

Detailed behavioral work is notably lacking. We know that parasitoids respond to airborne host chemicals, and some of these chemicals have been identified. But as far as I know, no behavioral tests (such as wind-tunnel experiments) or even theoretical studies have been undertaken to determine how a parasitoid uses such chemicals to find a host. Likewise, for close-range chemoreception there has been only one study of note that considers behavior as being substantially more than a bioassay tool (Waage 1978). Only a few instances of response variability due to previous conditions (conditioning) are known, but this is because the phenomenon has hardly been studied. The possibility exists that it occurs broadly. In such a young field there is a good opportunity for fruitful investigations at many levels.

From a practical standpoint, the mechanisms a parasitoid uses to locate its host can be expected to significantly influence its searching capacity, an important parameter of parasitoid effectiveness (Huffaker et al. 1976). It is at the level of host location that it may be possible to increase the effectiveness of parasitoids most advantageously through behavioral manipulation (Lewis et al. 1976). Also, a better understanding of olfactory responses of parasitoids to their hosts may lead to sampling procedures for adults in the field. Thus not only are there ample opportunities for additional research, but this research can be expected to contribute significantly to our ability to use natural enemies for applied purposes.

REFERENCES

Arthur, A. P. 1971. Associative learning by *Nemeritis canescens* (Hymenoptera: Ichneumonidae). Can. Entomol. 103:1137–1142.

Arthur, A. P., J. E. R. Stainer, and A. L. Turnbull. 1964. The interaction between *Orgilus obscurator* (Nees) (Hymenoptera: Braconidae) and *Temelucha interruptor* (Grav.) (Hymenoptera: Ichneumonidae), parasites of the pine shoot moth, *Rhyacionia buoliana* (Schiff.) (Lepidoptera: Olethreutidae). Can. Entomol. 96:1030–1034.

Bedard, W. D. 1965. The biology of *Tomicobia tibialis* (Hymenoptera: Pteromalidae) parasitizing *Ips confusus* (Coleoptera: Scolytidae) in California. Contrib. Boyce Thomps. Inst. Plant Res. 23:77–81.

Buleza, V. V. 1973. Poisk khozyaina u *Trissolcus viktorovi* yaitzeeda kapustnoi klopa. Zool. Zh. 52:1509–1513; through Biol. Abstr. 58, No. 2752.

Cade, W. 1975. Acoustically orienting parasitoids: Fly phonotaxis to cricket song. Science. 190:1312–1313.

Cardona, C., and E. R. Oatman. 1971. Biology of *Apanteles dignus* (Hymenoptera: Braconidae), a primary parasite of the tomato pinworm. Ann. Entomol. Soc. Am. 64:996–1007.

Carton, Y. 1971. Biologie de *Pimpla instigator* F. (Ichneumonidae, Pimplionae). 1. Mode de perception de l'hôte. Entomophaga 16:285–296.

Carton, Y. 1974. Biologie de *Pimpla instigator* (Ichneumonidae: Pimplionae). III. Analyse experimentale du processus de reconnaissance de l'hôte-chrysalide. Entomol. Exp. Appl. 17:265–278.

Corbet, S. A. 1971. Mandibular gland secretion of larvae of the flour moth, *Anagasta kuehniella,* contains an epideictic pheromone and elicits oviposition movement in a hymenopteran parasite. Nature (London) 232:481.

Corbet, S. A. 1973. Concentration effects and the response of *Nemeritis canescens* to a secretion of its host. J. Insect Physiol. 19:2119–2128.

DeLeon, D. 1935. The biology of *Coeloides dendroctoni* Cushman (Hymenoptera: Braconidae), an important parasite of the mountain pine beetle (*Dendroctonus monticolae* Hopk.). Ann. Entomol. Soc. Am. 28:411–424.

Dmoch, J., and Z. Rutkowska-Ostrowski. 1978. Host finding and host-acceptance mechanism in *Trichomalus perfectus* Walker (Hymenoptera, Pteromalidae). Bull. Acad. Polon. Sci., Ser. Sci. Biol. 26:317–323.

Gross, H. R., W. J. Lewis, R. L. Jones, and D. A. Nordlund. 1975. Kairomones and their use for management of entomophagous insects. III. Stimulation of *Trichogramma achaeae, T. pretiosum,* and *Microplitis croceipes* with host-seeking stimuli at time of release to improve their efficiency. J. Chem. Ecol. 1:431–438.

Hassell, M. P. 1968. The behavioral response of a tachinid fly (*Cyzenis albicans* (Fall.) to its host, the winter moth (*Operophtera brumata* (L.). J. Anim. Ecol. 37:627–639.

Hendry, L. B., Greany, P. D., and R. J. Gill. 1973. Kairomone mediated host-finding behavior in the parasitic wasp *Orgilus lepidus*. Entomol. Exp. Appl. 16:471–477.

Henson, R. D., S. B. Vinson, and C. S. Barfield. 1977. Ovipositional behavior of *Bracon mellitor* Say, a parasitoid of the boll weevil (*Anthonomus grandis* Boheman). III. Isolation and identification of natural releasers of ovipositor probing. J. Chem. Ecol. 3:151–158.

Herrebout, W. M. 1969. Some aspects of host selection in *Eucarcelia rutilla*. VIII. (Diptera: Tachinidae). Neth. J. Zool. 19:1–104.

Hsiao, T., F. G. Holdaway, and C. H. Chang. 1966. Ecological and physiological adaptations in insect parasitism. Entomol. Exp. Appl. 9:113–123.

Huffaker, C. B., F. J. Simmonds, and J. E. Laing. 1976. The theoretical and empirical basis of biological control. P. 41–78. *In* C. B. Huffaker and P. S. Messenger (eds.), Theory and Practice of Biological Control, Academic Press, New York.

Jacobi, E. F. 1939. Über Lebensweise, auffinden des Wirtes and Regulierung der Individuenzahl von *Mormoniella vitripennis* Walker. Arch. Neerl. Zool. Leiden 3:197–282.

Jones, R. L., W. J. Lewis, M. C. Bowman, M. Beroza, and B. A. Bierl. 1971. Host-seeking stimulant for parasite of corn earworm: Isolation, identification, and synthesis. Science 173:842–843.

Jones, R. L., W. J. Lewis, M. Beroza, B. A. Bierl, and A. N. Sparks. 1973. Host-seeking stimulants (kairomones) for the egg parasite, *Trichogramma evanescens*. Environ. Entomol. 2:593–596.

Kajita, H., and E. F. Drake. 1969. Biology of *Apanteles chilonis* and *A. flavipes* parasites of *Chilo suppressalis*. Mushi 42:163–179.

Kennedy, B. H. 1979. The effect of multilure on parasites of the European elm bark beetle, *Scolytus multistriatus*. Bull. Entomol. Soc. Am. 25:116–118.

Laing, J. 1937. Host-finding by insect parasites. I. Observations on the finding of hosts by *Alysia manducator, Mormoniella vitripennis,* and *Trichogramma evanescens*. J. Anim. Ecol. 6:298–317.

Lairachi, J., and J. Voegele. 1975. Modalities of host finding by two egg parasites of *Aelia germani* in Morocco: *Asolcus grandis* and *Ooencyrtus fecundus* (Hym., Scelionidae and Encyrtidae). Ann. Soc. Entomol. Fr. 11:541–549.

Lanier, G. N., M. C. Birch, R. F. Schmitz, and M. M. Furniss. 1972. Pheromones of *Ips pini*

(Coleoptera: Scolytidae). Variation in response among three populations. Can. Entomol. 104:1917–1923.

Leong, J. K. L., and E. R. Oatman. 1968. The biology of *Campoplex haywardi* (Hymenoptera: Ichneumonidae), a primary parasite of the potato tuber worm. Ann. Entomol. Soc. Am. 61:26–36.

Lewis, W. J. 1970. Life history and anatomy of *Microplitis croceipes* (Hymenoptera: Braconidae), a parasite of *Heliothis* spp. (Lepidoptera: Noctuidae). Ann. Entomol. Soc. Am. 63:67–70.

Lewis, W. J., and R. L. Jones. 1971. Substance that stimulates host-seeking by *Microplitis croceipes* (Hymenoptera: Braconidae), a parasite of *Heliothis* species. Ann. Entomol. Soc. Am. 64:471–473.

Lewis, W. J., A. N. Sparks, and L. M. Redlinger. 1971. Moth odor: A method of host-finding by *Trichogramma evanescens*. J. Econ. Entomol. 64:557–558.

Lewis, W. J., R. L. Jones, and A. N. Sparks. 1972. A host-seeking stimulant for the egg parasite *Trichogramma evanescens:* Its source and a demonstration of its laboratory and field activity. Ann. Entomol. Soc. Am. 65:1087–1089.

Lewis, W. J., R. L. Jones, D. A. Nordlund, and H. R. Gross, Jr. 1975. Kairomones and their use for management of entomophagous insects. II. Mechanisms causing increase in rate of parasitization by *Trichogramma* spp. J. Chem. Ecol. 1:349–360.

Lewis, W. J., R. L. Jones, H. R. Gross, and D. A. Nordlund. 1976. The role of kairomones and other behavioral chemicals in host finding by parasitic insects. Behav. Biol. 16:267–289.

Madden, J. 1968. Behavioural responses of parasites to the symbiotic fungus associated with *Sirex nictilio* F. Nature (London), 218:189–190.

Mangold, J. R. 1978. Attraction of *Euphasiopteryx ochracea, Corethrella* sp. and gryllids to broadcast songs of the southern mole cricket. Florida Entomol. 61:57–61.

Marsh, F. L. 1937. Biology of the ichneumonid, *Spilocryptus extrematis* Cresson. Ann. Entomol. Soc. Am. 30:40–42.

Mayer, K. 1934. Beitrage zur Sinnesphysiologie der Schlupfwespe *Nemeritis canescens* Grav. (Hym.: Ichneumonidae, Ophioninae). Arb. Physiol. Angew. Entomol. 1:245–248.

McKinney, T. R., and B. C. Pass. 1977. Olfactometer studies of host seeking in *Bathyplectes curculionis* Thoms. (Hymenoptera: Ichneumonidae). J. Kansas Entomol. Soc. 50:108–112.

Mitchell, W. C., and R. F. L. Mau. 1971. Response of the female southern green stink bug and its parasite, *Trichopoda pennipes,* to male stink bug pheromones. J. Econ. Entomol. 64:856–859.

Monteith, L. G. 1955. Host preferences of *Drino bohemica* Mesn. (Diptera: Tachinidae), with particular reference to olfactory responses. Can. Entomol. 87:509–530.

Monteith, L. G. 1956. Influence of host movement on selection of hosts by *Drino bohemica* Mesn. (Diptera: Tachinidae) as determined in an olfactometer. Can. Entomol. 88:583–586.

Monteith, L. G. 1958. Influence of host and its food plant on host-finding by *Drino bohemica* Mesn. (Diptera: Tachinidae) and interaction of other factors. Proc. Tenth Int. Congr. Entomol. 2:603–606.

Mudd, A., and S. A. Corbet. 1973. Mandibular gland secretion of larvae of the stored products pests *Anagasta kuehniella, Ephestia cautella, Plodia interpunctella* and *Ephestia elutella*. Entomol. Exp. App. 16:291–292.

Murr, L. 1930. Über den Geruchsinn der Mehmottenschluphwespe, *Habrobracon juglandis* zugleich Beitrag zum Orientierungsproblem. Z. Vgl. Physiol. 2:210–270.

Narayanan, E. S., and B. R. Subba Rao. 1955. Studies in insect parasitism. I–III. The effect of different hosts on the physiology, on the development and behaviour, and on the sex ratio of

Microbracon gelechiae Ashmead (Hymenoptera: Braconidae). Beitr. Entomol. 5:36–60.

Nettles, W. C., Jr., and M. L. Burks. 1975. A substance from *Heliothis virescens* larvae stimulating larviposition by females of the tachinid, *Archytas marmoratus*. J. Insect Physiol. 21:965–978.

Nordlund, D. A., W. J. Lewis, J. W. Todd, and R. B. Chalfant. 1977. Kairomones and their use for management of entomophagous insects. VII. The involvement of various stimuli in the differential response of *Trichogramma pretiosum* Riley to two suitable hosts. J. Chem. Ecol. 3:513–518.

Prokopy, R. J. and R. P. Webster. 1978. Oviposition deterring pheromone of *Rhagoletis pomonella*-kairomone for its parasitoid *Opius lectus*. J. Chem. Ecol. 4:481–494.

Rice, R. E. 1968. Observations on host selection by *Tomicobia tibialis* Ashmead (Hymenoptera: Pteromalidae). Contrib. Boyce Thompson Inst. 24:53–56.

Rice, R. E. 1969. Response of some predators and parasites of *Ips confusus* (Lec.) (Coleoptera: Scolytidae) to olfactory attractants. Contrib. Boyce Thompson Inst. 24:189–194.

Richerson, J. V., and J. H. Borden. 1972. Host finding by heat perception in *Coeloides brunneri* (Hymenoptera: Braconidae). Can. Entomol. 104:1877–1881.

Richerson, J. V., and C. J. DeLoach. 1972. Some aspects of host selection by *Perilitus coccinellae*. Ann. Entomol. Soc. Am. 65:834–839.

Roth, J. P., E. G. King, and A. C. Thompson. 1978. Host location behavior by the tachinid, *Lixophaga diatraeae*. Environ. Entomol. 7:794–798.

Ryan, R. B., and J. A. Rudinsky. 1962. Biology and habits of the douglas fir beetle parasite, *Coeloides brunneri* Viereck (Hymenoptera: Braconidae) in western Oregon. Can. Entomol. 94:748–763.

Sauls, C. E., D. A. Nordlund, and W. J. Lewis. 1979. Kairomones and their use for management of entomophagous insects. VIII. Effect of diet on the kairomonal activity of frass from *Heliothis zea* (Boddie) larvae for *Microplitis croceipes* (Cresson). J. Chem. Ecol. 5:363–369.

Schaffner, J. V., Jr. 1934. Introduced parasites of the brown-tail and gypsy (sic) moths reared from native hosts. Ann. Entomol. Soc. Am. 27:585–592.

Schmidt, G. T. 1974. Host-acceptance behavior of *Campolites sonorensis* toward *Heliothis zea*. Ann. Entomol. Soc. Am. 67:835–844.

Shorey, H. H. 1976. Animal Communication by Pheromones. Academic Press, New York.

Slobodchikoff, C. N. 1973. Behavioral studies of three morphotypes of *Therion circumflexum*. Pan-Pac. Entomol. 49:197–206.

Soper, R. S., G. E. Shewell, and D. Tyrrell. 1976. *Colcondamyia auditrix* nov. sp. (Diptera: Sacrophagidae), a parasite which is attracted by the mating song of its host, *Okanagana rimosa* (Homoptera: Cicadidae). Can. Entomol. 108:61–68.

Spradbery, J. P. 1968. The biology of *Pseudorhyssa sternata* Merrill (Hym., Icheumonidae), a cleptoparasite of siricid woodwasps. Bull. Entomol. Res. 59:291–297.

Spradbery, J. P., 1970a. The biology of *Ibalia drewseni* Borries, a parasite of siricid woodwasps. Proc. R. Entomol. Soc. London, Ser. A. 45:104–113.

Spradbery, J. P. 1970b. Host finding by *Rhyssa persuasoria*, an ichneumonid parasite of siricid woodwasps. Anim. Behav. 18:103–114.

Starks, K. J., and D. J. Schuster. 1974. Response of *Lysiphlebus testaceipes* in an olfactometer to a host and a non-host insect and to plants. Environ. Entomol. 3:1034–1035.

Sternlight, M. 1973. Parasitic wasps attracted by the sex pheromone of their coccid host. Entomophaga 18:339–342.

Thorpe, W. H., and F. G. W. Jones. 1937. Olfactory conditioning and its relation to the problem of host selection. Proc. R. Soc. London, Ser. B 124:56–81.

Ullyett, G. C. 1935. Notes on *Apanteles sesamilae* Cam., a parasite of the maize stalk-borer (*Busseola fiusca* Fuller) in South Africa. Bull. Entomol. Res. 26:253–262.

Ullyett, G. C. 1953. Biomathematics and insect population problems. Entomol. Soc. S. Afr. Mem. 2:1–89.

Van den Assem, J., and D. J. Kuenen. 1958. Host finding of *Choetospila elegans* Westw. (Hym., Chalcid) a parasite of *Sitophilus granarius* L. (Coleopt. Curcul.). Entomol. Exp. Appl. 1:174–180.

Viktorov, G. A., V. V. Buleza, E. P. Zinkevich, and S. B. Trofimov. 1975. Poisk khozyania u *Trissolcus grandis* i *Telenomus chloropus* yaitseedov vrednoi cherepashki. Zool. Zh. 54:922–927.

Vinson, S. B. 1968. Source of a substance in *Heliothis virescens* that elicits a searching response in its habitual parasite, *Cardiochiles nigriceps*. Ann. Entomol. Soc. Am. 61:8–10.

Vinson, S. B. 1975. Source of material in the tobacco budworm which initiates host-searching by the egg-larval parasitoid *Chelonus texanus*. Ann. Entomol. Soc. Am. 68:381–384.

Vinson, S. B., and W. J. Lewis. 1965. A method of host selection by *Cardiochiles nigriceps*. J. Econ. Entomol. 58:869–871.

Vinson, S. B., R. L. Jones, P. E. Sonnet, B. A. Bierl, and M. Beroza. 1975. Isolation, identification and synthesis of host-seeking stimulants for *Cardiochiles nigriceps*, a parasitoid of the tobacco budworm. Entomol. Exp. Appl. 18:443–450.

Vinson, S. B., R. D. Henson, and C. S. Barfield. 1976. Ovipositional behavior of *Bracon mellitor* Say (Hymenoptera: Braconidae), a parasitoid of the boll weevil (*Anthonomus grandis* Boh.). I. Isolation and identification of a synthetic releaser of oviposition probing. J. Chem. Ecol. 2:431–440.

Vinson, S. B., C. S. Barfield, and R. D. Henson. 1977. Oviposition behaviour of *Bracon mellitor*, a parasitoid of the boll weevil (*Anthonomus grandis*). II. Associative learning. Physiol. Entomol. 2:157–164.

Vinson, S. B., D. P. Harlan, and W. G. Hart. 1978. Response of the parasitoid *Microterys flavus* to the brown soft scale and its honey dew. Environ. Entomol. 7:874–878.

Waage, J. K. 1978. Arrestment responses of the parasitoid, *Nemeritis canescens*, to a contact chemical produced by its host, *Plodia interpunctella*. Physiol. Entomol. 3:135–146.

Walker, M. F. 1961. Some observations on the biology of the ladybird parasite, *Perilitus coccinellae* (Schrank) with special reference to host selection and recognition. Entomol. Mon. Mag. 97:240–244.

Weseloh, R. M. 1976. Behavioral responses of the parasite, *Apanteles melanoscelus*, to gypsy moth silk. Environ. Entomol. 5:1128–1132.

Weseloh, R. M. 1977. Effects on behavior of *Apanteles melanoscelus* females caused by modifications in extraction, storage, and presentation of gypsy moth silk kairomone. J. Chem. Ecol. 3:723–735.

Williams, J. R. 1951. The factors which promote and influence the oviposition of *Nemeritis canescens* Grav. (Ichneumonidae, Ophioninae). Proc. R. Entomol. Soc. London, Ser. A. 26:49–58.

Wilson, D. D., R. L. Ridgway, and S. B. Vinson. 1974. Host acceptance and oviposition behavior of the parasitoid *Campoletis sonorensis* (Hymenoptera: Ichneumonidae) Ann. Entomol. Soc. Am. 67:271–274.

CHAPTER SIX

HOST ACCEPTANCE BY PARASITOIDS

ALFRED P. ARTHUR

Agriculture Canada
Research Station Research Branch
Saskatoon, Saskatchewan
Canada

The increasingly extensive and successful use of biological control techniques in the control of insect pests has been limited by the great expense of mass producing biological agents and by our ignorance of the many stimuli involved in the whole field of parasitoid-host relationships. This chapter is confined to a detailed description of parasitoid-host acceptance. An adult parasitoid is a free-living insect. The female accepts another insect as her host by depositing eggs, or larvae, in, on, or near the host, which is eventually killed by the developing immature stages of the parasitoid.

During the last 10–15 years many new microanalytical chemical instrumentation and techniques have come into general use. These have aided in the discovery of stimuli, especially chemical stimuli, and assessment of their importance at many levels of parasitoid-host relationships. This knowledge should facilitate the increased effectiveness of biological control agents now that the long-term disadvantages to the continuous widespread use of persistent pesticides to the environment is realized.

Stimuli, especially chemicals, are now known to serve as cues for many species of parasitoids while they are searching for host habitats, for the host itself, or for ascertaining the acceptability of a particular host species. After finding a species of host that usually favors the development of her progeny, the female parasitoid must determine whether the individual is acceptable. Some of these individuals may be unsuitable due to inappropriate developmental stage, disease, death, or previous parasitization by another female, not necessarily of the same species.

The subject of host selection has been reviewed by Doutt (1959), and Vinson (1975a, 1976, 1977). The steps of host habitat finding, host finding, and host acceptance are progressive and overlap one another when many host-parasitoid relationships are described. To increase the understanding of the significance of the steps in host acceptance, some factors involved in host-habitat finding and host finding are briefly described here, although these topics have been discussed in Chapters 4 and 5. As indicated by Calvert (1973), host acceptance may itself be restricted by some factor found in earlier host habitat or host-finding procedures. Bragg (1974) defined host selection as the entire process of evaluation of host acceptability through tactile and chemical stimuli. He considered that oviposition is the actual acceptance of the egg, larva, pupa, or adult as a host by the adult parasitoid. Host acceptance behavior was divided into four definitive phases by Schmidt (1974): (1) encounter and examination, (2) thrusting with the ovipositor, (3) inserting the ovipositor, and (4) oviposition. The acceptance behavior of each parasitoid proceeds through each of the four phases, and if the sequence is interrupted at phase 2 or 3, it must begin again at phase 1 before oviposition can occur. Such interruption may be due to the aggressive behavior of a host or to the determination by the parasitoid of previous parasitization. Weseloh (1974) defined host recognition (= host acceptance) as the process whereby hosts are accepted or rejected for oviposition after contact has been made.

In earlier times, Thompson and Parker (1927) considered that parasitoid-host relationships were governed by instinct, which they defined as an operation suggestive of a judgment of desirability or suitability of a potential host, formulated by the insect without the data or the mental equipment required for producing a judgment in the true sense of the word. These authors concluded that the choice of a host suitable for parasitoid progeny was impossible to determine in terms of morphological or physiochemical properties, and that the hosts chosen were usually suitable.

Later, Salt (1935), working with *Trichogramma evanescens,* a polyphagous egg parasitoid, tried to express in scientific terms some of the criteria by which ovipositing females of this small egg parasitoid choose their hosts. He considered that if, after having drummed on an object with its antennae, a female tried to probe it with the ovipositor, she had accepted it as a potential host. He found that size was by far the most important criterion used by *Trichogramma* females during host acceptance. The females of the size he used accepted globular objects between 0.22 and 4.64 mm in diameter. If the objects were of irregular shape, they would be accepted, provided the smallest dimension could contain one adult parasitoid and the other dimensions were not more than four times as great. Shape was less important: spherical globules of mercury, eggs of insect, lumps of flour, seeds, glass beads, cylindrical rods of glass, rhombic crystals of calcium carbonate, particles of sand, and fragments of glass were all mounted and attempts were made to probe them. Texture was not important as long as the surface was not wet, odorous, or sticky, and was sufficiently firm to bear the parasitoid. Insect eggs not protruding above the surrounding surface were ignored. Because the *Trichogramma* females accepted the eggs of many insect species, which must have had different odors, Salt considered that odor was not an important cue to host acceptance by this species. In this well-planned series of experiments, the investigator showed that the criteria used in selecting a host by females of *Trichogramma* were often physical and definable. Therefore the theory of host selection by instinct as defined by Thompson and Parker (1927) is not valid for this species of parasitoid.

Many physical cues used by female parasitoids during host acceptance were recorded before detection of minute quantities of chemicals was possible. Salt (1937) showed that the internal and external presence of chemicals contaminating host eggs enabled the parasitoid to discriminate between eggs that were already parasitized and those that were not. *T. evanescens* females use pheromones to mark eggs that have been examined and/or parasitized, subsequently enabling females to discriminate between healthy and parasitized hosts as discussed in Chapter 8. Salt demonstrated that two processes were involved. The parasitoid was able to distinguish between clean hosts and hosts that had been merely walked on by another female of the same species. If hosts that had been walked on were washed in water, the parasitoid could no longer distinguish them externally from healthy hosts and attacked them. As soon as its ovipositor had penetrated into a parasitized egg, however, the parasitoid sensed that the egg was parasitized and usually withdrew immediately without laying another egg in the same host.

Jackson (1968) tentatively attributed discrimination of parasitized hosts by egg parasitoids to the presence of a pheromone injected into the host with the egg. This secretion is probably produced by the acid or alkaline gland of the female. It renders a parasitized *Caraphractus cinctus* egg unacceptable to all but novice females or experienced females long deprived of hosts. Often more than one type of cue (chemical and/or physical) is important in a single parasitoid-host relationship.

When the cues influencing various parasitoid-host relationships are being described, few generalities can be made. However, parasitoids attacking hosts in the same stage of development, or in the same microhabitat, often use similar cues.

CLASSIFICATION OF PARASITOIDS

Many insects are pests because they have been introduced accidentally without their primary parasitoids and predators. Most early biological control programs were based on returning to the country of origin of the pest, making large collections of the pest species, and rearing them under quarantine conditions to obtain the species of parasitoids that attack the pest. It soon became evident that all the species obtained in this way were not primary parasitoids, that is, insects that attacked only the pest species. Some species belonged to genera such as *Dibrachys, Gelis, Pteromalus, Habrocytus, Monodontomerus,* or *Perilampus,* which are known to attack both the pests and their parasitoids with equal frequency (Muesebeck et al. 1951). Naturally all individuals of this group known as secondary or hyperparasitoids, were killed, because their release would reduce the efficiency of the primary parasitoids.

Some complex behavioral relationships have been discovered between one species of parasitoid and another species or group that attack the same pests. For example, Arthur (1961) found that four species of larval parasitoids of the European pine shoot moth (*Rhyacionia buoliana*) had their parasitoid-host relationships interrelated. The larvae of this pest chew tunnels in newly developing pine shoots. The females of *Scambus hispae, Scambus tecumseh,* or *Exeristes comstockii* pierce the infested shoots with their ovipositors and puncture the shoot moth larvae, paralyzing them (Figure 6.1). They then partly withdraw their ovipositors and deposit their eggs in the tunnel on or near the inactivated host (Figure 6.2). A fourth larval parasitoid, *Eurytoma pini,* also emerges from pine buds infested with *R. buoliana* larvae. Early attempts to rear *E. pini* adults in the laboratory on active hosts failed because no oviposition occurred. Microscopic examination of buds from which *E. pini* adults had emerged showed that the immature stages of the other larval parasitoids were contained in many of them. Later females of *E. pini* were observed to oviposit on *R. buoliana* larvae that had been inactivated with heat and placed in hollowed-out buds, with or without immature stages of the other larval parasitoids. Apparently, these females will not accept active shoot moth larvae. These observations lead to the conclusion that females of *E. pini* would not accept shoot moth larvae as hosts in the field unless the larvae had been previously inactivated by females of other parasitoids. Because of this species host acceptance behavior, which does not add to the total parasitism of the pest, *E. pini* was classified as a cleptoparasitoid.

Cleptoparasitism was later defined by Spradbery (1969) as multiparasitism in which access to and paralysis of a host by one parasitoid species are essential before the host can be parasitized by a second parasitoid species (the cleptoparasitoid). This species upon eclosion destroys the egg or larva of the first parasitoid before feeding on the paralyzed primary host. Spradbery studied the host-parasitoid relationships of *Pseudorhyssa sternata* and other parasitoids of siricid woodwasps. He observed females of *P. sternata* and females of *Rhyssa persuasoria,* a primary parasitoid, together on siricid-infested logs. *P. sternata* stood by the primary parasitoid while it made drill shafts to deposit its eggs. After the primary parasitoid withdrew its ovipositor and moved away, *P. sternata* located the shaft and inserted its ovipositor to gain access to the host. *P. sternata*

FIGURE 6.1. Female of *Exeristes comstockii* ovipositing in an artificial pine bud containing larvae of *Galleria mellonella*. (From Arthur 1963. Courtesy of the Entomological Society of Canada.)

females were considered to be capable of finding the drill shaft by responding to the secretion of a kairomone from the vaginal gland of the primary parasitoid.

Clausen (1940) described the somewhat similar relationship for *Eurytoma monemae,* a gregarious external parasitoid of the mature larva of the oriental moth (*Monema flavescens*) within its cocoon, and for *Chrysis shanghaiensis* another parasite of the same host. In this relationship, the female of *E. monemae* is unable to penetrate the thick, hard, host cocoon but is attracted to the ovipositing female of *C. shanghaiensis*. After this parasitoid has cut the hole in the cocoon, has oviposited, and has plugged the hole with a spongy material, the

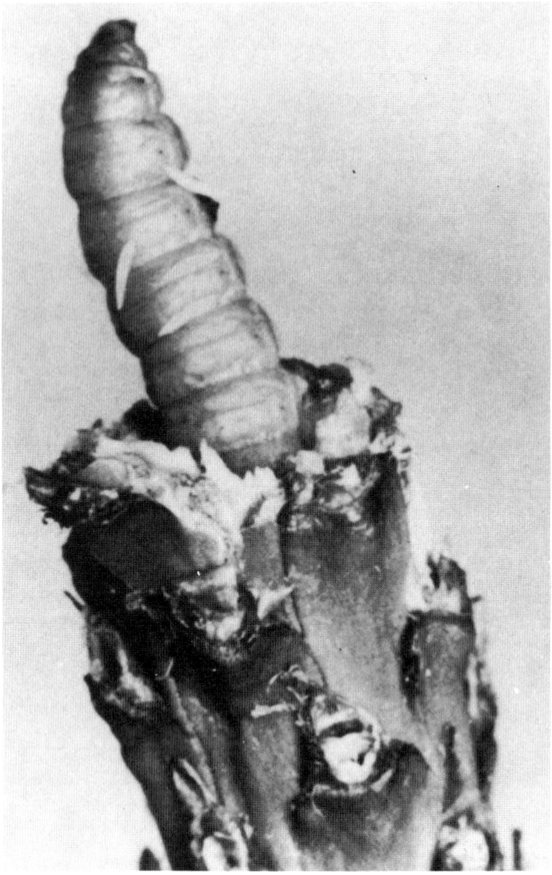

FIGURE 6.2. Eggs of *Exeristes comstockii* on paralyzed larvae of *Galleria mellonella*.

female of *E. monemae* thrusts her ovipositor through the plug and deposits her own eggs on the larva of *M. flavescens*.

An interesting ichneumonid-sarcophagid interaction with the gypsy moth (*Lymantria dispar*) as host was reported by Campbell (1963). Four species of ichneumonids stung and killed many more host pupae than they were able to parasitize successfully (as measured in number of ichneumonid offspring produced). The ratio between total number of hosts stung by ichneumonids and the number of offspring emerged was as high as 200–1. Many of these stings were probably made to obtain protein for egg development, as was described by Leius (1961), but some oviposition also occurred. Many puncture wounds made by the ichneumonids provided an avenue of entrance for larvae of *Sarcophaga aldrichi*. Of 114 pupae known to have been stung by ichneumonids, 52 produced sarcophagid progeny and only one ichneumonid was produced. Campbell collected many hosts known to have been stung by ichneumonids immediately after the attack. A second group of hosts, also known to have been attacked, were collected after having been exposed in the field for some time. About half the hosts from the field contained sarcophagids, but none of the hosts collected as soon as they were stung by the ichneumonids did. Field observations showed that sarcophagids associated with the gypsy moth usually larviposit on dead hosts and only rarely on living hosts. Sarcophagid maggots were observed working their way into wounds previously made by ichneumonids. This careful study indicated that the sarcophagids are usually cleptoparasitoids on gypsy moth pupae that have already been stung by ichneumonids.

Juillet (1960) and Arthur et al. (1964) studied the interrelations of two species of larval parasitoids of early instars of the European pine shoot moth in Ontario. Juillet (1960) observed that the females of *Orgilus obscurator* were very efficient searchers for shoot moth larvae. His observations showed that the females of the other parasitoid, *Temelucha interruptor,* were relatively inefficient searchers because they probed at random with their ovipositors and they also failed to locate infested areas. Dissection of more than 2000 shoot moth larvae showed only one *O. obscurator* per host. In the same lot, *T. interruptor* had laid one or two eggs per host; however, in up to 80% of the cases, one egg was laid in a host already parasitized by *O. obscurator*.

Arthur et al. (1964) showed that females of *T. interruptor* in the laboratory were strongly attracted to areas previously visited by females of *O. obscurator*. Such areas were strongly attractive even when no hosts were present. The females of *T. interruptor* examined with their antennae the areas contaminated by the females of *O. obscurator* and prodded them with their ovipositors. From the combined studies, females of *O. obscurator* were considered to mark their search trails and their oviposition sites with a pheromone to prevent superparasitism. Females of *T. interruptor* were perceived to use this chemical as a kairomone to aid them in locating and parasitizing shoot moth larvae. In this case, the progeny of *T. interruptor* (the more inefficient searcher) always survived when both species parasitized the same host. These studies indicated that insect parasitoids sometimes use chemical cues produced by other species of parasitoids to increase their own efficiency.

Turnbull and Chant (1961), on purely theoretical grounds, warned against the introduction of many species of parasitoids to combat a single host species without a thorough investigation of the characteristics of each species and the relationships among them. The discovery of cleptoparasitoids emphasizes the necessity for careful and detailed observations of the stimuli that influence host acceptance by parasitoids that attack the same pest.

PHYSICAL AND CHEMICAL STIMULI INFLUENCING HOST ACCEPTANCE

Size, Shape, and Surface Texture

Klomp and Teerink (1962) demonstrated that *Trichogramma* females examine and drum on a host egg with the antennae, to determine its size, before drilling it. The size of the host egg determines the number of parasitoid progeny that can successfully emerge from it. Experiments by Salt (1958) showed that there are certain limits beyond which a round object will be refused as a possible host by females of *Trichogramma*. He placed a small globule of mercury, smaller than a host egg, into a petri dish with a female *Trichogramma,* but the parasitoid was not attracted to it. Minute droplets of mercury were added to the globule until it reached a size acceptable to the parasitoid. The female then climbed the globule, examined it and attempted to pierce it with her ovipositor, sometimes for long periods. If droplets of mercury were added to the globule, the parasitoid no longer recognized it as a possible host because of its large size. Salt also found that the size limits of objects attacked varied with the size of females. Smaller females accepted objects of small diameter, which were rejected by larger females, although they would have been suitable hosts for their progeny (Salt 1940).

Price (1970) observed the behavior of *Pleolophus indistinctus,* an ichneumonid parasitoid of various sawfly cocoons. He found that that cocoon's roundness and length were important. The female parasitoid must be able to stand with all six legs on the cocoon, and also curl her antennae to tap the end for the host to be long enough to be accepted.

The behavioral relationships between a Ranunculus leaf mining fly, *Phytomyza ranunculi,* and its eulophid parasitoid *Kralochviliana* sp., were studied by Sugimoto (1977). He found that if a piece of colored paper, a ball of cotton, or a piece of nylon film replaced the natural host inside the mine, the parasitoid most often inserted its ovipositor into the cotton ball. The female apparently recognized the host, not by the stimuli produced by the host itself, but by the size of the swelling in the mine caused by the presence of the host, or a substitute host.

Schmidt (1974) investigated the host acceptance behavior of *Campoletis sonorensis* toward *Heliothis zea.* This author concluded that since the parasitoid deposits its egg in the hemocoele of its hosts, the heavily sclerotized integuments and thick subcuticular layers of fat of the large larvae may discourage successful insertion of the ovipositor and oviposition. Thus, large size is a limiting factor rather than a releasing stimulus.

The influence of shape, color, moisture content, state of the interior, surface texture, and size of primary parasitoids of the soft brown scale (*Coccus hesperidum*) attacked by the hyperparasitoid *Cheiloneurus noxius* was investigated by Weseloh (1971). He used artificial models placed within the scales to represent the primary parasitoid. A rounded or convex host model was discovered to be more acceptable than a flat or concave one, other properties having little influence on host discrimination by *C. noxius*.

Wilson et al. (1974) investigated physical and chemical stimuli associated with larvae of *Heliothis virescens* that influenced host acceptability by the parasitoid *C. sonorensis*. They found that shape had an appreciable effect: a straight cylindrical shape was more acceptable than a round or flat one, and size of host was less important to the oviposition process than these factors. A similar study by Burks and Nettles (1978) showed that shape was an important stimulus to females of *Eucelatoria* sp., a parasitoid of *H. virescens* larvae. The parasitoid females parasitized cuticles stretched over glass rods, approximately the natural shape of the host. However, they examined, but did not larviposit in a variety of flattened or semiflattened cuticles.

Laing (1937) found that the oval shape of muscoid fly puparia was important to their acceptance by the parasitoid *Nasonia vitripennis*. These observations were confirmed by Edwards (1954). He also found that a space must be present between the puparium wall and the pupa of the fly. This prerequisite was verified by Simmonds (1954) during studies with *Spalangia drosophilae*, a parasitoid of the puparia of *Drosophila*, and also by Wylie (1958) during studies with *N. vitripennis* a parasitoid of *Musca domestica*.

Griffiths (1961) studied the oviposition behavior of *Aptesis basizona*, a parasitoid of *Neodiprion sertifer* cocoons. The female mounts the cocoon and alternately taps it with her ovipositor and examines it with her antennae. The female then inserts her ovipositor for its full length into the cocoon. Later she inserts it less deeply. The host is apparently paralyzed by the first insertion. The egg is laid external to the pupa and is generally found loose within the cocoon or puparia. The host must be paralyzed because an active host could crush the egg against the inner cocoon or puparium wall.

Other parasitoids are stimulated to accept hosts by their surface texture. Vinson (1975b) concluded that a rough or sculptured surface is important or synergistic in eliciting host searching with the ovipositor by females of *Chelonus texanus* when attacking eggs of *H. virescens*, *Spodoptera exigua*, or *Spodoptera frugiperda*. Carton (1974) found that females of *Pimpla instigator* are stimulated to attack pupae within a paper tube more vigorously when the pupal surface is angular than when it is smooth.

Weseloh (1974) studied factors influencing host recognition by *Apanteles melanoscelus*, a parasitoid of young larvae of the gypsy moth. He showed that the hairiness of the host was an important stimulus that elicited examination and probing behavior by females of *A. melanoscelus*.

Movement

The movement of the host is an important cue in eliciting or preventing host acceptance by many species of parasitoids. Jackson (1968) studied the oviposition behavior of *C. cinctus,* a mymarid that parasitizes the eggs of Dytiscidae. She stated that *C. cinctus* will not parasitize an egg in which the host is ready to hatch, but ignores it or rejects it after a prick of the ovipositor. Nor will females of *Trichogramma* oviposit in a host egg that contains a moving embryo (Salt 1938). The acceptance of host eggs with advanced embryonic development probably would result in a loss of parasitoid progeny because the host would have hatched before the parasitoid could obtain sufficient food for development. Thus movement is a signal for unsuitability among egg parasitoids.

Van den Assem and Kuenen (1958) worked with *Choetospila elegans,* a parasitoid of *Sitophilus granarius,* that feeds inside whole kernels of grain. They noted that the parasitoid females carefully examined infested grains containing fourth instar larvae by drumming intensively with their antennae. After some time, the parasitoid turned around and oviposited in the grain. When the grain was opened, the parasitizad host larva was always near the surface. The female parasitoids did not oviposit on exposed larvae. The authors twice observed that a hollow grain was being inspected on its inside by two parasitoid adults. A third parasitoid walked on the grain and drummed vigorously finally ovipositing in the grain near the space occupied by the two parasitoids. The presence of the moving parasitoids within the grain had stimulated the third parasitoid to attack it from the outside.

The host acceptance behavior of females of *Coeloides brunneri,* a parasitoid of the Douglas fir beetle (*Dendroctonus pseudotsugae*) was studied by Ryan and Rudinsky (1962). They noted that parasitoid females located hosts by vibrations caused by their feeding rather than by odor. In the laboratory, the importance of vibrations was demonstrated when parasitoid females were induced to oviposit through a piece of isolated bark by scratching the undersurface gently with a pin.

Several workers have studied the parasitoid-host relationships of larvae that infest flour. Williams (1951) showed that excitation caused by intense host odor causes females of *Ventura* (*Nemeritis*) *canescens* to alight and search contaminated areas. When these areas are located, the parasitoid unsheaths its ovipositor and probes the flour. The recoil of the host when it is pricked further excites the parasitoid, which makes further frantic thrusts that increase its chance of penetrating the integument of the host and evoking oviposition.

Thorpe and Jones (1937) recorded that oviposition could be obtained in larvae of *Meliphora grisella,* a nonpreferred host of *V. canescens,* by contaminating these larvae with the odor of the normal host Mediterranean flour moth, *Anagasta kuehniella.* Oviposition was not elicited by dead hosts, although these were readily stabbed. Apparently, movement of the host when it is pricked was the final stimulus that induced the passage of the egg down the ovipositor. Fisher (1959) made similar observations on *Horogenes chrysostictos,* a parasitoid of *Ephestia sericarium*. This author concluded that the final stimulus to oviposition after penetration is probably the chemical condition of the host as perceived by sense organs at the tip of the ovipositor. These organs of discrimination have been described by Hawke et al. (1973).

The oviposition behavior of *Aphaereta pallipes,* a parasitoid of the onion maggot (*Hylemya antiqua*), was studied by Salkeld (1959). She found that females of *A. pallipes* were attracted to the host larva by the onion odor but ultimately found the host by a sense of touch. The parasitoid ignored immobile larvae, but apparently located by vibration larvae feeding within the onion, and oviposited in them through the onion tissue. Sugimoto (1977) found that artificially killed larvae of the Ranunculus leaf-mining fly were stung many times by its parasitoid *Kralochviliana* sp. without subsequent oviposition. Bragg (1974) made similar observations with freshly killed larvae of the artichoke plume moth (*Platyptilia carduidactyla*) offered to its parasitoid, *Phaeogenes cyriarae*.

Lloyd (1940) found that movement is necessary in larvae of the diamondback moth (*Plutella maculipennis*) before females of *Angitia cerophaga* or of *Apanteles plutellae* will attack and oviposit in them.

Females of *Perilampus hyalinus* were attracted and were stimulated to oviposit by the presence of feeding larvae of their host *Neodiprion swainei* (Tripp 1962). Whenever colonies of *N. swainei* larvae are disturbed, they jerk their heads up, more or less in unison, and exude a droplet of liquid from their mouths. This movement, probably originally a defense mechanism, now attracts the parasitoids, which deposit eggs within 30 cm of the sawfly colony. These eggs hatch, and the planidial first instar larvae stand erect, supported by the caudal sucker and balanced by the caudal stylets. The larvae remain motionless until disturbed by a nearby host larva, becoming attached to the moving host through circular head movement and later burrowing into the body cavity. Studies by Monteith (1955) indicated that females of *Drino bohemica* were also attracted by the movement of several species of sawfly larvae. He used a feather suspended in the arm of a Y-tube olfactometer to demonstrate that female parasitoids can be induced to probe a moving feather in the presence of sawfly odor (Monteith 1956). He concluded that if *D. bohemica* would probe something as different from a sawfly larva as a feather, movement must be an important stimulus.

Chemical Stimuli

Odor perceived at a distance through the antennal chemoreceptors of the female parasitoid is one of the most common cues used during host habitat and host location (see Chapters 4 and 5). Odors emitted by damaged leaves or plants (Arthur 1962, Camors and Payne 1972, McKenzie and Beirne 1972, Bragg

1974) attract parasitoids to the host-infested area and stimulate them to change from a random search to a host-seeking pattern. Some species of female tachinids, attracted by the odor of damaged leaves or needles, deposit microtype eggs on the plant that parasitize the host when they hatch in the larval intestine after ingestion (Simmonds 1944, Embree and Sisojevic 1965, Hassell 1968). Laing (1937) showed that females of *Alysia manducator* and of *N. vitripennis*, parasitoids of several carrion-feeding dipterous larvae and puparia, are attracted primarily by the odor of decaying meat. Bragg (1974) found that under laboratory conditions females of *P. cyriarae*, a parasitoid of *P. carduidactyla*, would oviposit into naked host larvae only if host plant juice was present; otherwise, the presence of the host plant was essential.

Stenobracon deesae, a parasitoid of certain lepidopterous borers in India, would not attack larvae of *Chilo zonellus* unless they were within tunnels in the host plant. Odor of the host plant apparently is also essential in this case (Narayanan and Chaudhuri 1954).

Many researchers (Ullyett 1935, Thorpe 1938, Bartlett 1941, Hsiao et al. 1966, Greany and Oatman 1972, Hendry et al. 1973, Nettles and Burks 1975, Lewis and Jones 1971, Henson et al. 1977, McKinney and Pass 1977, Roth et al. 1978) showed that females of other species of parasitoids are attracted by the odor of the frass of their hosts. Females of *Trichogramma* spp. are stimulated into an intensive search by host-seeking stimulants present in the scales left behind by ovipositing moths (Lewis et al. 1972, 1975a, 1975b, 1979).

Many parasitoids, led to the host by volatile odors, are stimulated to accept it by chemotactile stimuli on the cuticle or in the body, perceived through receptors on the antennae, tarsi, or ovipositor.

Herrebout (1969) found that the tachinid *Eucarcelia rutilla* will extrude its ovipositor and oviposit as soon as its front feet contact the normal cuticle of a *Bupalus piniarius* larva.

Vinson (1975b) investigated the source of the kairomone that stimulated females of the egg-larval parasitoid, *C. texanus* to accept as hosts, eggs of *H. virescens*. He found that an area containing any developing host egg, or ovariole tissue from the host, elicited strong antennal palpations, probing with the ovipositor, and oviposition as soon as the egg was found. He also discovered that extraction with Chlorox of the egg, or of egg material, rendered it unrecognizable as a host and negated oviposition activity. The kairomone was considered to be present in the cement binding the eggs together, or to the substrate.

Vinson and Lewis (1965) and Hays and Vinson (1971) found that oviposition stimuli were present in the cuticle of larvae of *H. virescens*. These kairomones could be extracted with alcohol or ether. Again, females of the parasitoid *Cardiochiles nigriceps* would examine, but would not oviposit on, a treated cuticle subsequently painted with the host-searching stimulant. A second chemical may be involved in host acceptance by this parasitoid. Other researchers have also studied the effects of cuticular extracts from *H. virescens*. The tachinid larval parasitoid *Eucelatoria* sp. was involved in the discovery that the kairomone that stimulates larviposition could be extracted from larval cuticles with a chloroform-methanol solution (2–1 v/v) (Burks and Nettles 1978). When this extract was applied to agar-filled extracted cuticles, the stimulating properties were partially restored. Natural levels of the reapplied stimulant gave the greatest larvipositional activity. When artificial hosts of the proper size and shape were formed and were coated with the stimulating extract, females of *Eucelatoria* sp. examined them carefully and repeatedly, without depositing maggots. These results indicate again that all chemicals responsible for larviposition were not present in the extract.

A third group of researchers (Wilson et al. 1974) also studied factors stimulating host acceptance by *C. sonorensis,* a larval parasitoid of *H. virescens*. They found that a chemical distributed throughout the host body provided a critical stimulus for oviposition. Further studies showed that acetone or methanol extracts of the host cuticle and frass produced the greatest activity.

Tucker and Leonard (1977) studied the role of kairomone in host acceptance by *Brachymeria intermedia,* a parasitoid of the gypsy moth. They extracted a kairomone from the pupal surface with *n*-hexane. Because washed pupae were not recognized as hosts, the authors concluded that the stimuli had been removed or had been reduced to a level below the threshold of activity. Other researchers (Weseloh 1974, Leonard et al. 1975) extracted from the cuticle of young gypsy moth larvae chemicals that elicited examination and probing behavior by females of *A. melanoscelus*.

Aphytis sp., an important parasitoid of the red citrus scale, *Aonidiella auranti,* in South Africa, examines its host with its feet and antennae (Quednau and Hubsch 1964). The chemical cue stimualting attack is a water-soluble substance present on the scale cover. The parasitoid has been found to examine disks of cellophane treated with the red scale extract. *Aphytis* sp. females also had to be in direct contact with the scale surface, as they could not find hosts covered with lens paper.

The chemicals influencing the secondary parasitoid *C. noxius* to deposit its eggs in the larvae of various primary parasitoids within the soft brown scale were investigated by Weseloh (1971) and by Weseloh and Bartlett (1971). Characteristics of the integument of unparasitized scales elicited the initiation of oviposition-related behavior, including probing by the ovipositor. The primary parasitoid was not detected until it was actually contacted by the ovipositor. Oviposition did not occur until substances, possibly proteins within the hemolymph of the primary parasitoid, were perceived by the ovipositor of the female of *C. noxius*.

Other kairomones that stimulate host acceptance by parasitoids have been discovered in special glands in the body of the host. Roth et al. (1978) reported that females of *Lixophaga diatraeae*, a parasitoid of the sugar cane borer (*Diatraea saccharalis*) larviposited on substances treated with extracts of the alimentary canal, but not on extracts from other parts of the host larva. Other tests showed that the primary stimulus for larviposition by *L. diatraeae* was detected by contact of the foretarsi of the parasitoid with host frass.

Kairomones that elicited host acceptance behavior by *C. sonorensis*, a parasitoid of the larvae of *H. zea*, were investigated by Schmidt (1974). He concluded that chemicals that elicit acceptance behavior are associated with larval frass, hemolymph, fat bodies, labial glands, and mandibular glands. Further tests seemed to indicate that mandibular glands may be the principal source of one or more kairomones.

Corbet (1971, 1973) found that the mandibular gland secretion of larvae of the Mediterranean flour moth contains a pheromone, which also acts as a kairomone because it elicits oviposition movements from its larval parasitoid, *V. canescens*.

Richerson and DeLoach (1972) showed that ovipositional stances and attacks by the parasitoid *Perilitis coccinellae*, were often increased by smearing coxal fluid from adult coccinellid hosts on metal and wooden models. This indicated that the coxal fluid contained an important kairomone for host acceptance by this species.

Factors stimulating oviposition by *Opius lectus* in apples infested by *Rhagoletis pomonella* were reported by Prokopy and Webster (1978). They found that several stimuli acted to retain the parasitoid on the apple, and to elicit antennal tapping and oviposition probes. These included unidentified fruit chemicals, characteristic fruit shape, size, and color; and an oviposition-deterring pheromone produced during oviposition in that apple by the first female apple maggots. This was the first demonstration that an oviposition-deterring pheromone produced by a phytophagous insect could act as a kairomone to one of its parasitoids.

Associative Learning and Host Acceptance

Ullyett (1936) considered that the host discrimination powers at the disposal of the female parasitoid were limited to instinct and to reactions to sensory impressions of shape, size, texture, and odor. Oviposition may be produced by a single stimulus, but more often it is caused by a chain of stimuli. An inherent instinct could provide the impulse for oviposition to occur in the substrate most suitable for the development of the progeny. However, the nutritional history of the adult female parasitoid can influence and stimulate its oviposition response in a similar manner. This influence, called "preimaginal conditioning," was examined by Thorpe and Jones (1937) and by Thorpe (1938, 1939) using larvae of *M. grisella* and *A. kuehniella* as hosts for *V. canescens*. The results showed that females reared on the preferred host *A. kuehniella* showed no response to the odor of *M.*

grisella, but those reared on the latter were conditioned to have an additional oviposition response to its odor. Such conditioning can take place either in the preimaginal stage, when the larva is feeding inside the host, or in the adult stage immediately on emergence (Thorpe 1938). This author concluded that conditioning is not a simple association between the effect of a given constituent of the environment (such as odor) and the eliciting of a single specialized response (oviposition). The association with a favorable environment in which the host itself occurs is also necessary.

Monteith (1955) concluded that cumulative conditioning does not occur in several unrelated groups of insects. The degree of conditioning for one particular host, or host tree, does not accumulate from one generation to the next, even by conditioning the same culture of insects to the same host, or tree, for a number of generations. He also found that olfactory stimuli frequently cause extrusion of the ovipositor, but contact with the host is necessary to induce probing and oviposition. Thorpe (1943) defined learning as a process that produces adaptive change in individual behavior as a result of experience. He did not expect it to be passed on from one generation to the next. Monteith (1963) showed that the number of parasitoids attracted to moving trays gradually increased when sawfly larvae exposed in trays were pushed repetitiously into cages containing females of *D. bohemica*. Once conditioned, these parasitoids continued to respond to tray movement for some time, even when the trays were empty.

Arthur (1966, 1967) has shown that females of the polyphagous pupal parasitoid *Itoplectis conquisitor* were capable of associative learning. They learned to associate tubes of various colors, sizes, and types of host shelter with the presence of hosts. This conditioned response was retained for 8–10 days after the conditioning period. As a result, Arthur (1967) concluded that each female probably makes its first host-searching flights more or less at random, especially when each generation is exposed to a different host species occupying different microhabitats. After having found and parasitized a few individuals of a species, the female learns to associate physical or chemical cues with that species of host. Hence, its parasitizing efficiency is greatly increased. These conclusions lead to the suggestion that the individual parasitoids should be first conditioned to a certain host, especially when a polyphagous parasitoid is to be released against a particular pest in a microhabitat. This may be done by placing the potential host on or in pieces of its food plant in the cage with the parasitoids in the laboratory, for variable periods before the parasitoids are transferred to the field. Parasitoids conditioned in this way would be in a searching-parasitizing behavior phase on release and would probably remain in the release area, thus increasing the chances of initial establishment on that host (see Chapter 8).

Arthur (1971) and Taylor (1974) demonstrated the occurrence of associative learning in the oligophagous parasitoid *V. canescens*. Arthur has shown that *V. canescens* could learn to associate the presence of hosts with the odor of a non-host-associated chemical, such as geraniol. Taylor (1974) concluded that females of *V. canescens* can learn to hunt in a novel environment. He also demonstrated that a model postulating the learning of two clues to host recognition fits better than a model postulating the learning of one. Possibly this explains why more than one type of stimulus is generally present in parasitoid-host relationships. Vinson et al. (1977), using *Bracon mellitor,* a larval parasitoid of the boll weevil (*Anthonomus grandis*), demonstrated that females of this parasitoid can learn associatively to respond to the antibiotic in hosts that have been reared on artificial diets. These authors warned that this rapid learning ability can complicate attempts to isolate kairomones determined genetically from other cues that have been learned. They said that care should be taken in experiments to use only females that have not been exposed to artificially reared hosts, during kairomone tests. Van Lenteren and Bakker (1975) found that learning was important in relation to discrimination by the parasitic wasp, *Pseudeucoila bochei.* Only female parasitoids that had been exposed to unparasitized hosts appeared able to refrain from ovipositing in parasitized hosts, when they were given a choice.

Host Acceptance of Artificial Diets

The ovipositor has been recognized for some time as a sensitive chemical probe for female parasitoids. Dethier (1947), King and Rafai (1970), Ganesalingam (1972), and Hawke et al. (1973) have described some of the sensory receptors present on the ovipositors of various species. Great progress toward the identification of chemical stimuli that elicited host acceptance by the parasitoid *I. conquisitor* was made by Arthur et al. (1969 and 1972) and by Hegdekar and Arthur (1973). In the laboratory, females of *I. conquisitor* were attracted to paper tubes, which they examined with the antennae and probed with the ovipositor. If pupae were placed in these tubes, they became parasitized. Eggs were also deposited in Parafilm® tubes containing pupal hemolymph of the greater wax moth (*Galleria mellonella*), but not in distilled water or in saline solution (Figure 6.3). Once it was known that *I. conquisitor* females would oviposit in *G. mellonella* hemolymph, it was realized that host movement was not essential for this process, and efforts were made to determine what fraction of the hemolymph stimulated oviposition. Extracted hemolymph was passed through a Sephadex® column. Oviposition occurred in a distinct fraction of the hemolymph that contained hexoses and approximately 19 common amino acids.

Parallel studies showed that *I. conquisitor* would oviposit in a neutralized (pH 7.0) mixture of 17 amino acids used by Bracken (1965) in a diet for adult *E. comstockii*. A few eggs were also laid in five neutralized solutions of individual amino acids, alanine and serine being the most active. A standard solution containing serine (0.5 M) and arginine (0.05 M) was selected for comparison to

FIGURE 6.3. Females of *Itoplectis conquisitor* ovipositing into a Parafilmfl tube containing one of various oviposition stimulants.

the oviposition results of other solutions. Various inorganic ions and sugars were added to this solution. Of these, magnesium chloride (0.25 M) and trehalose (1.0 M) elicited the laying of the greatest number of eggs. As a result, various combinations of chemicals were tested to stimulate oviposition (Table 6.1). Solutions containing three different mixtures of amino acids and magnesium chloride at three concentrations were significantly better than those containing serine, arginine, or trehalose, with or without magnesium chloride. They were also significantly better for stimulating oviposition than a solution of hemolymph of *G. mellonella* diluted with water (1–2 v/v). These studies confirmed that the oviposition of *I. conquisitor* is easily influenced by chemicals found within prospective hosts. The data presented above in the section on chemical stimuli indicated that this situation is widespread among parasitoid-host relationships.

These findings are important for the development of mass propagation methods for parasitoids and should have great economic significance. The creation of a proper artificial medium requires a thorough knowledge not only of

TABLE 6.1. The Average Number of Eggs Deposited by Females of *Itoplectis conquisitor* Exposed to Solutions Containing Various Combinations of Oviposition Stimulants

Chemical and Concentration (M)							Average Number of Eggs Bioassay[a]
MgCl 0.025	Arginine 0.05	Serine 0.5	Isoleucine 0.065	Methionine 0.065	Leucine 0.065	Trehalose 1.0	
+	+						0 B[b]
	+					+	0.33 B
		+				+	1.0 B
+		+					2.6 B
	+	+					16.6 A
+		+	+				28.3 A
+		+		+			30.6 A
+		+			+		40.0 A
	+	+				+	40.3 C
+	+	+				+	52.6 C
+	+	+					54.3 C
+	+	+		+			78.0 D
+	+	+	+				80.0 D
+	+	+			+		95.6 D
Wax moth hemolymph and water (diluted 1−2)							50.6 C

(From Arthur et al. 1972. Courtesy of the Entomological Society of Canada).

[a] Calculated from three bioassays containing two tubes each. Each tube was exposed in two cages of 15 females for a total of 2 hours.

[b] Yields designated with the same letter are not statistically different at the 5% level but those designated by different letters are different, as determined by the Mann-Whitney U-test (Siegel 1956).

the nutritional requirements for *I. conquisitor* (Yazgan and House 1970, Yazgan 1972), but also of the requirements for host acceptance. Recently House (1978), using a Parafilm® tube developed by Arthur et al. (1969), reported the first *in vitro* oviposition and rearing of a parasitic insect on an artificial host consisting of an encapsulated synthetic medium.

Progress in the development of an acceptable artificial host for the egg parasitoid *Trichogramma californicium* has been reported by Rajendram and Hagen (1974). These authors reported oviposition in parafilm droplets containing physiological saline, Neisenheimer's saline solution, and some mixtures of amino acids, but not in droplets containing distilled water. More recently Hoffman et al. (1975) reported that *Trichogramma* females readily oviposited viable eggs in artificial eggs filled with hemolymph from *H. zea*.

CONCLUSION

Host acceptance is the third phase (following host habitat location and host finding) in the process of host selection by parasitoids. During this phase, the female parasitoids may use physical and/or chemical cues in host recognition. These cues include host size, shape, surface texture, and movement, and the presence of volatile or nonvolatile chemicals. The cues are now known to be either instinct determined or learned by association.

Our knowledge of the cues involved in host acceptance will facilitate the development of inexpensive mass rearing techniques. Their use before and during parasitoid release will greatly increase the chances of initial establishment and increased parasitism at the release area.

ACKNOWLEDGMENTS

I thank T. H. Stovell who photographed Figures 6.1–6.3 at the Canada Agriculture, Research Institute, Belleville, Ontario (closed in 1972); and G. R. F. Davis and J. F. Doane of the Canada Agriculture, Research Station, Saskatoon, Saskatchewan, for editing the manuscript.

REFERENCES

Arthur, A. P. 1961. The cleptoparasitic habits and the immature stages of *Eurytoma pini* Bugbee (Hymenoptera: Chalcidae), a parasite of the European pine shoot moth, *Rhyacionia buoliana* (Schiff.) Lepidoptera: Olethreutidae). Can. Entomol. 93:655–660.

Arthur, A. P. 1962. Influence of host tree on abundance of *Itoplectis conquisitor* (Say) (Hymenoptera: Ichneumonidae), a polyphagous parasite of the European pine shoot moth, *Rhyacionia buoliana* (Schiff.) (Lepidoptera: Olethreudidae). Can. Entomol. 94:337–347.

Arthur, A. P. 1963. Life histories and immature stages of four ichneumonid parasites of the European pine shoot moth *Rhyacionia buoliana* (Schiff.) in Ontario. Can. Entomol. 95:1078–1091.

Arthur, A. P. 1966. Associative learning in *Itoplectis conquisitor* (Say) (Hymenoptera: Ichneumonidae). Can. Entomol. 98:213–223.

Arthur, A. P. 1967. Influence of position and size of host shelter on host-searching by *Itoplectis conquisitor* (Hymenoptera: Ichneumonidae). Can. Entomol. 99:877–886.

Arthur, A. P. 1971. Associative learning by *Nemeritis canescens* (Hymenoptera: Ichneumonidae). Can. Entomol. 103:1137–1141.

Arthur, A. P., J. E. R. Stainer, and A. L. Turnbull. 1964. The interaction between *Orgilus obscurator* (Nees) (Hymenoptera: Braconidae) and *Temelucha interruptor* (Grav.) (Hymenoptera: Ichneumonidae), parasites of the pine shoot moth, *Rhyacionia buoliana* (Schiff.) (Lepidoptera: Olethreutidae). Can. Entomol. 96:1030–1034.

Arthur, A. P., B. M. Hegdekar, and L. Rollins. 1969. Component of the host haemolymph that induces oviposition in a parasitic insect. Nature (London) 223:966–967.

Arthur, A. P., B. M. Hegdekar, and W. W. Batsch. 1972. A chemically defined synthetic medium that induces oviposition in the parasite *Itoplectis conquisitor* (Hymenoptera: Ichneumonidae). Can. Entomol. 104:1251–1258.

Bartelett, K. A. 1941. The biology of *Metagonistylum minense* Tns., a parasite of the sugar cane borer. Puerto Rico Experiment Station, U.S. Department of Agriculture Bulletin No. 40.

Bracken, G. K. 1965. Effects of dietary components on fecundity of the parasitoid *Exeristes comstockii* (Cress) (Hymenoptera: Ichneumonidae). Can. Entomol. 97:1037–1041.

Bragg, D. E. 1974. Ecological and behavioral studies of *Phaeogenes cyriarae:* Ecology, host specificity; searching and oviposition; and avoidance of superparasitism. Ann. Entomol. Soc. Am. 67:931–936.

Burks, M. L., and W. C. Nettles, Jr. 1978. *Eucelatoria* sp.: Effects of cuticular extracts from *Heliothis virescens* and other factors on oviposition. Environ. Entomol. 7:897–900.

Calvert, D. J. 1973. Experimental host preferences of *Monoctonus paulensis* (Hymenoptera: Braconidae), including a hypothetical scheme of host selection. Ann. Entomol. Soc. Am. 66:28–33.

Camors, F. B., Jr., and T. L. Payne. 1972. Response of *Heydenia unica* (Hymenoptera: Pteromalidae) to *Dendroctonus frontalis* (Coleoptera: Scolytidae) pheromones and a host-tree terpene. Ann. Entomol. Soc. Am. 65:31–33.

Campbell, R. W. 1963. Some ichneumonid-sarcophagid interactions in the gypsy moth *Porthetria dispar* (L.) (Lepidoptera: Lymantriidae). Can. Entomol. 95:337–345.

Carton, Y. 1974. Biologie de *Pimpla instigator* (Ichneumonidae: Pimplinae). III. Analyse expérimentale du processus de reconnaissance de l'hôte-chrysalide. Entomol. Exp. Appl. 17:265–278.

Clausen, C. P. 1940. Entomophagous Insects, McGraw-Hill, New York.

Corbet, S. A. 1971. Mandibular gland secretion of larvae of the flour moth, *Anagasta kuehniella*, contains an epideictic pheromone and elicits oviposition movements in a hymenopteran parasite. Nature (London) 232:481–484.

Corbet, S. A. 1973. Concentration effects and the response of *Nemeritis canescens* to a secretion of its host. J. Insect Physiol. 19:2119–2128.

Dethier, V. G. 1947. The response of hymenopterous parasites to chemical stimulation of the ovipositor. J. Exp. Zool. 105:199–207.

Doutt, R. L. 1959. The biology of parasitic Hymenoptera. Annu. Rev. Entomol. 4:161–182.

Edwards, R. L. 1954. The host-finding and oviposition behavior of *Mormoniella vitripennis* (Walker) (Hym., Pteromalidae), a parasite of muscoid flies. Behaviour. 7:88–112.

Embree, D. G., and P. Sisojevic. 1965. The bionomics and population density of *Cyzenis albicans* (Fall.) (Tachinidae: Diptera) in Nova Scotia. Can. Entomol. 97:631–639.

Fisher, R. C. 1959. Life history and ecology of *Horogenes chrysostictos* Gmelin (Hymenoptera: Ichneumonidae), a parasite of *Ephestia sericarium* Scott (Lepidoptera, Phycitidae). Can. J. Zool. 37:429–446.

Ganesalingam, V. K. 1972. Anatomy and histology of the sense organs of the ovipositor of the ichneumonid wasp, *Devorgilla canescens*. J. Insect Physiol. 18:1857–1867.

Greany, P. D., and E. R. Oatman. 1972. Analysis of host discrimination in the parasite *Orgilus lepidus* (Hymenoptera: Braconidae) Ann. Entomol. Soc. Am. 65:377–383.

Griffiths, K. J. 1961. The life history of *Aptesis basizona* (Grav.) on *Neodiprion sertifer* (Geoff.) in Southern Ontario. Can. Entomol. 93:1005–1010.

Hassell, M. P. 1968. The behavioral response of a tachinid fly *(Cyzenis albicans* (Fall.) to its host, the winter moth *(Operophtera brumata* (L.) J. Anim. Ecol. 37:627–639.

Hawke, S. D., R. D. Farley, and P. D. Greany. 1973. The fine structure of sense organs in the ovipositor of the parasitic wasp, *Orgilus lepidus* Muesebeck. Tissue Cell 5:171–184.

Hays, D. B., and S. B. Vinson. 1971. Acceptance of *Heliothis virescens* (F.) (Lepidoptera, Noctuidae) as a host by the parasite *Cardiochiles nigriceps* Viereck (Hymenoptera, Braconidae). Anim. Behav. 19:344–352.

Hegdekar, B. M., and A. P. Arthur. 1973. Host hemolymph chemicals that induce oviposition in the parasite *Itoplectis conquisitor* (Hymenoptera: Ichneumonidae). Can. Entomol. 105:787–793.

Hendry, L. B., P. D. Greany, and F. J. Gill. 1973. Kairomone mediated host-finding behavior in the parasitic wasp *Orgilus lepidus*. Entomol. Exp. Appl. 16:471–477.

Henson, R. D., S. B. Vinson, and C. S. Barfield. 1977. Oviposition behavior of *Bracon mellitor* Say, a parasitoid of the boll weevil (*Anthonomus grandis* Boh.). III. Isolation and identification of natural releasers of oviposition probing. J. Chem. Ecol. 3:151–158.

Herrebout, W. M. 1969. Some aspects of host selection in *Eucarcelia rutilla* Vill. (Diptera: Tachinidae). Neth. J. Zool. 19:1–104.

Hoffman, J. D., C. M. Ignoffo, and W. A. Dickerson. 1975. *In vitro* rearing of the endoparasitic wasp. *Trichogramma pretiosum*. Ann. Entomol. Soc. Am. 68:335–336.

House, H. L. 1978. An artificial host encapsulated synthetic medium for *in vitro* oviposition and rearing the endoparasitoid *Itoplectis conquisitor* (Hymenoptera: Ichneumonidae). Can. Entomol. 110:331–333.

Hsiao, T. H., F. G. Holdaway, and H. C. Chiang. 1966. Ecological and physiological adaptations in insect parasitism. Entomol. Exp. Appl. 9:113–123.

Jackson, D. J. 1968. Observations on the female reproductive organs and the poison apparatus of *Caraphractus cinctus* Walker (Hymenoptera: Mymaridae). Zool. J. Linn. Soc. 48:59–81.

Juillet, J. A. 1960. Immature stages, life histories, and behavior of two hymenopterous parasites of the European pine shoot moth, *Rhyacionia buoliana* (Schiff.) (Lepidoptera: Olethreutidae). Can. Entomol. 92:342–346.

King, P. E., and J. Rafai. 1970. Host discrimination in a gregarious parasitoid *Nasonia vitripennis* (Walker) (Hymenoptera: Pteromalidae). J. Exp. Biol. 53:245–254.

Klomp, H., and B. J. Teerink. 1962. Host selection and number of eggs per oviposition in the egg-parasite *Trichogramma embryophagum* Htg. Nature (London) 195:1020–1021.

Laing, J. 1937. Host finding by insect parasites. 1. Observations on the finding of hosts by *Alysia manducator, Mormoniella vitripennis* and *Trichogramma evanescens*. J. Anim. Ecol. 6:298–317.

Leius, K. 1961. Influence of food on fecundity and longevity of adults of *Itoplectis conquisitor* (Say) (Hymenoptera: Ichneumonidae). Can. Entomol. 93:771–780.

Leonard, D. E., B. A. Bierl, and M. Beroza. 1975. Gypsy moth kairomones influencing behavior of the parasitoids *Brachymeria intermedia* and *Apanteles melanoscelus*. Environ. Entomol. 4:929–930.

Lewis, W. J., and R. L. Jones. 1971. Substance that stimulates host seeking by *Microplitis croceipes* (Hymenoptera: Braconidae), a parasite of *Heliothis* species. Ann. Entomol. Soc. Am. 64:471–473.

Lewis, W. J., R. L. Jones, and A. N. Sparks. 1972. A host-seeking stimulant for the egg parasite *Trichogramma evanescens:* Its source and a demonstration of its laboratory and field activity. Ann. Entomol. Soc. Am. 65:1087–1089.

Lewis, W. J., R. L. Jones, D. A. Nordlund, and A. N. Sparks. 1975a. Kairomones and their use for management of entomophagous insects: I. Evaluation for increasing rates of parasitism by *Trichogramma* spp. in the field. J. Chem. Ecol. 1:343–347.

Lewis, W. J., R. L. Jones, D. A. Nordlund, and H. R. Gross, Jr. 1975b. Kairomones and their use for management of entomophagous insects. II. Mechanisms causing increase in rate of parasitization by *Trichogramma* spp. J. Chem. Ecol. 1:349–360.

Lewis, W. J., M. Beevers, D. A. Nordlund, H. R. Gross, Jr., and K. S. Hagen. 1979. Kairomones and their use for management of entomophagous insects. IX. Investigations of various kairomone-treatment patterns for *Trichogramma* spp. J. Chem. Ecol. 5:673–680.

Lloyd, D. C. 1940. Host selection by hymenopterous parasites of the moth *Plutella maculipennis* Curtis. Proc. R. Soc. London, Ser. B 128:451–484.

McKenzie, L. M., and B. P. Beirne. 1972. A grape leafhopper, *Erythroneura ziczac* (Homoptera: Cicadelliae) and its mymirid (Hymenoptera) egg-parasite in the Okanagan Valley, British Columbia. Can. Entomol. 104:1229–1233.

McKinney, T. R., and B. C. Pass. 1977. Olfactometer studies of host seeking in *Bathyplectes curculionis* Thoms. (Hymenoptera: Ichneumonidae). J. Kansas Entomol. Soc. 50:102–112.

Monteith, L. G. 1955. Host preferences of *Drino bohemica* Mesn. (Diptera: Tachinidae) with particular reference to olfactory responses. Can. Entomol. 87:509–530.

Monteith, L. G. 1956. Influence of host movement on selection of hosts by *Drino bohemica* Mesn. (Diptera: Tachinidae) as determined in an olfactometer. Can. Entomol. 88:583–586.

Monteith, L. G. 1963. Habituation and associative learning in *Drino bohemica* Mesn. (Diptera: Tachinidae). Can. Entomol. 95:418–426.

Muesebeck, C. F. W., K. V. Krombein, and H. K. Townes. 1951. Hymenoptera of America north of Mexico. Department of Agriculture Monograph No. 2.

Narayanan, M. A., and R. P. Chaudhuri. 1954. Studies on *Stenobracon deesae* (Cam.), a parasite on certain lepidopterous borers of graminaceous crops in India. Bull. Entomol. Res. 45:647–659.

Nettles, W. C., Jr., and M. L. Burks. 1975. A substance from *Heliothis virescens* larvae stimulating larviposition by females of the tachinid, *Archytas marmoratus*. J. Insect Physiol. 21:965–978.

Price, P. W. 1970. Trail odors: Recognition by insects parasitic on cocoons. Science 170:546–547.

Prokopy, R. J., and R. P. Webster. 1978. Oviposition deterring pheromone of *Rhagoletis pomonella*, a kairomone for its parasitoid *Opius lectus*. J. Chem. Ecol. 4:481–494.

Quednau, F. W., and H. M. Hubsch. 1964. Factors influencing the host-finding and host-acceptance pattern in some *Aphytis* species (Hymenoptera: Aphelinidae). S. Afr. J. Agric. Sci. 7:543–554.

Rajendram, G. F., and K. S. Hagen. 1974. *Trichogramma* oviposition into artificial substrates. Environ. Entomol. 3:399–401.

Richerson, J. V., and C. J. DeLoach. 1972. Some aspects of host selection by *Perilitus coccinellae*. Ann. Entomol. Soc. Am. 65:834–839.

Roth, J. P., E. G. King, and A. C. Thompson. 1978. Host location behavior by the tachinid, *Lixophaga diatraeae*. Environ. Entomol. 7:794–798.

Ryan, R. B., and J. A. Rudinsky. 1962. Biology and habits of the Douglas fir beetle parasite *Coeloides brunneri* Viereck (Hymenoptera, Braconidae), in western Oregon. Can. Entomol. 94:748–763.

Salkeld, E. H. 1959. Notes on anatomy, life history and behaviour of *Aphereta pallipes* (Say) (Hymenoptera: Braconidae), a parasite of the onion maggot, *Hylemia antiqua* (Meig.) Can. Entomol. 91:93–97.

Salt, G. 1935. Experimental studies on insect parasitism. III. Host selection. Proc. R. Soc. London, Ser. B 117:413–435.

Salt, G. 1937. The sense used by *Trichogramma* to distinguish between parasitized and unparasitized host. Proc. R. Soc. London, Ser. B 122:57–75.

Salt, G. 1938. Experimental studies in insect parasitism. VI. Host suitability. Bull. Entomol. Res. 29:223–246.

Salt, G. 1940. Experimental studies in insect parasitism. VII. The effects of different hosts on the parasite *Trichogramma evanescens* Westw. (Hym., Chalcidoidea) Proc. R. Entomol. Soc. London, Ser. A 15:81–124.

Salt, G. 1958. Parasite behavior and the control in insect pest. Endeavour July:145–148.

Schmidt, G. T. 1974. Host acceptance behavior of *Campoletis sonorensis* toward *Heliothis zea*. Ann. Entomol. Soc. Am. 67:835–844.

Siegel, S. 1956. Nonparametric statistics for the behavioral sciences. McGraw-Hill, New York.

Simmonds, F. J. 1944. The propagation of insect parasites on unnatural hosts. Bull. Entomol. Res. 35:219–226.

Simmonds, F. J. 1954. Host finding and selection by *Spalangia drosophilae* Ashm. Bull. Entomol. Res. 45:527–537.

Spradbery, J. P. 1969. The biology of *Pseudorhyssa sternata,* a cleptoparasite of siricid woodwasps. Bull. Entomol. Res. 59:291–297.

Sugimoto, T. 1977. Ecological studies on the relationship between the Ranunculus leaf mining fly, *Phytomyza ranunculi* Schrank (Diptera: Agromyzidae) and its parasite, *Kralochviliana* sp. (Hymenoptera: Eulophidae) from the viewpoint of spatial structure. Analysis of searching and attacking behaviors of the parasite. Appl. Entomol. Zool. 12:87–103.

Taylor, R. J. 1974. Role of learning in insect parasitism. Ecol. Monogr. 44:89–104.

Thompson, W. R., and H. L. Parker. 1927. The problem of host relations with special reference to entomophagous parasites. Parasitology 19:1–34.

Thorpe, W. H. 1938. Further experiments on olfactory conditioning in a parasitic insect. The nature of the conditioning process. Proc. R. Soc. London Ser. B 126:370–397.

Thorpe, W. H. 1939. Further studies on preimaginal olfactory conditioning in insects. Proc. R. Soc. London, Ser. B 127:424–433.

Thorpe, W. H. 1943. Types of learning in insects and other arthropods. Part 1. Br. J. Psychol. 33:220–234.

Thorpe, W. H., and F. G. W. Jones. 1937. Olfactory conditioning in a parasitic insect and its relation to the problem of host selection. Proc. R. Soc. London, Ser. B 124:56–81.

Tripp, H. A. 1962. The biology of *Perilampus hyalinus* Say (Hymenoptera: Perilampidae), a primary parasite of *Neodiprion swainei* Midd. (Hymenoptera: Diprionidae) in Quebec, with descriptions of the egg and larval stages. Can. Entomol. 94:1250–1270.

Tucker, J. E., and D. E. Leonard, 1977. The role of kairomones in host recognition and host acceptance behavior of the parasite *Brachymeria intermedia*. Environ. Entomol. 6:527–531.

Turnbull, A. L., and D. A. Chant. 1961. The practice and theory of biological control of insects in Canada. Can. J. Zool. 39:697–753.

Ullyett, G. C. 1935. Notes on *Apanteles sesamiae* Cam. A parasite of the maize stalk-borer (*Busseola frisca* Fuller) In South Africa. S. Afr. Bull. Entomol. Res. 26:253–262.

Ullyett, G. C. 1936. Host selection by *Microplectron fuscipennis* Zett. (Chalcididae: Hymenoptera). Proc. R. Soc. London, Ser. B 120:253–291.

van den Assem, J., and D. J. Kuenen. 1958. Host finding of *Choetospila elegans* Westw. (Hym. Chalcid) a parasite of *Sitophilus granarius* L. (Coleopt. Curcul.) Entomol. Exp. Appl. 1:174–180.

van Lenteren, J. C., and K. Bakker. 1975. Discrimination between parasitized and unparasitized hosts in the parasitic wasp *Pseudeucoila bochei:* A matter of learning. Nature (London) 254:417–419.

Vinson, S. B. 1975a. Biochemical coevolution between parasitoids and their hosts. P. 14–48. *In* P. W. Price (ed.), Evolutionary Strategies of Parasitic Insects and Mites. Plenum Press, New York.

Vinson, S. B. 1975b. Source of material in the tobacco budworm which initiates host-searching by the egg-larval parasitoid, *Chelonus texanus*. Ann. Entomol. Soc. Am. 68:381–384.

Vinson, S. B. 1976. Host selection by insect parasitoids. Annu. Rev. Entomol. 21:109–133.

Vinson, S. B. 1977. Behavioral chemicals in the augmentation of natural enemies. P. 237–279. *In* R. L. Ridgway and S. B. Vinson (eds.), Biological Control by Augmentation of Natural Enemies. Plenum Press, New York.

Vinson, S. B., and W. J. Lewis. 1965. A method of host selection by *Cardiochiles nigriceps*. J. Econ. Entomol. 58:869–887.

Vinson, S. B., and C. S. Barfield, and R. D. Hensen. 1977. Oviposition behavior of *Bracon mellitor*, a parasitoid of the boll weevil *(Anthonomus grandis)*. II. Associative learning, Physiol. Entomol. 2:157–164.

Weseloh, R. M. 1971. Influence of primary (parasite) hosts on host selection of the hyperparasite *Cheiloneurus noxius* (Hymenoptera: Encyrtidae). Ann. Entomol. Soc. Am. 64:1233–1236.

Weseloh, R. M. 1974. Host recognition by the gypsy moth larval parasitoid, *Apanteles melanoscelus,* Ann. Entomol. Soc. Am. 67:583–587.

Weseloh, R. M., and B. R. Bartlett. 1971. Influence of chemical characteristics of the secondary scale host on host selection behavior of the hyper-parasite *Cheiloneurus noxius* (Hymenoptera: Encyrtidae). Ann. Entomol. Soc. Am. 64:1259–1264.

Williams, J. R. 1951. The factors which promote and influence the oviposition of *Nemeritis canescens* Grav. (Ichneumonidae, Ophionidae). Proc. R. Entomol. Soc. London, Ser. A 26:49–58.

Wilson, D. D., R. L. Ridgway, and S. B. Vinson. 1974. Host acceptance and oviposition behavior of the parasitoid *Campoletis sonorensis* (Hymenoptera: Ichneumonidae). Ann. Entomol. Soc. Am. 67:271–274.

Wylie, H. G. 1958. Factors that affect host-finding by *Nasonia vitripennis* (Walk.) (Hymenoptera: Pteromalidae). Can. Entomol. 90:597–608.

Yazgan, S. 1972. A chemically defined synthetic diet and larval nutritional requirements of the endoparasitoid *Itoplectis conquisitor* (Hymenoptera). J. Insect Physiol. 18:2123–2141.

Yazgan, S., and H. L. House. 1970. A hymenopterous insect, the parasitoid *Itoplectis conquisitor,* reared axenically on a chemically defined synthetic diet. Can. Entomol. 102:1304–1306.

CHAPTER SEVEN

PREY SELECTION

PATRICK D. GREANY

AR–SEA–USDA
Insect Attractants, Behavior and Basic Biology Research Laboratory
Gainesville, Florida

KENNETH S. HAGEN

Division of Biological Control
Department of Entomology
University of California at Berkeley
Albany, California

Allelochemics play a significant role in prey selection by many arthropod predators of insects and mites. Along with physical stimuli, chemical signals often mediate prey finding and prey recognition. These chemical messengers generally fall into two categories: kairomones and allomones. Most commonly, prey-produced kairomones have been implicated as attractants, arrestants, or phagostimulants. In a few cases, however, predators produce substances that attract prey or otherwise exploit prey chemical communication systems. These substances may therefore be regarded as allomones, since they are interspecific chemical communicating agents that benefit the emitter rather than the receiver (Nordlund and Lewis 1976). Defensive secretions produced by potential prey to deter predators also play a role in the chemical ecology of predators, but this topic has been reviewed by Blum (1978) and is not considered here. Reviews by Wilson (1971) and Weaver (1978) also include much relevant information on the chemical ecology of insect predators.

Predators that use prey-produced kairomones are discussed first, followed by a description of predacious insects and spiders that produce allomones to assist them in capturing their prey. Efforts to use allelochemics to enhance predator utilization in biological control are described briefly in conclusion.

USE OF KAIROMONES IN PREY SELECTION

Kairomones are recognized to be important to the adults and/or immature stages of many predators, including insect and mites. Kairomones assist some predators in prey finding and recognition, but for other species, kairomones mediate prey recognition alone, with prey finding being accomplished by other means. Examples of these uses of kairomones by representative insect and mite predators are presented below.

Predacious Insects

Chrysopidae

The common green lacewing *(Chrysopa carnea)* is a good example of a predator that is predacious only in the larval stage (Figure 7.1); the adults obtain nutrients from pollen and honeydew (Hagen et al. 1976a). Green lacewing adults can be attracted to proteinaceous artificial honeydews (Hagen et al. 1971), which promote both feeding and oviposition. Similar responses of green lacewing adults to natural honeydew ensure availability of prey for the larvae when they emerge, so that the adult females play an important role in prey finding by selecting a suitable environment for the larvae before ovipositing.

Studies to define the attractive agent in artificial honeydews implicated the amino acid tryptophan as the attractant source (Hagen et al. 1976a). Because of the low volatility of tryptophan it was hypothesized that attraction was probably due to a tryptophan degradation product. This was tested by van Embden and Hagen (1976), who found that lacewings induced to fly while tethered in an olfactometer responded positively to a volatile tryptophan oxidation product, indole acetaldehyde. The attractiveness of this substance in the field has not yet been confirmed. *Chrysopa* adults also have been found to respond to chemical attractants such as methyl eugenol and terpenyl acetate (reviewed by Hagen et al. 1976a) and caryophyllene (Flint et al. 1979). These responses are suspected to relate to prey habitat finding.

While the studies on attraction of *C. carnea* to honeydew help explain the means used by these predators to find an appropriate site for oviposition, actual prey finding is left to the larvae. Fleschner (1950) found no evidence for perception of citrus red mites *(Panonychus citri)* at a distance by *C. carnea* larvae. He stated that actual contact with a prey was required before perception occurred. In one of his experiments, however, he separated the *Chrysopa* larvae from mites by muslin cloth, and noted that the larvae probed through the cloth and attacked the mites, which suggests that the larvae may have some ability to detect prey at very short range. He also found that *Chrysopa* larvae would probe through the

FIGURE 7.1. Green lacewing larva attacking an aphid.

cloth into sites that had previously held prey, but from which the prey had been removed, indicating that the prey may leave behind a kairomone.

More recently, Lewis et al. (1977b) found that scales from ovipositing *Heliothis zea* moths stimulated *C. carnea* larvae to search intensively in the vicinity of the scales. They suggested that this response was elicited by a kairomone associated with the moth scales. In addition, Nordlund et al. (1977) found evidence for a kairomone associated with the *H. zea* eggs themselves. Thus, *C. carnea* larvae respond in a manner similar to that of the parasitoid *Trichogramma evanescens* by intensifying their search upon encountering kairomones associated with their hosts (Lewis et al. 1977a). Fleschner (1950) observed that *C. carnea* larvae increased their rate of turning upon encountering a prey, a strategy that would be well suited to discovery of additional prey of species that tend to be clumped, such as aphids. Bänsch (1966) observed similar behavior by the larvae of *Chrysopa vulgaris*.

Hagen et al. (1976b) suggested that the composition of the cuticle of prey may be important in eliciting the biting or sucking response of predators that seem to find their prey by moving about until they directly contact them. The influence of the internal composition of the prey on feeding by *C. carnea* larvae might be ascertained by modifying the encapsulated diet technique developed by Hagen and Tassan (1965) for this species. Instead of providing a completely nutritious medium, defined solutions of individual amino acids or other agents could be encapsulated and tested as phagostimulants. Arthur et al. (1972) successfully used this approach to test the ovipositional response of the parasitoid *Itoplectis conquisitor* to specific biochemicals within surrogate hosts.

Syrphidae and Cedidomyidae

Some syrphid species are attracted to artificial honeydew (Hagen et al. 1971, Ben Saad and Bishop 1976) and to a substance Ben Saad and Bishop (1976) termed "aphid juice," but these substances did not promote oviposition by the syrphid species in their studies. However, Bombosch and Volk (1966) demonstrated that aphid odor did stimulate oviposition by *Syrphis corollae*.

Prey finding by syrphid larvae apparently is accomplished by detection of the prey by direct contact during random searching of the foliage rather than by perception of prey at a distance (Bänsch 1966). On the other hand, larvae of the cecidomyid *Aphidoletes aphidimyza* reportedly perceive aphids over a short distance by detecting the aphids' odor or that of their honeydew (Wilbert 1974).

Coccinellidae and Malachiidae

A few studies on prey selection by coccinellids and malachiids have implicated plant-produced compounds in habitat and prey finding. For example, Kesten (1969) found that adults of the coccinellid *Anatis ocellata* find their usual prey, the pine aphid *(Schizolachnus pineti)*, by responding initially to aromatic substances from pine needles. Flint et al. (1979) found that the malachiid *Collops vittatus*, a common predator in cotton fields, was attracted to the terpenoid caryophyllene oxide; they interpreted this as a habitat-finding mechanism, since the compound is produced by cotton foliage.

Artificial honeydew in the form of Wheast® plus sucrose serves to arrest movement of a variety of immature coccinellids (including *Hippodamia convergens, Coccinella novemnotata*, and *Coleomegilla maculata*) that contact it in the course of their movements, but it is not attractive to the adult beetles (Hagen et al. 1971, Nichols and Neel 1977).

In contrast, adults of *Coccinella septempunctata* have been found to use olfactory cues to find aphids; but rather than responding to aphid honeydew or odors produced by the aphids themselves, these beetles use the pheromone trails of aphid-tending ants as a means of finding the aphids (A. P. Bhatkar, personal communication).

There is no evidence that coccinellid larvae detect their prey at a distance. Instead, prey appear to be perceived only on direct contact (Fleschner 1950,

Banks 1957, Dixon 1959, Kaddau 1960, Bansch 1966, Rowlands and Chapin 1978). Whether prey recognition is mediated by chemical or tactile stimuli is not known, but it seems likely that contact chemical stimuli are involved. Hagen et al. (1976b) suggested that the composition of the cuticular waxes of the prey may be involved in acceptance, since the paraffin coating of artificial diets developed by Hagen and Tassan (1965) stimulated biting and piercing by the coccinellid larvae, perhaps by mimicking natural feeding stimulants.

Coccinellid larvae, like chrysopid larvae, often modify their searching behavior after encountering prey, switching from fairly straight-line or gently curved paths to highly convoluted routes (Rowlands and Chapin 1978, and references therein). Marks (1977) noted this type of klinokinetic behavior after prey discovery by *C. septempunctata* larvae. He also found that plants previously searched unsuccessfully were recognized by detection of a pheromone secreted from the anal disk during searching. Since he found that *C. septempunctata* larvae were unable to detect aphids by sight or smell, he suggested that recognition of previously searched areas would minimize wasteful energy expenditure. This behavior is akin to that observed for many other entomophagous and phytophagous insects that have been found to mark already exploited sites by using epideictic pheromones (Chapters 9 and 10).

It seems quite likely that prey recognition by highly prey specific coccinellid species such the vedalia beetle *(Rodolia cardinalis)* is mediated by chemical factors perceived on direct contact with the prey. Contact chemical stimuli could include internal as well as external factors. Hagen (unpublished) unsuccessfully attempted to improve artificial diets encapsulated in paraffin for acceptance by larvae of the vedalia beetle by incorporating waxes from the cuticle of cottonycushion scales *(Icerya purchasi)* into the paraffin. Here the issue of the physical aspect of the presentation comes into question also. For example, Weseloh (1976, 1977) found that the kairomone associated with silk from the gypsy moth *(Lymantria dispar)* stimulates the parasitoid *Apanteles melanoscelus* only if it is presented as an impregnated fiber, even a cotton thread. Since the covering of the cottonycushion scale is quite highly textured, it might be necessary to stimulate not only the chemical composition of the cuticle, but the texture as well.

Anthocoridae and Lygaeidae

Apparently little research has been performed on the prey selection behavior of the minute pirate bugs (Anthocoridae). Dixon and Russel (1972) studied the prey-searching behavior of an aphid-feeding *Anthocoris* species and found that they recognize their prey by touching them with the antennae. Whether recognition is mediated by contact or by chemical or physical stimuli was not determined.

The lygaeid *Geocoris punctipes* is considered by Crocker (1977) to use its antennae and vision to detect prey. This investigator found by extirpating the distal half of the antennae by *Geocoris* adults that their ability to detect *Trichopulsia ni* eggs was markedly impaired, but they still actively preyed on *T. ni* larvae, apparently detecting them visually. Again, it was not established whether prey recognition using the antennae was a response to chemical or physical stimuli associated with the *T. ni* eggs.

Predacious Pentatomids

Two recent studies on predacious pentatomids implicate kairomones in prey selection. Norman Marston of the U.S. Department of Agriculture Biological Control of Insects Research Laboratory, Columbia, Missouri, found that spined soldier bugs *(Podisus maculiventris)* orient to soybean plants infested by larvae of *T. ni,* presumably because of a chemical stimulus either from the larvae themselves or from a chemical liberated from the wounded plant tissue (personal communication).

McLain (1979) found that three species of predatory pentatomids all follow terrestrial trails made by a variety of caterpillars. He demonstrated that artificial trails could be drawn on paper by using aqueous solutions of either frass or hemolymph from *T. ni* larvae and found that these trails were readily followed by the predators.

Although we were unable to find other examples of responses by predators to kairomone trails produced by feeding insects, it seems likely that a wide array of hunting-type predators, such as carabids and cicindelids, also use kairomone trails to help locate their prey.

"Eavesdropping" by Bark Beetle and Termite Predators

One of the more fascinating uses of semiochemicals in prey selection is that of "eavesdropping" by a variety of bark beetle predators (Table 7.1). These predators use either the aggregating pheromones of their prey, scolytid bark beetles, or a combination of the pheromones plus host tree terpenes as kairomones to indicate the presence of their prey. Several parasitoids of bark beetles and other insects also have been found to use the sex pheromones of their hosts as kairomones to indicate the probable presence of their hosts (Vinson 1977).

Similarly, scouts of the ant *Megaponera foetens,* an obligate predator of termites, detect their prey by using kairomones from the termites that are believed to serve as pheromones within the termite community (Longhurst and Howse

TABLE 7.1. "Eavesdropping" Bark Beetle Predators

Predator	Prey	Attractant	Reference
		COLEOPTERA	
Cleridae			
Thanasimus undatulus	*Dendroctonus rufipennis*	Frontalin	Kline et al. (1974), Dyer (1975)
Thanisimus dubius	*Dendroctonus frontalis*	Frontalin	Vité and Williamson (1970)
Thanisimus formicarius and *T. rufipes*	*Ips typographus*	Synthetic pheromone	Bakke and Kvamme (1978)
Enoclerus lecontei	*Dendroctonus brevicomis*	Host tree terpenes	Pitman and Vité (1971)
Enoclerus lecontei	*Ips confusus*	Host tree terpenes	Rice (1969)
Enoclerus lecontei	*I. confusus*	Synthetic pheromone	Wood et al. (1968)
Trogostidae (= Ostomidae)			
Temnochilia virescens	*Dendroctonus brevicomis*	Exobrevicomin	Pitman and Vité (1971)
Temnochilia virescens	*Ips confusus*	Host tree terpenes plus *n*-heptane	Rice (1969)
		DIPTERA	
Dolichopodidae			
Medetera bistriata	*Dendroctonus frontalis*	Frontalin + α-pinene + verbenone or verbenol	Williamson (1971)
Medetera bistriata	Scolytid spp.	D-α-Pinene [a]	Fitzgerald and Nagel (1972)

[a] Adults use as oviposition stimulant; larvae orient to prey entrance holes using this agent.

1978). Recruited workers follow a pheromone trail laid by the scout ant back to the termite nest, which they then pillage.

Bee Wasps

One of the classic studies of insect behavior was performed by Niko Tinbergen (1972 and references therein) on the prey-seeking behavior of a "bee wasp" *(Philanthus triangularum)*. Bee wasps, or "bee wolfs" as they are also known, prey on honey-bees *(Apis mellifera)* and are of some economic importance to bee-keepers in Europe because they kill more bees than are required to provision the nest (Clausen 1940).

Tinbergen found that initial detection of the bee was by visual means and that final recognition of the captured bee was based on chemical stimuli. One of Tinbergen's first tests was to extract a honeybee in ether, let it dry, and test its acceptability to the bee wasps by suspending it by a thread in a habitat frequented by bee wasps. He found that the wasps would approach extracted bees but would not strike at them. If he then imparted the usual scent of bees to an extracted bee by rubbing it against a normal honeybee, it was again acceptable for attack. He also found that syrphid flies, which mimic bees visually, were approached by the bee wasps but were not attacked unless they were made to smell like bees by rubbing them against honeybees. After this, they were attacked and even stung. Thus, this study demonstrated the role of both vision and chemoreception in prey selection by bee wasps.

Predacious Aquatic Insects

It would seem that many predacious aquatic insects would use kairomones to detect and locate their prey, but we discovered only one published account. Tinbergen (1936, 1951, 1965) found that the feeding response of the predacious diving beetle *Dytiscus marginalis* is elicited by the scent of its prey, which includes tadpoles and small fish as well as insects. He found that in spite of having well-developed eyes, these beetles do not react at all to visual stimuli when capturing prey. He determined this by the following experiments.

First, Tinbergen placed a tadpole in a test tube near a hungry beetle in an aquarium. The beetle did not react at all, even though the tadpole was clearly visible. Next, he placed the tadpole in a small cheesecloth bag near the beetle, effectively hiding the tadpole from sight but allowing chemicals to escape. This time the beetle reacted vigorously, either by grabbing the bag with the forelegs and chewing it up, or, if it happened to swim past under the bag, by immediately diving to the bottom and swimming around in irregular searching movements below. Tinbergen did not demonstrate a similar response with an insect prey, but the beetles likely use kairomones from insects in a similar way.

Predacious Mites

Representatives of at least three families of predacious mites have been found to use kairomones as an aid in prey finding. Jalil and Rodriguez (1970a) found that the uropodid *Fuscuropoda vegetans,* an egg predator of the housefly *(Musca domestica),* is attracted to the odor of housefly eggs. Similarly, they found that the macrochelid *Macrocheles muscaedomesticae,* which preys on housefly eggs and parasitizes housefly adults, is attracted to housefly eggs and adults by olfactory stimuli (Jalil and Rodriguez 1970b). In both cases the responses are mediated by olfactory receptors on the fore tarsi. Farish and Axtell (1966) noted that sensory structures on the palps of *M. muscaedomesticae* appear to detect contact chemical stimuli associated with adult houseflies. Jalil and Rodriguez (1970b) also found that *M. muscaedomesticae* were attracted to low concentrations of ammonia and skatole, which may assist them in habitat finding, since their prey are often found in manure, which emits ammonia and skatole as a result of protein degradation.

Hislop et al. (1978) reported that the phytoseiid *Amblyseius fallacis* uses kairomones deposited by the two-spotted spider mite *(Tetranychus urticae)* and by the European red mite *(Panonychus ulmi).* They found that leaf disks previously infested by prey mites and from which the mites had been removed were searched much more thoroughly than never-infested leaf disks. They also noted that the two-spotted mites deposited stronger arrestant cues than the red mites. Further tests indicated that the kairomone was associated with the silk left by the prey mites and that it could be extracted in methanol and reapplied to fresh disks without loss of activity.

Research on the phytoseiid *Metaseiulus (= Typhlodromus) occidentalis* by Marjorie Hoy and Jan Smilanick of the University of California at Berkeley (personal communication) showed that this predator also appears to detect kairomones from several spider mite prey species, including *P. ulmi, P. citri, T. urticae, Tetranychus pacificus, Eotetranychus willamettei,* and *Bryobia rubioculus.* Of these prey, the arrestment effect was strongest when the predators contacted sites previously infested by *T. urticae* and *T. pacificus.*

Another phytoseiid, *Phytoseiulus persimilis,* which feeds largely on the eggs and immature stages of certain tetranychids, is reported to recognize prey by using contact chemoreceptors and possibly mechanoreceptors on the pedipalps (Jackson and Ford 1973, Jackson 1975).

USE OF ALLOMONES IN PREY SELECTION

As noted, many predators locate and recognize their prey by responding to prey-produced kairomones, but some predators reverse this process and produce allomones that attract prey or otherwise exploit the chemical communication systems of the prey. When the substances resemble those used intraspecifically by the prey, the phenomenon is known as aggressive chemical mimicry (Eberhard 1977).

Myrmecophilous Beetles

The relationship of myrmecophilous staphylinid beetles and the ants they parasitize and prey on provides an outstanding example of aggressive chemical mimicry. Research by Hölldobler (1971) has shown that these beetles are able to communicate in the chemical and mechanical languages used by their ant hosts. He found that the beetles, as exemplified by *Atelmes pubicollis,* apparently mimic the pheromones of the host ants. By this means, they are not only admitted to the ants' nests, but their larvae are cared for and reared by the ants as well.

Hölldobler discovered that the *Atelmes* adults find the ant nests by detecting the nest odor and move upwind to its source. They then induce worker ants to allow them to enter the ant nest by producing secretions from "appeasement" and "adoption" glands in the abdomen. These secretions suppress the aggressive behavior of ants and even induce the ants to carry the beetles directly into the brood chambers. Beetle larvae produced there obtain food by preying on ant larvae as well as by cannibalizing other beetle larvae. In addition, the beetle larvae possess glands that secrete allomones that induce the worker ants to feed them by trophyllaxis.

Reduviidae

Two species of assassin bugs have been found to produce allomones that are attractive to their prey. Weaver et al. (1975) showed that the bug *Apiomerus pictipes* attracts stingless bees *(Trigona fulviventris)* by using allomones that mimic the pheromones of the bees themselves. Similarly, Jacobson (1911) showed that the assassin bug *Ptilocerus ochraceus* produces a glandular secretion from ventral trichomes that initially attracts ants. Ants that feed upon the secretion are soon tranquilized and finally paralyzed, which renders them easy prey.

Bolas Spiders

Another use of aggressive chemical mimicry by a predator is displayed by a bolas spider, *Mastophora* sp. (Eberhard 1977). This spider derives its name from its unique prey-capturing technique. The spider spins a web that consists of a sticky ball suspended on the end of a short vertical thread. The spider hands the thread from one front leg, and swings the ball at passing insects. When the ball sticks to an insect, the spider descends the line, paralyzes the prey, and feeds.

Eberhard discovered that nearly all the prey captured by mature female *Mastophora* sp. were male noctuid moths either of *Spodoptera frugiperda* (the fall armyworm) or of a species of *Leucania*. He concluded that these spiders attract their prey with a volatile substance that mimics the female sex pheromone of *S. frugiperda*.

This behavior represents a rather extreme degree of specialization for a predator. As pointed out by Eberhard, newly emerged bolas spiders are only a fraction of the size of adult male moths, which suggests that the small spiders must use either some other attractant for smaller prey or some other trapping technique. At any rate, this is an interesting example of the use of allelochemics in prey selection.

GENERALIZATIONS AND POTENTIAL APPLICATIONS

The examples cited suggest that many arthropod predators of insects and mites probably use allelochemics at some point during prey finding. Some predators, such as myrmecophilous beetles, appear to rely quite heavily on allelochemics, whereas others, such as bee wasps, may use them in conjunction with visual stimuli. It is possible that very general predators such as the European praying mantis *(Mantis religiosa)* may use visual and tactile stimuli exclusively, but even these may use chemical cues to select an appropriate habitat.

The fact that a predator is a general feeder does not preclude use of chemical cues for prey finding. This is illustrated by the example of the predacious diving beetle, which appears to use chemoreception almost exclusively in detecting prey. In a case like this, the predator might respond to very common biochemicals such as amino acids, whereas highly specialized predators such as those attacking scolytid beetles may respond to a much narrower spectrum of chemicals.

The findings on attraction of aphid predators to artificial honeydew are already being applied. Hagen and Bishop (1978) reviewed research on use of artificial honeydews to attract, arrest, and stimulate egg production in various chrysopids, coccinellids, and syrphids. As they point out, this approach has been of demonstrable value in reducing aphid populations in crops of cotton, bell peppers, and alfalfa.

An unexploited potential application of kairomones for augmentation of predators is in increasing the acceptability of artificial food substances to be used in lieu of natural prey for rearing predators. Artificial foodstuffs might be used either in mass rearing programs or in the field, to conserve populations of natural enemies during temporary periods of low prey density or to build up predator populations prior to pest population increases. This approach would be especially helpful in rearing predators that recognize their prey and begin to feed only after detecting phagostimulant kairomones.

Other uses of allelochemics as tools for predator augmentation will probably become evident as integrated pest management becomes increasingly sophisticated.

REFERENCES

Arthur, A. P., B. M. Hegdekar, and W. W. Batsch. 1972. A chemically defined synthetic medium that induces ovipos8tion in the parasite *Itoplectis conquisitor* (Hymenoptera: Ichneumonidae). Can. Entomol. 104:1251–1258.

Bakke, A., and T. Kvamme. 1978. Kairomone response by the predators *Thanasimus formicarius* and *Thanasimus rufipes* to the synthetic pheromone of *Ips typographus*. Norw. J. Entomol. 25:41–43.

Banks, C. J. 1957. The behavior of individual coccinellid larvae on plants. Br. J. Anim. Behav. 5:12–24.

Bänsch, R. 1966. On prey-seeking behaviour of aphidophagous insects. P. 123–128, *In* I. Hodek (ed.), Ecology of Aphidophagous Insects. Academia, Prague.

Ben Saad, A. A., and G. W. Bishop. 1976. Effect of artificial honeydews on insect communities in potato fields. Environ. Entomol. 5:453–457.

Blum, M. S. 1978. Biochemical defenses of insects. P. 466–513. *In* M. Rockstein (ed.), Biochemistry of Insects. Academic Press, New York.

Bombosch, S., and S. Volk. 1966. Selection of oviposition site by *Syrphus corollae* F. P. 117–119, *In* I. Hodek (ed.), Ecology of Aphidophagous Insects, Academia, Prague.

Clausen, C. P. 1940. Entomophagous Insects. McGraw-Hill, New York.

Crocker, R. L. 1977. Components of the feeding niches of *Geocoris* spp. (Hemiptera: Lygaeidae). Unpublished Ph.D. Dissertation, University of Florida, Gainesville.

Dixon, A. F. G. 1959. An experimental study of the searching behavior of the predatory coccinellid beetle *Adalia decempunctata* (L.). J. Anim. Ecol. 28:259–281.

Dixon, A. F. G., and R. J. Russel. 1972. The effectiveness of *Anthocoris nemorun* and *A. confusus* (Hemiptera: Anthocoridae) as predators of the sycamore aphid, *Drepanosiphum platanoides*. I. Searching behavior and the incidence of predation in the field. Entomol. Exp. Appl. 15:35–50.

Dyer, E. D. A. 1975. Frontalin attractant in stands infested by the spruce beetle, *Dendroctonus rufipennis* (Coleoptera: Scolytidae). Can. Entomol. 107:979–988.

Eberhard, W. G. 1977. Aggressive chemical mimicry by a bolas spider. Science 198:1173–1175.

Farish, D. J., and R. C. Axtell. 1966. Sensory functions of the palps and first tarsi of *Macrocheles muscaedomesticae* (Acarina: Macrochelidae), a predator of the housefly. Ann. Entomol. Soc. Am. 59:165–170.

Fitzgerald, T. D., and W. P. Nagel. 1972. Oviposition and larval bark-surface orientation of *Medetera aldrichii* (Diptera: Dolichopodidae): Response to a prey-liberated plant terpene. Ann. Entomol. Soc. Am. 65:328–330.

Fleschner, C. A. 1950. Studies on the searching capacity of the larvae of three predators of the citrus red mite. Hilgardia 20:233–265.

Flint, H. M., S. S. Salter, and S. Walters. 1979. Caryophyllene: An attractant for the green lacewing. Environ. Entomol. 8:1123–1125.

Hagen, K. S., and R. L. Tassan. 1965. A method of providing artificial diets to *Chrysopa* larvae. J. Econ. Entomol. 58:999–1000.

Hagen, K. S., E. F. Sawall, Jr., and R. L. Tassan. 1971. The use of food sprays to increase effectiveness of entomophagous insects. Proc. Tall Timbers Conf. Ecological Animal Control by Habitat Management 2:59–81.

Hagen, K. S., P. D. Greany, E. F. Sawall, Jr., and R. L. Tassan. 1976a. Tryptophan in artificial honeydews as a source of an attractant for adult *Chrysopa carnea*. Environ. Entomol. 5:458–468.

Hagen, K. S., S. Bombosch, and J. A. McMurtry. 1976b. The biology and impact of predators. P. 92–142, *In* C. B. Huffaker and P. S. Messenger (eds.), Theory and Practice of Biological Control, Academic Press, New York.

Hagen, K. S., and G. W. Bishop. 1978. Use of supplemental foods and behavioral chemicals to increase the effectiveness of natural enemies. *In* D. W. Davis, J. A. McMurtry, and S. C. Hoyt (eds.), Biological Control and Insect Management. California Agriculture Experiment Station publication (in press).

Hislop, R. G., N. Alaves, and R. J. Prokopy. 1978. Kairomone of *Tetranychus urticae* (Acarina: Tetranychidae) influencing host-searching behavior of its predator, *Amblyseius fallacis* (Acarina: Phytoseiidae). J. N.Y. Entomol. Soc. 86:297.

Hölldobler, B. 1971. Ants and their guests. Sci. Am. 224:86–93.

Jackson, G. J. 1975. Chaetotaxy and setal morphology of the palps and first tarsi of *Phytoseiulus persimilis* A.-H. (Acarina: Phytoseiidae). Acarology. 16:583–595.

Jackson, G. J., and J. B. Ford. 1973. The feeding behavior of *Phytoseiulus persimilis* (Acarina: Phytoseiidae), particularly as affected by certain pesticidies. Ann. Appl. Biol. 75:165–171.

Jacobson, E. 1911. Biological notes on the hemipteran *Ptilocerus ochraceus*. Tidschr. Entomol. 54:175–179.

Jalil, M., and J. G. Rodriguez. 1970a. Biology of and odor perception by *Fuscuropoda vegetans* (Acarina: Uropodidae), a predator of the housefly. Ann. Entomol. Soc. Am. 63:935–938.

Jalil, M., and J. G. Rodriguez. 1970b. Studies of behavior of *Macrocheles muscaedomesticae* (Acarina: Macrochelidae). Ann. Entomol. Soc. Am. 63:738–744.

Kaddou, I. 1960. The feeding behavior of *Hippodamia quinquesignata* (Kirby) larvae. Univ. California Publ. Entomol. 16:181–232.

Kesten, U. 1969. Morphologie und Biologie von *Anatis ocellata* (L.). Z. Angew. Entomol. 63:412–445.

Kline, L. M., R. F. Schmitz, J. A. Rudinsky, and M. M. Furniss. 1974. Repression of spruce beetle (Coleoptera) attraction by methylcyclohexanone in Idaho. Can. Entomol. 106:485–491.

Lewis, W. J., R. L. Jones, D. A. Nordlund, and H. R. Gross. 1977a. Kairomones and their use for management of entomophagous insects. P. 455–469, *In* Int. Symp. Troph. Rel. Insects. Tours, 1976. Colloques Int. CNRS 265.

Lewis, W. J., D. A. Nordlund, H. R. Gross, Jr., R. L. Jones, and S. L. Jones. 1977b. Kairomones and their use for management of entomophagous insects. V. Moth scales as a stimulus for predation of *Heliothis zea* (Boddie) eggs by *Chrysopa carnea* Stephens larvae. J. Chem. Ecol. 3:483–487.

Longhurst, C., and P. E. Howse. 1978. The use of kairomones by *Megaponera foetens* (Hymenoptera: Formicidae) in the detection of its termite prey. Anim. Behav. 26:1213–1218.

Marks, R. J. 1977. Laboratory studies of plant searching behavior by *Coccinella septempunctata* L. larvae. Bull. Entomol. Res. 67:235–241.

McLain, K. 1979. Terrestrial trail following by three species of predatory stink bugs. Florida Entomol. 62:152–154.

Nichols, P. R., and W. W. Neel. 1977. The use of food Wheast as a supplemental food for *Coleomegilla maculata* (Degeer) (Coleoptera: Coccinellidae) in the field. Southwest. Entomol. 2:102–106.

Nordlund, D. A., and W. J. Lewis. 1976. Terminology of chemical releasing stimuli in intraspecific and interspecific interactions. J. Chem. Ecol. 2:211–220.

Nordlund, D. A., W. J. Lewis, R. L. Jones, H. R. Gross, Jr., and K. S. Hagen. 1977. Kairomones and their use for management of entomophagous insects. VI. An examination of the kairomones for the predator *Chrysopa carnea* Stephens at the oviposition sites of *Heliothis zea* (Boddie). J. Chem. Ecol. 3:507–511.

Pitman, G. B., and J. P. Vité. 1971. Predator-prey response to western pine beetle attractants. J. Econ. Entomol. 64:402–404.

Rice, R. E. 1969. Response of some predators and parasites of *Ips confusus* (LeConte) (Coleoptera: Scolytidae) to olfactory attractants. Contrib. Boyce Thompson Inst. 24:189–194.

Rowlands, M. L. J., and J. W. Chapin. 1978. Prey searching behavior in adults of *Hippodamia convergens* (Coleoptera: Coccinellidae). J. Georgia Entomol. Soc. 13:309–315.

Tinbergen, N. 1936. Eenvoudige Proeven over de Zintuigfuncties van Larve en Imago van de geelgerande Watertor. Levende Nat. 41:225–235.

Tinbergen, N. 1951. The Study of Instinct. Oxford University Press, London.

Tinbergen, N. 1965. Animal Behavior. Time-Life books, New York.

Tinbergen, N. 1972. The Animal in Its World, Vol. I. Field Studies. Harvard University Press, Cambridge, MA.

van Embden, H. F., and K. S. Hagen. 1976. Olfactory reactions of the green lacewing, *Chrysopa carnea,* to tryptophan and certain breakdown products. Environ. Entomol. 5:469–473.

Vinson, S. B. 1977. Behavioral chemicals in the augmentation of natural enemies. P. 237–379, *In* R. L. Ridgway and S. B. Vinson (eds.), Biological Control by Augmentation of Natural Enemies. Plenum Press, New York.

Vité, J. P., and D. L. Williamson. 1970. *Thanasimus dubius:* Prey perception. J. Insect Physiol. 16:233–239.

Weaver, E. C., E. T. Clarke, and N. Weaver. 1975. Attractiveness of an assassin bug to stingless bees. J. Kansas Entomol. Soc. 48:17–18.

Weaver, N. T. 1978. Chemical control of behavior. P. 391–418. *In* M. Rockstein (ed.), Biochemistry of Insects, Academic Press, New York.

Weseloh, R. M. 1976. Behavioral responses of the parasite *Apanteles melanoscelus* to gypsy moth silk. Environ. Entomol. 5:1128–1132.

Weseloh, R. M. 1977. Effects on behavior of *Apanteles melanoscelus* females caused by modifications in extraction, storage and presentation of gypsy moth silk kairomone. J. Chem. Ecol. 3:723–735.

Wilbert, H. 1974. Die Wahrnehmung von Beute Durch die Eilarven von *Aphidoletes aphidimyza* (Cecidomyiidae). Entomophaga 19:173–181.

Williamson, D. L. 1971. Olfactory discernment of prey by *Medetera bistriata* (Diptera: Dolichopodidae). Ann. Entomol. Soc. Am. 64:586–588.

Wilson, E. O. 1971. The Insect Societies. Harvard University Press, Cambridge, MA.

Wood, D. L., L. E. Browne, W. D. Bedard, P. E. Tilden, R. M. Silverstein, and J. O. Rodin. 1968. Response of *Ips confusus* to synthetic sex pheromones in nature. Science 159:1373–1374.

CHAPTER EIGHT

EMPLOYMENT OF KAIROMONES IN THE MANAGEMENT OF PARASITOIDS

HARRY R. GROSS, JR.

AR–SEA–USDA
Southern Grain Insects Research Laboratory
Tifton, Georgia

The importance of allelochemics in host selection, habitat selection, host location, and host acceptance by parasitoids has been reviewed in detail in Chapters 4–6. In this chapter I expand on the aspects that deal specifically with kairomone employment and attempt to formulate kairomone use strategies for enhancing the performance of both host-generalist and host-specific parasitoids in the management of economic lepidopteran species.

During the past decade, scientists have become increasingly aware of the importance of allelochemics, particularly kairomones, to the performance of parasitoids. Accumulating information has advanced us from an immediate past in which it was generally believed that parasitoid efficiency was solely dependent on the ability of these organisms to randomly encounter hosts, while being limited by the supply of available ova. There is now general agreement among behaviorists that the host-finding efficiency of most well-adapted parasitoids, within favorable habitats, is determined primarily by their frequency of encounter with allelochemics (emanating from the host, host by-products, nonhost organisms associated with the host habitat, host plants, by-products from damaged host of the target species, etc.) and by inherent parasite capacity to orient via allelochemics to high host probability sites. Similarly, there has been modification of the belief that the efficiency of parasitoids in a given environment is innately fixed and cannot be improved beyond predictable limits that, more frequently than not, rendered unsatisfactory control of the target species.

These changes in perception of the potential of kairomones for behavioral management of parasitoids have resulted from studies made to (1) identify the sources of these materials and the behavioral responses they evoke (Jones et al. 1971, 1973, Lewis and Jones 1971, Lewis et al. 1972, Vinson 1968, Greany et al. 1977, Hendry et al. 1973, and others), (2) consider the effect of host food on their level of activity (Sauls et al. 1979, Roth et al. 1978), and (3) with increasing frequency, elucidate their chemical identity (Chapter 12). The use now made of this accumulating information could determine whether it enters the literature as another interesting curiosity of the Class Insecta or becomes as much a part of insect pest management strategies as insecticides are today.

In fact, the potential for employment of kairomones appears overwhelmingly favorable. Likewise, the desire and demand for their utilization, particularly in conjunction with inundative releases of entomophagous insects, is gaining increasing support.

HOST GENERALISTS: *TRICHOGRAMMA*

For decades, *Trichogramma* spp. have been recognized as generalist hymenopteran egg parasitoids with great potential for managing lepidopteran pests, primarily because of their commonness of occurrence among diverse habitats, their seasonal predictability, and their periodic high rates of parasitization, particularly in association with high host densities in insecticide-free agroecosystems (Oatman and Platner 1971, Sanford 1976, Stinner 1977, and others). Perspectives for employment of *Trichogramma* spp. were greatly enhanced by the discovery that rates of parasitization could be increased via the application of kairomones (Lewis et al. 1972) and by the development of methods of mass producing the parasite on alternate hosts and at competitive costs (Morrison et al. 1978).

Trichogramma spp. are noted for many assets, but predictability of performance following inoculative or inundative releases is not one of them. Reasons offered for the species' erratic performance have included factors such as high sensitivity to insecticide, low vigor of parasitoids reared on alternate hosts, failure to select the appropriate species or strain, inadequate rates of release, unfavorable climatic conditions, detrimental plant characteristics, and low host densities. It is only recently that suboptimal quantities and poor distribution of kairomones have become suspect.

Laing (1937) was the first to demonstrate that *Trichogramma evanescens* perceived an odor left by adult moths. However, the significance of this finding apparently went unnoticed for several decades until Lewis and Jones (1971), working with *T. evanescens,* affirmed that moth odor of *Heliothis zea* increased the host finding and rates of parasitization of *H. zea* eggs. Subsequently, Lewis et al. (1972) identified moth scales as the source of the mediator that stimulates the host-seeking response by *T. evancescens,* and Jones et al. (1973) found that

C_{23} (tricosane) present in *H. zea* moth scales was the most active hydrocarbon in eliciting significant orientation and stimulating parasitism by *T. evanescens*. Since these discoveries, much information has accumulated regarding the role of kairomones in the host-seeking behavior of *Trichogramma* spp. (Lewis et al. 1975a, 1975b, 1976, 1979, Gross et al. 1975, and others), most of which suggest that kairomones serve to improve interception, retention, egg distribution, and efficiency of *Trichogramma* spp.

Application Strategies: Prerelease Stimulation

The difficulty of retaining released parasitoids in target areas is perhaps the most critical limitation repeatedly noted by investigators during field releases. Approaches offered for improving retention and colonization of entomophages have included physiological and developmental acclimation (Force 1967, Stinner 1977), prerelease feeding and/or parasitization of hosts (Gross et al. in press), provision of supplemental resources (Rabb et al. 1976, Hagen et al. 1970), and partial removal of the wings (Ignoffo et al. 1977).

Findings by Gross et al. (1975) affirmed that the innate tendency of released *Trichogramma achaeae* and *Trichogramma pretiosum* to disperse upon release could be overridden by exposing them, at time of release, to kairomones extracted from *H. zea* moth scales. When tricosane (Jones et al. 1973) and the *H. zea* moth scale extract (MSE) were used as sign stimuli (releasers) for *T. achaeae* and *T. pretiosum*, respectively, at time of release from laboratory containers, rates of parasitization on *H. zea* eggs were significantly increased, compared with rates for unstimulated parasitoids (Gross et al. 1975). These findings have far-reaching implications and may play a significant role in developing strategies for parasitoid releases. The current trend of investigators evaluating inundative releases of *Trichogramma* spp. is the uniform distribution of parasitized eggs over target areas. However, it may be that a packaging method whereby the parasitoids are exposed to kairomones upon release will prove best for on-site retention. Zil'Berg et al. (1976) have already developed a paper sphere encapsulation system for *Trichogramma* spp. parasitized eggs. A device mounted on the rear of a tractor punctures a hole (0.8 – 1 m) in the capsule wall and distributes the eggs over the field. The tendency for dispersal could be substantially reduced by the use of this device or a similar one (Figure 8.1), if the inner surface or neck of the container were treated with an appropriate kairomone, to expose the emerging parasitoids as they are released.

Kairomone Distribution Patterns

The application of kairomones to the agroecosystem, in combination with parasitoid releases, offers probably the greatest immediate potential for enhanc-

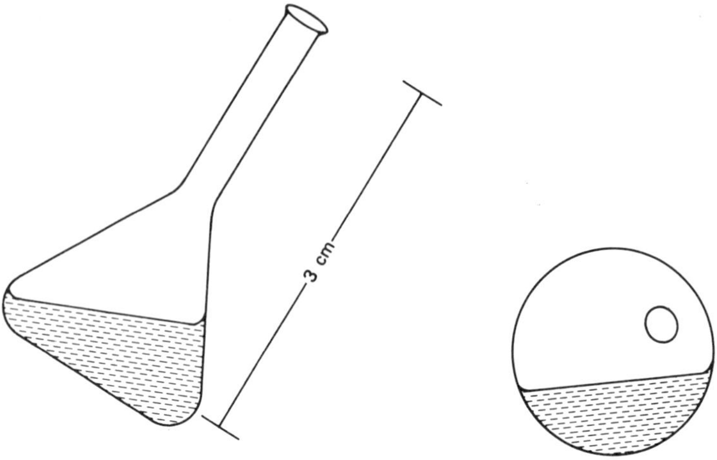

FIGURE 8.1. Packaging methods for field distribution and prelease host-seeking stimulation of *Trichogramma* spp.

ing the consistency of performance and increasing rates of parasitization by *Trichogramma* spp. Therefore, after the selection of the appropriate kairomone, the greatest immediate need is the determination of distribution patterns that retain maximum numbers of parasitoids and reinforce host-seeking behavior within target areas.

Lewis et al. (1975b) examined the effect of increasing diameters of tricosane-treated filter paper on the ability of *T. achaeae* to locate and parasitize eggs of *H. zea*. Rates of parasitization improved as the treated area was expanded, and the highest rates of parasitization were obtained with total surface treatments. Total surface treatments of potted peas also yielded higher rates of parasitization than treatments of restricted locations and no treatment. From these studies, basic assumptions regarding *Trichogramma*-kairomone relationships were drawn, including the following: (1) *Trichogramma* were not attracted to thc host-seeking stimuli but rather were arrested by it and stimulated to search, via antennal contact with kairomone-treated surfaces, also (2) the most efficient distribution of kairomones in arresting and retaining the greatest number of parasitoids was a broadcast or blanket application.

These hypotheses were also field tested and apparently reaffirmed (Lewis et al. 1975b). However, the lack or information about the natural field distribution of moth scales from *Heliothis* spp. had, in fact, caused a bias in favor of the blanket application. Subsequent studies revealed that *H. zea* eggs naturally oviposited on cotton received higher rates of parasitization by *T. pretiosum* than did eggs from laboratory-reared insects placed on cotton. This difference was not due to lessened acceptability of laboratory eggs (Gross et al. unpublished data) but rather seemed to be attributable to concentrations of kairomone on the substrate surrounding naturally oviposited eggs.

A laboratory universe was therefore developed to permit a further critical analysis of *Trichogramma* behavior in response to kairomone distributions (Lewis et al. 1979). On butcher paper sheets measuring 117.5 × 30.4 cm, a 7.6-cm diameter area immediately surrounding the host egg was sprayed with MSE to create a simulated oviposition site (SOS), so-called spot treatments. When blanket MSE treatments were compared to no MSE using the above-described assay system, the percentage parasitization by *T. pretiosum* was higher on host eggs placed on the solid MSE. However, host eggs placed on the laboratory universe with blanket MSE were significantly less parasitized than host eggs exposed to the SOS treatment (Table 8.1). Field studies with soybeans also were conclusively in favor of the SOS treatment. Subsequent laboratory observations revealed that blanket MSE (Figure 8.2a) served to retain *T. pretiosum* but prevented them from concentrating search activity at sites where there was a high probability of finding hosts. Rates of parasitization by *T. pretiosum* exposed to SOS treatments (Figure 8.2b) were significantly higher because search was intensified only in the vicinity of the host.

Since no data yet suggest that *Trichogramma* spp. are capable of orienting via kairomones to sites of oviposition, it is likely that many parasitoids in crops containing low to intermediate host densities or hosts with clumped distributions disperse prior to detection of the first oviposition site. The need, therefore, was for a supplemental distribution of kairomone (Figure 8.2c) that would intermittently reinforce host seeking by *Trichogramma* without interfering with the innate klinotaxis of the species. Anakrom-U® diatomaceous earth particles (40–50 mesh) impregnated with MSE were therefore tested as simulated moth scales (SMS) on laboratory universes previously described. When *T. pretiosum*

TABLE 8.1. Response of *T. pretiosum* Exposed to *H. zea* Moth Scale Extract (MSE) on an Expanded Laboratory Universe[a,b]

Treatment	Parasitization (%)[c]
Solid	23.6
Control	14.0
Solid	18.0
SOS[d]	33.8

[a] Butcher paper 117 x 30 cm.
[b] Two vials of 5 *T. pretiosum* females released per test.
[c] Means significantly different ($P = 0.05$) as measured by the paired t test.
[d] SOS = simulated oviposition site.

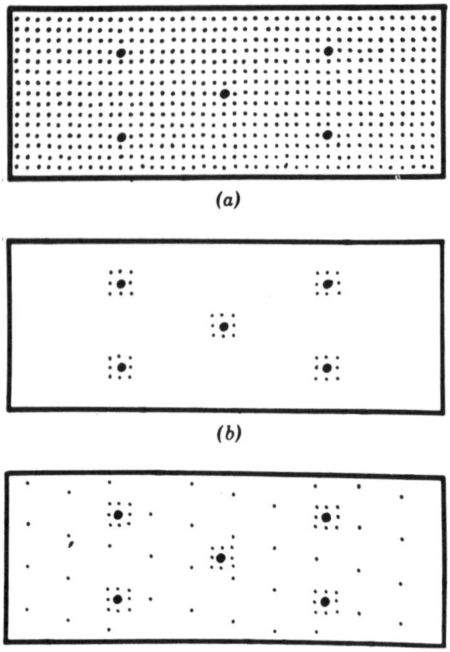

(a)

(b)

(c)

FIGURE 8.2. Simulated distribution of kairomones (small dots) and host eggs (large dots) and their effect on retention and efficiency of *Trichogramma* spp.: (a) high retention, low efficiency; (b) low retention, high efficiency; and (c) high retention, high efficiency.

were exposed to *H. zea* eggs on SOSs (Table 8.2), they parasitized significantly more eggs on universes containing MSE-treated SMS than on universes with the untreated SMS (Lewis et al. 1979).

The predictability of response by *T. pretiosum* to kairomone distributions was revealed by studies conducted on plots of cotton (30.4 × 30.4 m) in the San Joaquin Valley, California, where no evidence of naturally occurring *Heliothis* or *Trichogramma* spp. populations was detected (Gross et al. unpublished data) (Table 8.3). Kairomones (MSE) were evaluated in conjunction with released *T. pretiosum* against introduced *H. zea* eggs. The *H. zea* host eggs in all plots were spaced evenly over the canopy at about 1-m intervals. MSE treatments were as follows: (1) SOS around eggs, (2) SOS around eggs and a second SOS equidistant between two eggs, and (3) no SOS around eggs. Eggs collected from plots receiving treatment 1 were 70.6% parasitized; eggs from untreated plots (treatment 3) were 4.2% parasitized. Thus, *T. pretiosum* were obviously strongly dependent on the presence of kairomones at oviposition sites. However, plots

TABLE 8.2. Response of *T. pretiosum* Exposed to Simulated Moth Scales (SMS) Treated with *H. zea* Moth Scale Extract (MSE) on an Expanded Laboratory Universe[a,b]

Treatment	Parasitization (%)[c]
SOS[d] + control SMS	27.2
SOS + treated SMS	43.9

[a] Butcher paper 117.5 x 30.4 cm.
[b] Two vials of 5 *T. pretiosum* females released per test.
[c] Means significantly different ($P = 0.05$) as measured by the paired t test.
[d] SOS = simulated oviposition site.

containing an SOS between eggs were only 48.0% parasitized. Apparently, the efficiency of host finding by *Trichogramma* was impaired by the presence of SOSs that were not associated with the host. The need for distributions of kairomone that encourage unimpeded movement by *Trichogramma* between high host probability areas was thus reemphasized.

Effect of Host Densities

Unpublished findings by Gross et al. affirmed that rates of parasitization by released *T. pretiosum* on eggs of *H. zea* on cotton increased with host density. Apparently, the reduced rates of parasitization by *Trichogramma* commonly associated with low host densities result primarily from loss of the parasite from target areas after it has failed to encounter hosts during a defined period of search. Mean retention time (time to attempted dispersal) of *T. pretiosum* stimulated at a release site with MSE but receiving no additional stimulation was

TABLE 8.3. Response of *T. pretiosum* Exposed to *H. zea* Moth Scale Extracts in the Presence or Absence of Single or Multiple Simulated Oviposition Sites

Treatment	Parasitization (%)[a,b]
SOS[c] around eggs	70.6
SOS around eggs and 1 SOS equidistant between eggs	48.0
No SOS around eggs	4.2

[a]40,000 *T. pretiosum* per 0.4 ha in cotton.
[b] Plots 30.4 x 30.4 m in cotton, San Joaquin Valley, California.
[c] SOS = simulated oviposition site.

1.2 minutes on a laboratory substrate (Gross et al. in press). Mean retention time of *T. pretiosum* similarly stimulated and then permitted to parasitize an *H. zea* egg was approximately six times greater. The act of parasitization was thus unequivocally the primary reinforcer of host seeking, followed secondarily by kairomone stimulation. Opportunities for employment of kairomones and/or supplemental hosts to extend the time of host seeking appear to be favorable when low to intermediate host densities are encountered. In other words, the less dense the host, the greater the need for applied host-seeking stimuli; the higher the density, the less the need for additional stimuli. Also, the effectiveness of supplemental kairomones in manipulating the behavior of *Trichogramma* is limited by the density of naturally occurring moth scales. As moth populations in favored host crops approach peak densities in late season, the environment becomes saturated with naturally occurring moth scales, and the effect of any artificially introduced kairomone becomes negligible, as shown by Gross et al. (unpublished data) in Table 8.4.

From these observations we can hypothesize the scheduling of search-inductive components for *Trichogramma* to effect optimum performance (Figure 8.3). The low natural host densities common in early season would require the employment of supplemental host (SH), kairomones (K), and released *Trichogramma* (T). As host densities increase, only K and T would be introduced; in late season, as natural K accumulates, only T would be released. Therefore, the manipulation of *Trichogramma* should not consist of regular periodic application of multiple components SH, K, and T, but intermittent scheduling of the components in response to the demands of a particular environment. Moreover, the intermittent early-season applications of SH could induce in-field cycling of naturally occurring T that would then be present to assist in the control of later generations of target species. This concept was discussed in depth by Knipling and McGuire (1968).

TABLE 8.4. Response of *T. pretiosum* Exposed to Distributions of *H. zea* Moth Scale Extract (MSE) Applied in Late Season (Tifton, GA, 1976)

	Parasitization (%) [a]		
Treatment	Sept. 10	Sept. 16	Sept. 17
SOS [b] around eggs	48.9	64.1	57.8
SOS around eggs and 1 SOS equidistant between eggs	47.1	64.4	57.3
SOS around eggs and a blanket MSE treatment	47.4	64.1	53.6

[a] 50,000 *T. pretiosum* per 0.4 ha; plots, 4 rows × 7.3 m in soybeans; *H. zea* host egg placed per 0.9 m of row.
[b] SOS = simulated oviposition site.

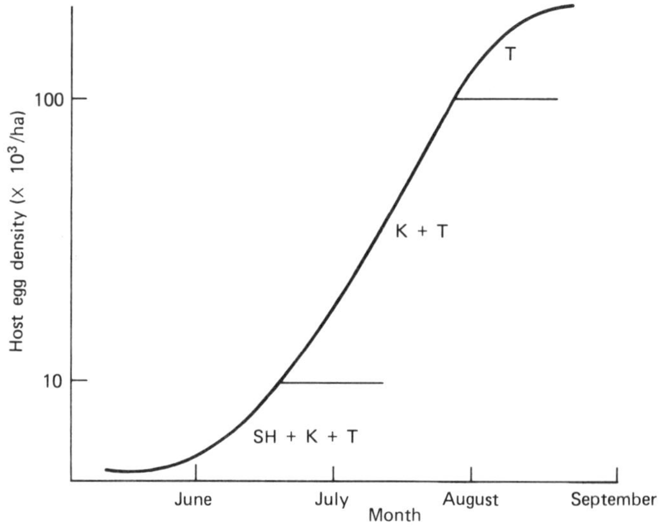

FIGURE 8.3. Hypothetical scheduling of supplemental hosts (SH) Kairomones (K), and *Trichogramma* (T) based on host density of target species.

HOST SPECIALISTS

A diagram representing the host selection process (Figure 8.4) developed by Lewis et al. (1975b) illustrates strategies with kairomones that might be useful in manipulating host-specific parasitoids. The S_3 stimuli identified there represent primary mediators associated with the investigation of host trails within habitats. They offer the best opportunity for management. Moreover, host-specific parasitoids, unlike host-generalist species such as *Trichogramma*, are likely to respond to multiple S_3 stimuli. Thus, strategies for host-specific parasitoids will probably be quite different from those for host generalists. For example, 13-methylhentriacontane (S_{3a}) was identified by Jones et al. (1971) as the host-seeking stimuli in *H. zea* larval frass that induced antennation by *Microplitis croceipes*. (It is detected only by antennation.) Larval frass of *H. zea* larvae was, therefore, used by Gross et al. (1975) as prerelease stimuli (releasers) to induce host seeking in *M. croceipes*. As a result, the tendency of dispersal upon release was overcome, and the rate of parasitization on *H. zea* larvae was increased over that of unstimulated parasitoids. Recent studies (personal observations) affirmed that another kairomone (S_{3b}) was detectable by *M. croceipes* in flight at distances to about 0.5 m. It is not unreasonable to assume that yet another kairomone (S_{3c}) might be detectable in flight at even greater distances.

Host-specific kairomones that induce antennation of substrates would appear to have limited potential for application, as noted by Vinson (1977). The

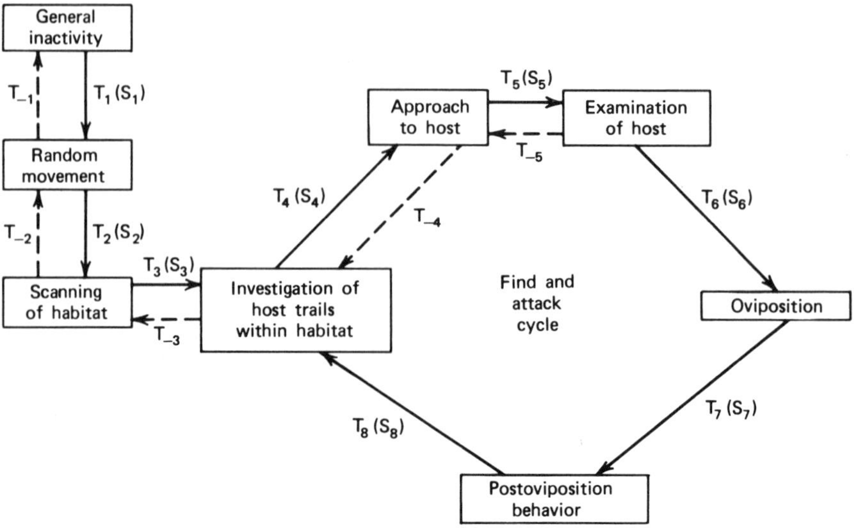

FIGURE 8.4. Basic sequence of host-finding activities by females of parasitic insects: T_1 to T_8 and T_{-1} to T_{-4}, transitions among the indicated behavioral acts; $S_{1\text{-}6}$, stimuli releasing the indicated behavioral patterns; S_2, olfactory, visual, and physical cues associated with host plants on other habitats; S_3, primarily chemical cues from frass, moth scales, and decomposition products associated with the presence of host insects. S_4 olfactory, visual, auditory, and other chemical or visual cues from host insect. S_3 and S_6, olfactory, tactile, and auditory, cues, or a combination of these, from individual host. (Originally published in Lewis et al 1975b, P. 356.)

application of such kairomones detectable by parasitoids at close range to the host would serve only to intensify search uniformly and without regard for the actual distribution of the host. On the other hand, a volatile (S_{3c}) kairomone that was detectable in flight at some distance from the source could be broadcast (Figure 8.5) to arrest immigrant parasitoids from random search. Then, interception by short-range stimuli (S_{3b}) would ultimately orient them to sources of contact kairomone (S_{3a}) and to the host. The need, therefore, is to find and identify kairomones that are detectable by parasitoids at intermediate to long distances, and for this, bioassays must first be developed. Observations of the responses of *M. croceipes* to aerially detectable kairomones placed within standard sticky traps (designed for use with sex pheromones) revealed that their precise movements dismissed their blundering into such traps.

Olfactometer studies have not been any more encouraging. Although advances are likely to come slowly, the development of behavioral management strategies for host-specific parasitoids may eventually yield the more efficient and predictable control of many economic lepidopteran species attacking row crops.

FIGURE 8.5. Hypothetical application of S_{3c} kairomones for the management of host-specific parasitoids.

USE OF KAIROMONES FOR SELECTION OF EXOTIC PARASITOIDS

Despite the best efforts of biological control specialists, the ability to identify parasitoids that will successfully colonize target host species and control them effectively remains totally unpredictable. The use of kairomones appears to offer an opportunity for increasing the frequency of successful colonizations. If a parasitoid is inherently incapable of responding to kairomones associated with the intended host, there is little chance for its successful establishment on that host. Assaying introduced parasitoids against kairomones from their intended host would provide an immediate indication of the inherent capacity of that parasitoid to respond. Substantial saving in time and funds for rearing, maintenance, and field evaluation of parasitoids is probable if such a strategy is adopted.

CONCERNS AND EXPECTATIONS

The use of kairomones in the field to enhance the efficiency of parasitoids is not without controversy (Knipling 1976, Vinson 1977). A major question involves the possible impact of widespread kairomone use on the natural balance and distribution of parasitoids and their hosts in the total ecosystem. The concern is that parasitoids may be pulled into a kairomone-treated area to the detriment of surrounding areas.

It seems reasonable to assume that parasitoids whose appetitive needs are met and whose host-seeking behavior is being reinforced by success in parasitizing target species would have minimum tendency to disperse from a productive agroecosystem. Increasing the number of parasitoids in a kairomone-treated area is dependent on their interception from a random search phase. Those individuals are, as described by Hagen et al. (1976) for *Chrysopa carnea* Steph. adults, "looking for a home." Those investigators observed the response of adult *C. carnea* to the kairomone tryptophan and found that only the individuals that were in a random search phase were attracted and intercepted. Individuals functioning productively in nearby environments did not respond. I see no reason to suspect that parasitoids in agroecosystems containing adequate host densities would respond differently.

Also in question is whether retention of parasitoids in kairomone-treated ecosystems of limited host resources would be detrimental to the seasonal population dynamics of the parasitoids. Just as productive environments would not be likely to lose parasitoids to a kairomone-treated area, a kairomone-treated environment would not retain parasitoids that are not reinforced by parasitization of hosts. Stimulation to search without periodic benefit of parasitization results in a return of parasitoids to a random movement and ultimate dispersal from kairomone-treated environments with low host densities.

The concerns addressed are but few of innumerable questions that must be resolved before kairomones can be successfully employed on a large scale to control the behavior of parasitoids. However, motivation of behaviorists to decipher and utilize the complex chemical relationships that link parasitoids and their hosts will continue to lie with the envisioned rewards of a novel and enduring technology for the management of insect pests.

REFERENCES

Force, D. C. 1967. Genetics in the colonization of natural enemies for biological control. Ann. Entomol. Soc. Am. 60:722–729.

Greany, P. D., J. H. Tumlinson, D. L. Chambers, and G. M. Boush. 1977. Chemically-mediated host finding by *Biosteres* (*Opius*) *longicaudatus*, a parasitoid of tephritid fruit fly larvae. J. Chem. Ecol. 3:189–195

Gross, H. R., Jr., W. J. Lewis, and R. L. Jones. 1975. Kairomones and their use in the management of entomophagous insects. III. Stimulation of *Trichogramma achaeae*, *T. pretiosum*, and *Microplitis croceipes* with host-seeking stimuli at time of release to improve their efficiency. J. Chem. Ecol. 1:431–438.

Gross, H. R., Jr., W. J. Lewis, and R. L. Jones. 1975. Kairomones and their use for management of entomophagous insects. III. Stimulation of *Trichogramma achaeae*, *T. pretiosum*, and *Microplitis croceipes* with host-seeking stimuli at time of release to improve their efficiency. J. Chem. Ecol. 1:431–438.

Hagen, K. S., E. F. Sawall, Jr., and R. L. Tassan. 1970. The use of food sprays to increase effectiveness of entomophagous insects. Proc. Tall Timbers Conf. Ecological Animal Control by Habitat Management 10:59–81.

Hagen, K. S., P. D. Greany, E. F. Sawall, Jr., and R. L. Tassan. 1976. Tryptophan in artificial honeydews as a source of an attractant for adult *Chrysopa carnea*. Environ. Entomol. 5:458–468.

Hendry, L. B., P. D. Greany, and R. J. Gill. 1973. Kairomone-mediated host-finding behavior in the parasitic wasp *Orgilus lepidus*. Entomol. Exp. Appl. 16:471–477.

Ignoffo, C. M., C. Garcia, W. A. Dickerson, G. T. Schmidt, and K. D. Biever. 1977. Imprisonment of entomophages to increase effectiveness: Evaluation of a concept. J. Econ. Entomol. 70:292–294.

Jones, R. L., W. J. Lewis, M. C. Bowman, M. Beroza, and B. A. Bierl. 1971. Host-seeking stimulant for parasite of corn earworm: Isolation, identification, and synthesis. Science 173:842–843.

Jones, R. L., W. J. Lewis, M. C. Bowman, M. Beroza, B. A. Bierl, and A. N. Sparks. 1973. Host-seeking stimulants (kairomones) for the egg parasite *Trichogramma evanescens*. Environ. Entomol. 2:593–596.

Knipling, E. F., and J. U. McGuire. 1968. Population models to appraise the limitations and potentialities of *Trichogramma* in managing host insect populations. U.S. Department of Agriculture Technical Bulletin No. 1387.

Knipling, E. F. 1976. Role of pheromones and kairomones for insect suppression systems and their possible health and environmental impacts. Environ. Health Perspect. 14:145–152.

Laing, J. 1937. Host-finding by insect parasites. I. Observations on finding of host by *Alysis manducator, Mormoniella vitripennis,* and *Trichogramma evanescens*. J. Anim. Ecol. 6:298–317.

Lewis, W. J., and R. L. Jones. 1971. Substance that stimulates host-seeking by *Microplitis croceipes,* a parasite of *Heliothis* species. Ann. Entomol. Soc. Am. 64:471–473.

Lewis, W. J., R. L. Jones, and A. N. Sparks. 1972. A host-seeking stimulant for the egg parasite *Trichogramma evanescens:* Its source and a demonstration of its laboratory and field activity. Ann. Entomol. Soc. Am. 65: 1087–1089.

Lewis, W. J., R. L. Jones, D. A. Nordlund, and A. N. Sparks. 1975a. Kairomones and their use for management of entomophagous insects. I. Evaluation for increasing the rates of parasitization by *Trichogramma* spp. in the field. J. Chem. Ecol. 1:343–347.

Lewis, W. J., R. L. Jones, D. A. Nordlund, and H. R. Gross, Jr. 1975b. Kairomones and their use for management of entomophagous insects: II. Mechanisms causing increase in rate of parasitization by *Trichogramma* spp. J. Chem. Ecol. 1:349–360.

Lewis, W. J., R. L. Jones, H. R. Gross, Jr., and D. A. Nordlund. 1976. The role of kairomones and other behavioral chemicals in host finding by parasitic insects. Behav. Biol. 16:267–289.

Lewis, W. J., M. Beevers, D. A. Nordlund, H. R. Gross, Jr., and K. S. Hagen. 1979. Kairomones and their use for management of entomophagous insects. IX. Investigations of various kairomone-treatment patterns for *Trichogramma* spp. J. Chem. Ecol. 5:673–680.

Morrison, R. K., S. L. Jones, and J. D. Lopez. 1978. A unified system for the production and preparation of *Trichogramma pretiosum* for field release. Southwest. Entomol. 3:62–68.

Oatman, E. R., and G. R. Platner. 1971. Biological control of the tomato fruitworm, cabbage looper, and hornworms on processing tomatoes in southern California, using mass releases of *Trichogramma pretiosum*. J. Econ. Entomol. 64:501–506.

Rabb, R. L., R. E. Stinner, and R. van den Bosch. 1976. Conservation and augmentation of natural enemies. P. 433–544. *In* C. B. Huffaker and P. S. Messenger (eds.), Theory and Practice of Biological Control. Academic Press, New York.

Roth, J. P., E. G. King, and A. C. Thompson. 1978. Host location behavior by the tachinid *Lixophaga diatraeae*. Environ. Entomol. 7:794–798.

Sanford, J. W. 1976. Sugarcane borer: Seasonal distribution and fate of eggs deposited on sugarcane in Lousiana. J. Georgia Entomol. Soc. 11:332–334.

Sauls, C. E., D. A. Nordlund, and W. J. Lewis. 1979. Kairomones and their use for management of entomophagous insects. VIII. Effect of diet on the kairomonal activity of frass from *Heliothis zea* (Boddie) larvae for *Microplitis croceipes* (Cresson). J. Chem. Ecol. 5:363–369.

Stinner, R. E. 1977. Efficacy of inundative releases. Annu. Rev. Entomol. 22:515–531.

Vinson, S. B. 1968. Source of a substance in *Heliothis virescens* that elicits a searching response in its habitual parasite, *Cardiochiles nigriceps*. Ann. Entomol. Soc. Am. 61:8–10.

Vinson, S. B. 1977. Behavioral chemicals in the augmentation of natural enemies. P. 237–279. *In* R. L. Ridgway and S. B. Vinson (eds.), Biological Control by Augmentation of Natural Enemies. Plenum Press, New York.

Zil' Berg, L. P., A. S. Abashkin, and B. P. Adashkevich. 1976. Capsules for the distribution of *Trichogramma*. Zashch. Rast. 5:27.

SECTION III

ROLE AND SIGNIFICANCE OF PHEROMONES

CHAPTER NINE

HOST DISCRIMINATION BY PARASITOIDS

JOOP C. VAN LENTEREN

Department of Ecology
Zoological Laboratory
University of Leiden
Leiden
The Netherlands

George Salt, the first researcher who intensively and, in my opinion, properly analyzed host discrimination, wrote (1958):

> It is naive to suppose that natural enemies can be effectively used for control with inadequate knowledge of their biological nature, especially their physiology and behaviour. . . . The behaviour of insect parasites plays a great part in fixing their values as pest controls; and it follows that we must know their behaviour before we can use them effectively.

I know that not all workers in the field of biological pest control think this way. Several colleagues from California consider it unimportant to study parasitoid behavior in great detail before releases are made (see van Lenteren 1980). During my experience in biological control I have become convinced that detailed observations may play an important part in developing a good pest control program. Of course, it is not a matter of "never or always": in some situations the necessary basic research may be limited, whereas in others a long research period is essential if the application of a natural enemy is to be successful.

Vinson and Weseloh (Chapters 4 and 5) described how parasitoids locate their hosts, and Arthur (Chapter 6) discussed why a parasitoid may reject or accept a host. Arthur's chapter dealt mainly with what I would describe as *host selection,* that is, selection between hosts of different stages and/or species. *Host discrimination* is usually defined as the ability of a parasitoid to distinguish unparasitized from parasitized hosts.

In this chapter I first present a brief historical review of the development of ideas on host discrimination. Then I discuss how different opinions about the ability to discriminate arose, and also some possible causes of superparasitism (the situation that arises when wasps of one species deposit more eggs in or on a host than can develop in that host). Because host discrimination has often been analyzed improperly, some experiments are described by which the presence of host discrimination can easily be ascertained. Later I deal with the ways the parasitoid may mark the host and how it can perceive such a mark. My final topics are the biological significance of host discrimination for the parasitoid and the importance of host discrimination for biological pest control and population dynamics.

HISTORICAL REVIEW

Until 1900 it had generally been supposed that a parasitoid selects its hosts in the best interest of its offspring and deposits eggs only in individuals that are unparasitized. Most entomologists apparently accepted the idea that insects have a proper instinct for avoiding mistakes. Howard (1897) observed superparasitism and wrote that "such mistakes are, of course, much more likely to occur during such times of extraordinary multiplication than when a species is normally abundant." Howard realized that superparasitism was abnormal and occurred only when the parasitoid density was high relative to host density. Thus the relation between superparasitism and lack of suitable hosts had already been discovered at the turn of the century, although it was neglected or unnoticed by most of the pre–World War II workers of the next half-century.

In 1909 Fiske and Thompson wrote: "A considerable amount of evidence has been accumulated during the past three years in the laboratory which tends strongly to support the contention that the prescience [i.e. instinct] of the female parasite [of several species] is insufficient to enable her to distinguish parasitized and unparasitized hosts for her progeny." Fiske (1910) tried to define this phenomenon (the condition that results "when an individual host is attacked by two or more species of primary parasites, or by one species more than once") and presented examples of parasitoids that superparasitized. He further proved that the result of "ovipositing more than once" in one host might be greatly to the disadvantage of the parasitoid if it resulted in the elimination of supernumerary or all parasitoids, the weakening of the emerging parasitoids, or the premature death of the host.

On the basis of these data about superparasitism, Fiske (1910) stated that wasps of several species are unable to discriminate. He reasoned that: "The prevalence of superparasitism depends entirely upon whether or not the female is gifted with a prescience which will enable her to select healthy hosts for her offspring.'' And also: ''Presupposition that she possesses this instinct is equivalent to the denial of the existence of superparasitism.'' The idea expressed in this last sentence appears to have caused many problems for later researchers. It was implicitly or explicitly used to prove that a certain parasitoid does not distinguish healthy hosts from parasitized hosts *because* sometimes more than one parasitoid egg per host was found. The way Fiske described superparasitism gives me the feeling that he saw host discrimination as an ''all or nothing'' phenomenon: either complete avoidance of superparasitism or random oviposition.

Fiske (1910) was the first to analyze egg distributions mathematically, as is clear from the following quotation: ''Total absence of any such instinct [host discrimination] makes the prevalence of superparasitism wholly dependent upon and governed by the law of chance.'' He made predictions about egg distributions to be expected from a perfectly discriminating parasitoid and from one that lays its eggs at random, and he analyzed the (field) data then available. These studies led him to conclude that superparasitization of the degree he found could be caused only by random oviposition (see, however, the discussion of main causes of different opinions, below).

Thompson (1924) supported Fiske's ideas and developed together with Deltheil a formula to calculate the percentage of parasitism on the basis of random oviposition. Thompson was interested in such a formula because in preceding papers he had used population models in which the parasitoids could discriminate; but on the basis of the available data, he became more and more convinced that there was no reason at all to suppose that parasitoids have such a capability. Therefore he wanted to include in his model a parasite that oviposits at random.

The whole situation changed when Salt in the 1930s started to study egg distributions from field and laboratory work. He described host discrimination as: ''the ability to discriminate between healthy hosts and those already parasitized'' (Salt 1934). His field data (own research and data from other people) and laboratory work showed that the parasitoid species involved could distinguish, at least to a certain degree, between parasitized and unparasitized hosts. Both Salt (1934) and Walker (1937) analyzed field data mathematically with essentially the same formula that was used by Thompson (1924) and proved that the theory of random distribution was not applicable to all parasitoids. Somewhat later Salt (1937) showed, on the basis of laboratory research, several important facts. First, the host *Sitotroga cerealella* is marked externally and internally by the parasitoid *Trichogramma evanescens*. Second, the act of discrimination may take place at two moments: (1) during external examination with the antennae and/or (2) during internal examination with the ovipositor. Third, superparasitization does occur, but only after the parasitoid has had to stay with parasitized hosts for a long period, that is, has failed to contact any unparasitized hosts. Even if the parasitoid provisionally accepts an already parasitized host, this does not usually result in an oviposition: the parasitoid stabs her ovipositor into the host for a short while and abandons it without laying an egg.

Salt's most essential statement was that avoidance of superparasitization involves two distinct faculties: (1) the ability to discriminate between healthy and parasitized hosts, and (2) the ability to refrain from oviposition when suitable hosts are not available (Salt, 1934). Many later researchers mixed up these two separate faculties. Usually this led them to draw incorrect conclusions. Generally, superparasitism occurs not because of absence of the ability to discriminate or failure to exercise it, but because of a failure of restraint.

At the end of his 1934 article Salt concluded that the faculty of host discrimination seems to be widespread among parasitic Hymenoptera. Decades later (Salt, 1961) he confirmed that conclusion, and many other students of parasitoids have expressed the same idea (e.g. Clausen 1940, Doutt 1959, Askew 1971, DeBach 1974, Huffaker and Messenger 1976, Ridgway and Vinson 1977). Salt does not expect discrimination to occur among parasitic Diptera. Askew (1971) recently wrote that discrimination has not so far been recognized among parasitic Diptera. However, there are very few data on the oviposition behavior of this group.

I estimate that the capability of discrimination has now been established in 150–200 species of hymenopterous parasitoids, spread over all families. However, many scientists still feel that the hymenopterous parasitoids they have studied do not discriminate (for references, see next section). In several cases inadequate data were used; in others the data were all right but the wrong conclusions were drawn, generally because of the improper use of definitions.

If we study all recent information, can we determine whether host discrimination occurs in most hymenopterous parasitoids? If it does, why do many researchers maintain the opposite opinion?

MAIN CAUSES OF DIFFERENT OPINIONS

Wrong Data

One of the most common errors in discrimination studies is the basing of a worker's conclusions on field data. Usually these data are the only ones available and there is no possibility of doing laboratory research. The following procedure is almost standard: host are collected at random in the field, they are dissected and parasitoid eggs are counted, and data from all dissections are added together. If a random distribution of parasitoid eggs or larvae among the hosts is found, it is concluded that the parasitoid cannot distinguish parasitized from unparasitized hosts (e.g. Fiske 1910, Varley 1941, Schröder 1974, Jørgensen 1975, Carl 1976).

In no case, however, should results from several sites be added together. These sites may differ in the ratio of hosts to parasitoids, resulting in different degrees of parasitization that cannot be leveled out because of the limited migratory power of host and/or parasitoids. Although the egg distributions at each of these different sites may all be nonrandom, they will usually have different means. A summation of such nonrandom distributions may automatically give a random distribution through the combination of these individually nonrandom distributions (for an example, see van Lenteren et al. 1978).

It is not permissible to draw a conclusion about the absence of host discrimination based on data collected in the way described above, because one should at least know:

1 The number of suitable hosts that were available to the parasitoids in the period during which they were searching for hosts.

2 The actual number and the kind of contacts between host and parasitoids, since only in this way can data on acceptance and/or rejection of hosts encountered be obtained.

3 The condition of the parasitoids (e.g. freshly emerged, inexperienced, experienced, number and moment of preceding ovipositions); the importance of this knowledge will become clear when the causes of superparasitism are treated (p. 158).

This means that a proper field study of host discrimination will be very laborious, and in many cases even impossible to carry out. Raw field data can only sometimes be used to indicate the existence of host discrimination. We may conclude that Salt was very lucky to find a nonrandom egg distribution in hosts collected in the field, since it appears from the literature that very often erroneous conclusions are drawn because of the general unsuitability of such data.

Wrong Definitions and/or Terminology

Another common error is that, despite the availability of good data, the researchers use the term "host discrimination" carelessly. Salt (1934) clearly distinguished between the ability to discriminate and the ability to refrain from oviposition. Several workers did not know this, forgot about it, or mixed up the two abilities and simply concluded after the observation of superparasitism that the parasitoid species involved could not discriminate or that discrimination had broken down, varied, and so on (e.g. Fiske, 1910; Beling, 1932, Fulton 1933, Lloyd 1940, Caudri 1941, Ullyett 1943, 1945, 1949, Jenni 1947, 1951,

Venkatraman 1964, Force and Messenger 1965, Rogers and Hassell 1974, Schröder 1974, Jørgensen 1975, Rogers 1975, Carl 1976, Lawrence et al. 1978).

In many cases of authors who concluded that "their" parasitoid could not discriminate, I discovered from their own data that the superparasitization appeared to be caused not by a failure of discrimination but by a failure of restraint. The egg distributions given by these authors were uniformly found to differ significantly from random distributions, although superparasitism was common. This means that eggs were laid mainly in hosts with the lowest number of parasitoid eggs, hence that the parasitoid can discriminate (Bakker et al. 1967, 1972). To avoid mistakes about whether the ability of host discrimination is present, it would be advisable to extend the meaning of the term "host discrimination" to include the ability to distinguish hosts with different numbers of eggs (van Lenteren et al. 1978).

Wrong Arguments

Some authors have correct data available and use the proper terms, but conclude on the basis of wrong arguments that discrimination does not occur. An example: Gerling and Schwartz (1974) state that superparasitism in the parasitoid *Telenomus remus* is quite common and conclude that the parasitoid therefore can not discriminate. They argue that this parasitoid does not need to discriminate because the host lays hundreds of eggs in each batch and the number of hosts available for immediate oviposition by an individual parasitoid usually exceeds the capacity of the female to oviposit several times. To me, this argument is nonsense. First, the parasitoid may revisit parasitized hosts during her stay and, second, conspecific females may visit the same egg cluster afterwards. Furthermore, one of the figures these authors present shows that oviposition is certainly not random, and finally the superparasitism they found can almost certainly be attributed to a strong tendency to oviposit and/or to the parasitoid's not yet having learned to discriminate (see below).

An important part of the differences in opinion is surely caused by one or more of these three errors.

FACTORS THAT MAY CAUSE SUPERPARASITISM

Based on the foregoing discussion, we can safely assume that many parasitoid species are incorrectly put on the list of "nondiscriminators." Students of discrimination are likely to be deceived by the occurrence of high degrees of superparasitism, and therefore I decided several years ago to study factors that may cause superparasitism in the parasitoid *Pseudeucoila bochei* ovipositing in larvae of *Drosophila* species (see van Lenteren 1976) (Figure 9.1).

Possible causes of superparasitism were suggested as follows:

1 A female lays more than one egg per oviposition.
2 A female does not recognize hosts parasitized by other females.
3 A female lays a second egg within the period needed for building up a factor that causes avoidance of superparasitism.
4 Two or more females lay eggs simultaneously in one host.
5 A female's tendency to oviposit increases when she encounters only parasitized hosts for a long period and she will then lay eggs in these hosts.

During the research, a sixth possible cause was discovered:

6 A female has not yet learned to discriminate.

All except the last cause can be found in the literature on host discrimination as possible causes for superparasitism. Especially the fifth cause is frequently seen, because most students of parasitoids take for granted that causes 1–4 never or only very infrequently play a role; that is:

1 Solitary parasitoids do not lay more than one egg per oviposition.
2 They do recognize hosts parasitized by conspecific females.
3 A mark is recognizable immediately after the first parasitization.
4 Simultaneous ovipositions do not occur frequently.

Unimportant Factors

After my experiments on the importance of the factors outlined above, I can agree with the students who expected the first two possible causes to be unimportant. *Pseudeucoila* females do not lay more than one egg at an oviposition, and this result was also found for several other solitary parasitoids (e.g. Lloyd 1938, Caudri 1941, Simmonds 1956, Hays and Vinson 1971, Greany and Oatman 1972). Some authors mention that occasionally two eggs are laid per oviposition by a solitary parasitoid (e.g. Beling 1932, Wilbert 1964). The recognition by *Pseudeucoila* of hosts parasitized by conspecific females is perfect, and the same was found for a number of other parasitoids (e.g. Chamberlin and Tenhet 1926, Hill 1926, Rosenberg 1934, Lloyd 1935, 1956, Salt 1937, Ayyar 1941, Ullyett 1949, Jackson 1966, King and Rafai 1970, Rabb and Bradley 1970, Fisher 1971).

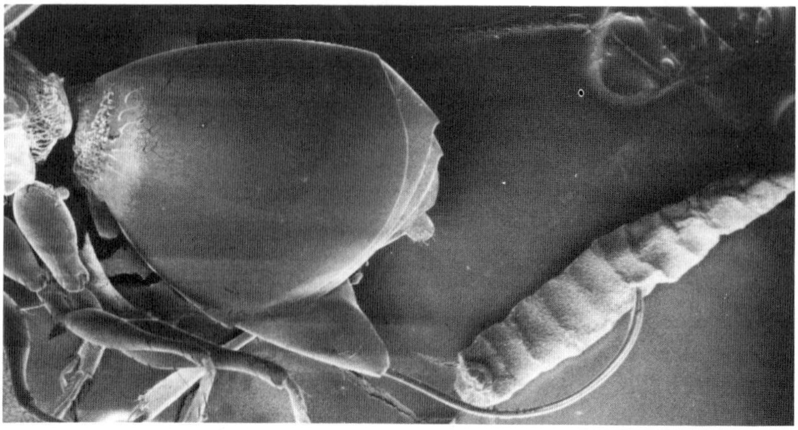

FIGURE 9.1. *Pseudeucoila bochei* female killed while laying an egg in its host, a larva of *Drosophila melanogaster*. (Scanning-electron micrograph, 46 ×, by the author.)

The third factor may, in some parasitoid species, cause superparasitizations. The wasp *Pseudeucoila* superparasitized a larva of *Drosophila* if she returned to a host within 70 seconds after the first oviposition. Figure 9.2 shows that some second contacts within this 70-second period led to ovipositions. If the wasp stabbed the host for the second time close to the spot where she had oviposited, that host would be rejected even if the second contacts occurred within 20 seconds. The two ovipositions that occurred after 50 seconds took place at the other end of the host. The relationship between distance between successive contacts and probability of superparasitization at the second or later contacts indicates that the marking pheromone has to spread over the host's body and that this takes at most 70 seconds. Rogers (1972) also noticed that it takes some time after oviposition before the host is recognizable as being parasitized. Wylie (1971) said that the distance between the sites at which a host is pierced plays a role.

Through the fourth factor, which is in fact a special case of the third, superparasitism may be caused in *Pseudeucoila*, but only if the parasitoid density is high relative to host density. Simultaneous ovipositions are also known to occur in other parasitoids (e.g. Howard 1897, Hase 1925, Ullyett, 1936, 1949, Wylie 1958, Wilson 1961, Jackson 1966).

A Female's Tendency to Oviposit

The fifth factor has not been analyzed frequently, although many authors mention it as the main cause for superparasitism. A multitude of terms is used as substitutes for relevant data. Phrases such as "a breakdown of the ability to

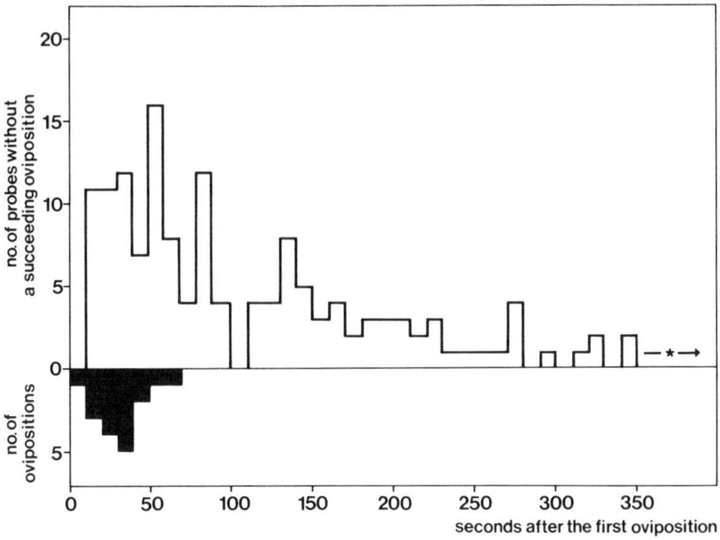

FIGURE 9.2. Number of probes without a succeeding oviposition (open columns) and number of ovipositions (solid columns) after the first oviposition (per 10-second period); asterisk indicates the period from 350 to 1800 seconds after the first oviposition, when 56 probes were observed, and no ovipositions.

discriminate," "increased desire to get rid of the eggs," "a varying degree of discrimination," "urge to oviposit," "a high drive to oviposit," "the need to oviposit is great," and "a great pressure to oviposit," are used to describe the cause of the loss of the ability to refrain from ovipositing in parasitized hosts (e.g. Ullyett 1936, 1943, 1949, Walker 1937, Doutt and DeBach 1964, Chacko 1969, Schröder 1974, Jørgensen 1975, Carl 1976, and many others). It was again Salt (1934) who properly described how superparasitism could be caused by an "urge to oviposit." He wrote:

> Towards the end of the experiment, when most of the available hosts had been found and the parasite came to host after host it had already parasitized, it became more and more "apathetic" and would move along the hosts without paying much attention to them, but on coming to one not previously attacked it seemed suddenly to become "alert", then "excited", and soon oviposited in it. At this stage of the experiments long periods intervened while the parasite roamed in the barren areas of the dish or stood cleaning itself.

Caltagirone (1977, personal communication) found the same kind of behavior in the parasitoid *Phanerotoma flavitestacea:* first many ovipositions were seen, then more and more probings without oviposition followed; finally, if the parasitoid and the parasitized hosts were confined for a long period, after an interval with probings only some superparasitizations would follow, and all would be interrupted by several probes. In our laboratory we found very much the same phenomena in the parasitoids *P. bochei* (van Lenteren 1976), *Encarsia formosa* (van Lenteren et al. 1976), *Aphidius matricariae* ('t Hart et al. 1978), *Aphytis lingnanensis, Aphytis coheni,* and *Aphytis melinus* (van Lenteren et al. 1977, unpublished data), *Pachycrepoideus vindemiae* (Figure 9.3) (Nell 1976, unpublished data), *Asobara tabida* (van Aphen 1977, unpublished data), *Opius pallipes* and *Dacnusa sibirica* (Figure 9.4) (Dijkgraaf 1978, unpublished data). Salt's observations that "the parasite roamed in the barren areas of the dish" indicates that the parasitoid apparently moves away from a spot where most hosts are parasitized. Indeed, all the above-mentioned parasitoids we tested leave the experimental setup, if open, as soon as most of the hosts have been parasitized.

The data in Figure 9.5 illustrate this behavior. When a female wasp arrives at a site where unparasitized hosts are present, she will start ovipositing. In the control situation (Figure 9.5*a*), where every parasitized host was substituted by a fresh unparasitized one, the parasitoid continued ovipositing for several hours. The wasp only rarely leaves the patch, and if she does leave, it is only for short periods. In the experiments in which parasitized hosts were not replaced (Figure 9.5*b*), the number of ovipositions per unit of time decreased strongly. The number of probes increased and after a number of probes in succession, the wasp

FIGURE 9.3. *Pachyrepoideus vindemiae* laying an egg in *Drosophila* pupa; this parasitoid applies an external and an internal mark during oviposition. (Photograph by H. W. Nell.)

FIGURE 9.4. *Dacnusa sibirica* laying an egg in a leaf miner larva; this parasitoid marks its host internally during oviposition (Photograph by the author.)

left the patch and tried to get out of the experimental arena. After a period of standing still and preening her body, the wasp returned to the patch, resumed searching for hosts, and probed frequently. Sometimes an oviposition occurred in an already parasitized host. Then she again left the patch, and the cycle could start anew. The importance of this behavior is discussed on page 172.

Learning to Discriminate

The last possible cause of superparasitism, the necessity to learn to discriminate, needs some more attention. Before I started to study the causes of superparasitism in *P. bochei*, the aspect of learning had not occurred to me, although previously we had found that parasitoids of several species need to experience their oviposition behavior before the act is performed properly (Samson-Boshuizen et al. 1974, van Lenteren 1976, Nell et al. 1976, 't Hart et al. 1978). During the analysis of the importance of the first five factors, we discovered that all together they did not cause sufficient superparasitizations to explain the results in earlier experiments. We then realized that experience might also be necessary before proper host discrimination could be performed.

In a series of experiments we offered parasitized and/or unparasitized hosts to inexperienced females (i.e. females that had never oviposited) for half an hour,

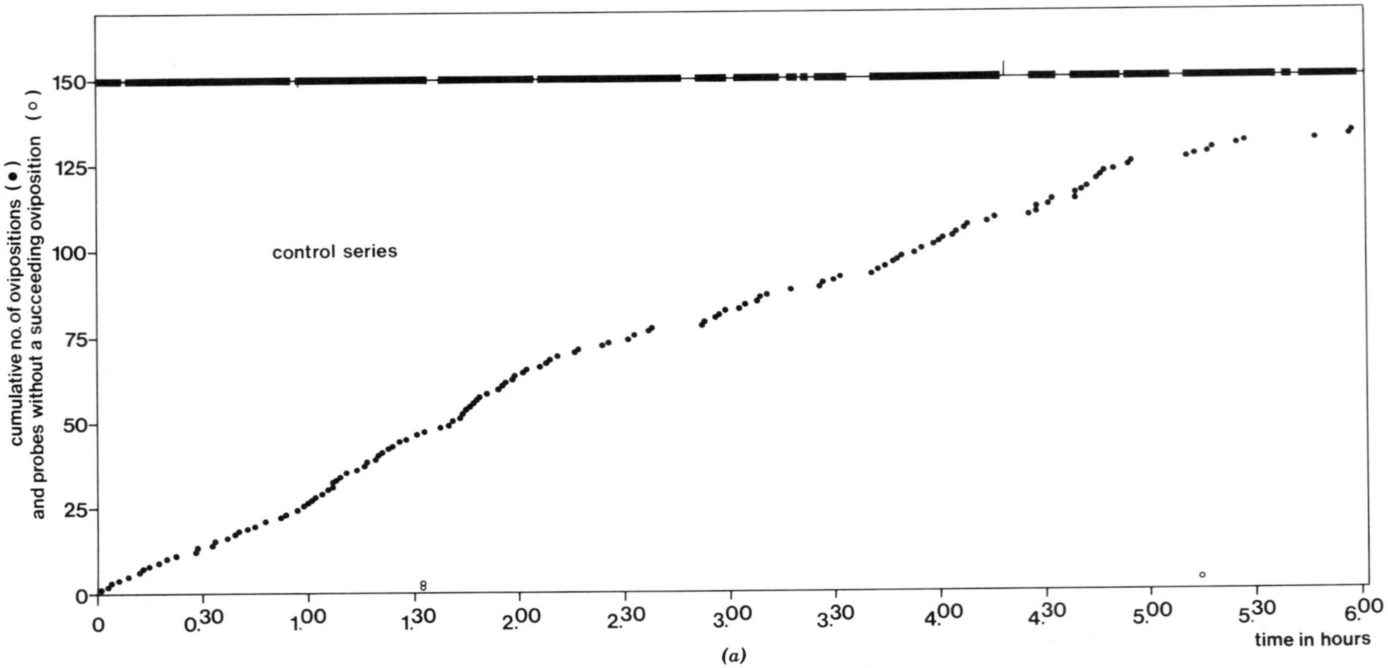

FIGURE 9.5. Cumulative number of ovipositions and probes without a succeeding oviposition per minute. Top of each graph shows time spent on host medium (thick horizontal bars), time spent outside host medium (thin-horizontal bars), and attempts to escape from the experimental setup (vertical lines). *(a)* Replacement of parasitized hosts by unparasitized hosts.

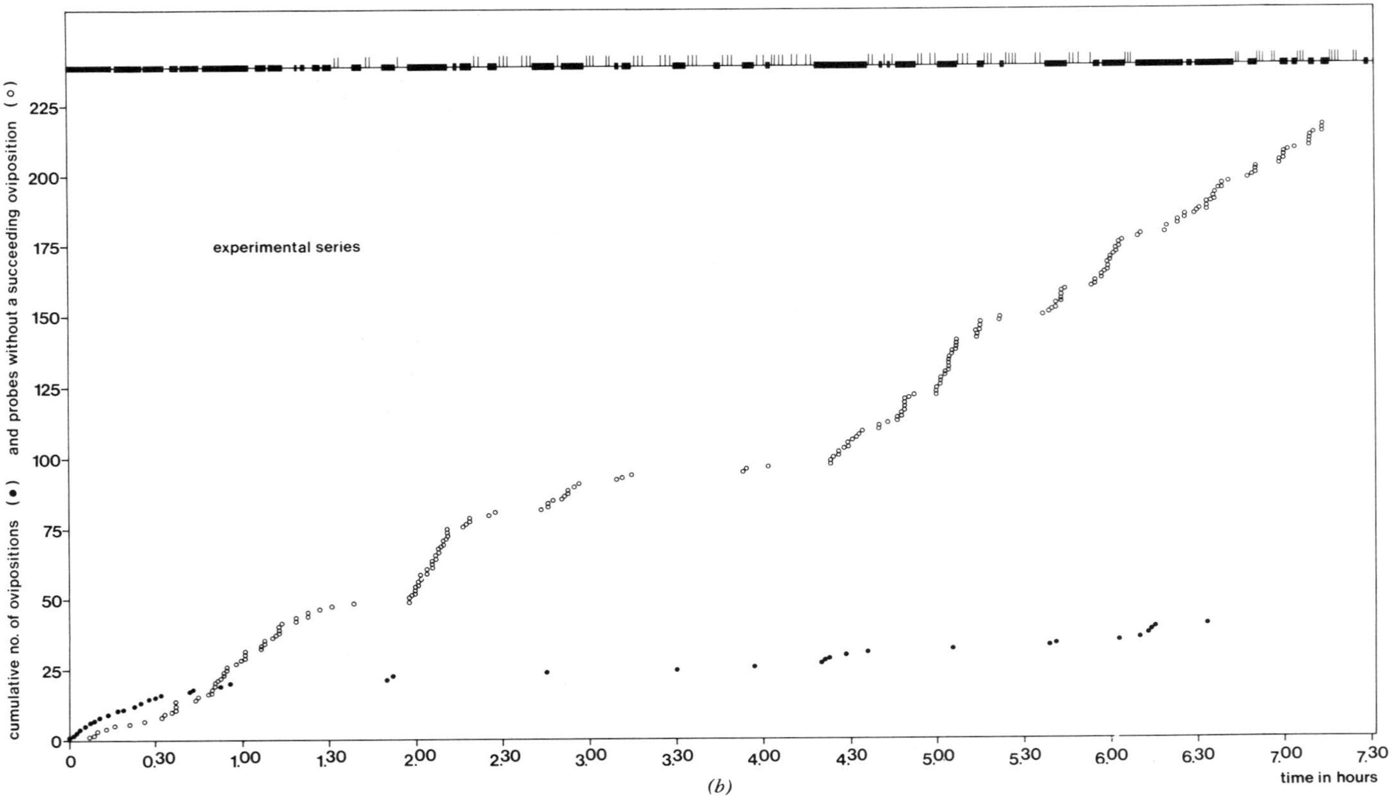

FIGURE 9.5. *(b)* No replacement of parasitized hosts.

three to five times in succession (Figures 9.6*a*–*c*). In the first series only parasitized hosts were offered; the wasps accepted these hosts in most cases, though some rejections occurred. In the control series (Figure 9.6*b*) inexperienced females were introduced to unparasitized hosts. These wasps accepted most hosts, but also rejected some. These series show that inexperienced wasps accept parasitized hosts as easily as they accept unparasitized ones.

In another series (Figure 9.6*c*) inexperienced wasps were presented alternately with unparasitized and parasitized hosts. Four out of five wasps laid hardly any eggs in parasitized hosts after a first period with unparasitized ones. Female number 5 was a slow learner; she accepted most of the parasitized hosts the first time they were presented, but rejected most of them the next time. The change in behavior (viz. rejecting instead of accepting parasitized hosts after contacts with unparasitized hosts), which is a result of experience, can be interpreted as "learning" according to Thorpe's (1956) definition: "Learning is that process which manifests itself by adaptive changes in individual behavior as a result of experience." That the change in behavior is adaptive is clear because only one parasitoid develops from each superparasitized host (for more details see van Lenteren and Bakker 1975, and van Lenteren 1976).

In a few cases data in the literature indicate that experience is involved in proper host discrimination (Jackson 1966, 1969, Rabb and Bradley 1970). Van

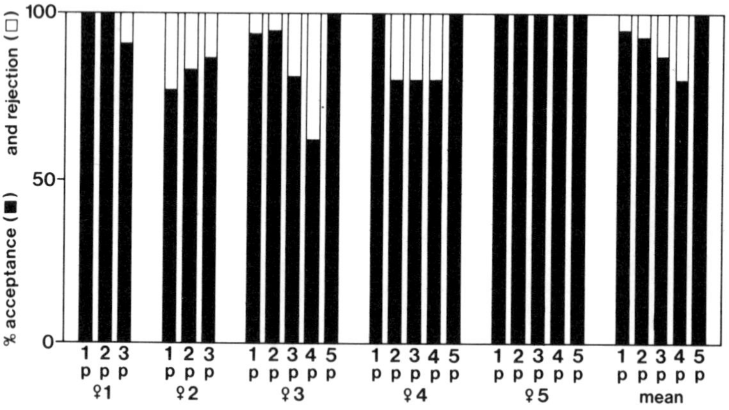

(*a*)

FIGURE 9.6. (*a*) Percentage acceptance and rejection of once-parasitized (*p*) hosts offered to inexperienced wasps for three to five 30-minute periods. (*b*) Percentage acceptance and rejection of unparasitized (*u*) hosts offered to inexperienced wasps for 5 30-minute periods. (*c*) Percentage acceptance and rejection of unparasitized and once-parasitized hosts offered in succession to inexperienced wasps for three or four 30-minute periods.

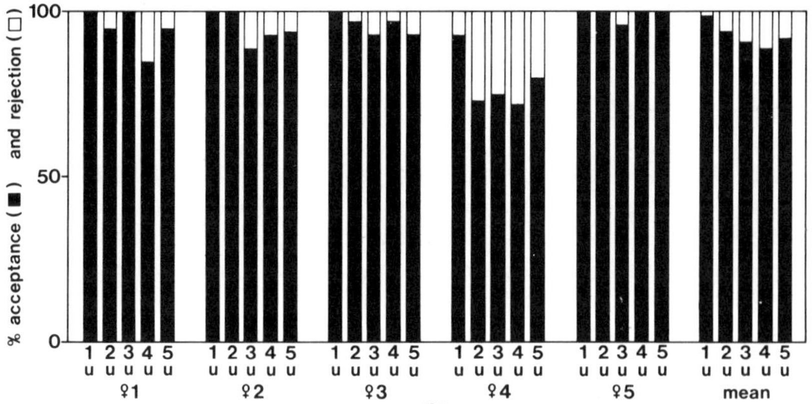

(b)

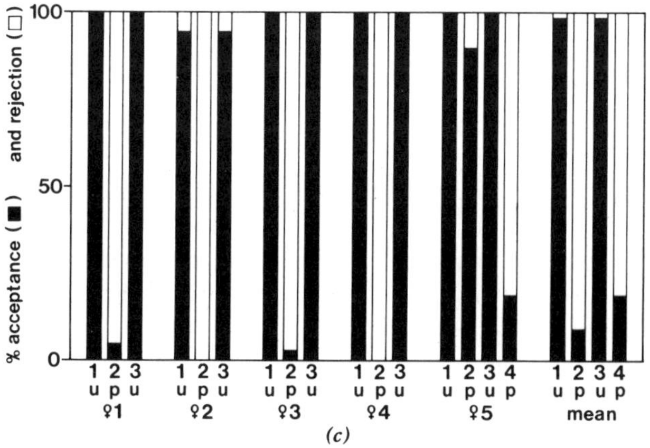

(c)

FIGURE 9.6. *(Continued)*

Alphen (1979, personal communication), Klomp et al. (1980), and Salt (1979, personal communication) told me that data from their experiments with *A. tabida* (Braconidae), *T. evanescens* (Trichogrammatidae), and *Trichogramma embryophagum* (Trichogrammatidae), respectively, can be interpreted as a result of learning to discriminate. Together, these few but clear data show that inexperienced females may superparasitize many hosts until they encounter unparasitized hosts and learn to discriminate. This casts a heavy shadow over many data from laboratory research, either because the condition (experienced-inexperienced) of the parasitoids that were used is not known or, if the condition was known, because the parasitoids were inexperienced during the first part of the experiment.

HOW TO TEST HOST DISCRIMINATION

As I have explained, it is difficult to test for the presence of the ability to discriminate in the field. Observations of the parasitoid's behavior are usually not possible in the field during long periods, and field samples provide an unreliable basis for analysis. A series of simple tests can be used to study host discrimination in the laboratory. If it is at all possible, the behavior of the parasitoids toward parasitized and unparasitized hosts should be observed (the direct tests). The test that gives the most information about the presence and the mechanism of host discrimination gives the following test.

Offer a batch of unparasitized hosts to a wasp; record the sequence, time interval, kind, and number of contacts with each host the wasp meets; do not replace the parasitized hosts. If sufficient hosts are presented, the parasitoid will start ovipositing. After a while she will oviposit in unparasitized hosts and probe parasitized ones and finally, when all hosts are parasitized, only probing will be performed (see Figure 9.5*b*). If the experiment lasts long enough, the wasp will eventually try to leave the patch with hosts or will superparasitize. A parasitoid that does not discriminate will lay an egg at almost every encounter with a host, will show no increasing number of probings, and will not leave the patch after all hosts have been parasitized once.

Sometimes it may be useful and sufficient to test host discrimination in an indirect way, although some of the risks connected with analyzing field data also apply to these tests (indirect tests). The following test proved to be a good one. Arrange for the parasitoid to be able to emigrate from the place where she was introduced—for example, a central box with four "escape" channels connected with other boxes containing many unparasitized hosts. In the central box a certain number of hosts is offered, and the wasp is allowed to oviposit as long as she likes. Afterward the degree of parasitization is determined by dissection. This setup allows one to ascertain whether host discrimination has occurred (the egg distribution shows this) and if so, whether the parasitoid reacted to contacts with once-parasitized hosts by emigrating.

Several other tests can be found in Bakker et al. (1967, 1972), van Lenteren (1976), and van Lenteren et al. (1978).

THE WAY THE PARASITOID MARKS THE HOST AND PERCEIVES THE MARK

To be able to discriminate, the parasitoid must be able to cause a change in the host after parasitization *and* be able to detect this change, as well.

The Way of Marking

Host marking can be accomplished in several ways. Some parasitoids put marking pheromones into or onto the host preceding, during, or after oviposition. Some examples: *Nasonia vitripennis* applies a pheromone into the host, pupae of *Musca domestica*, prior to oviposition (Wylie 1965, 1970, 1971). *Caraphractus cinctus* injects a marking pheromone into the host, eggs of *Agabus bipustulatus*, during oviposition (Jackson 1966, 1969). *Asolcus basalis* applies an external mark to the host, eggs of *Nezara viridula*, after oviposition (Wilson 1961).

In other cases the condition of the host may change because of parasitization: it may die or become paralyzed, it may develop in a different way, or it may produce a substance in reaction to the introduction of a foreign material (i.e. the parasitoid egg or fluids—not the marking pheromone itself—that are injected during the oviposition), or it may simply show an oviposition hole where body juices are exuding. *Microplectron fuscipennis* is said to react to, among other stimuli, the presence of a large parasitoid larva in its host, the pupae of *Diprion polytomum* (Ullyett 1936). *Spalangia drosophilae* avoids ovipositing in its hosts, pupae of several dipteran species, when cessation of heartbeat or death of the host occurs because of preceding parasitization (Simmonds 1956). *Aphelinus semiflavus* is reluctant to accept its hosts, larvae of *Rhopalomyzus ascolonicus*, for parasitization when they contain a hole, made by the ovipositor at a previous contact, where substances are oozing out (Wilbert 1964).

Sometimes not the host itself but only its surroundings are marked by the wasp (Chrystal 1930, Salt 1937, DeBach 1944, Price 1970, Greany and Oatman 1972, van Lenteren and Bakker 1978, Waage 1979), and here an interesting parallel between the marking of already visited sites by parasitoids and by phytophagous insects as discussed by Prokopy in Chapter 10 can be seen. The substrate marks can occur alone or may be additional to the mark applied to the host. Presence of these pheromones results in increased efficiency in finding spots with unparasitized hosts.

For the parasitoid *O. pallipes*, which lays her eggs in larvae of the tomato leafminer, *Liriomyza bryoniae*, we found that besides marking the host, a pheromone is applied to the leaf in which the mine is present (J. Kooloos, unpublished data). A second visit to the same leaf does not last as long (on average 104.9 ±78.7 seconds, $n = 15$) as the first visit (on average 282.8 ±228.6 seconds, $n = 12$), and on the leaves that had been visited before, the parasitoid less often found the host (in 15% of 20 visits) than on unvisited leaves (in 38% of 21 visits).

Only speculations exist about the source of production and the nature of these external marking pheromones (e.g. Greany and Oatman 1972); also no data are available about the period that these pheromones remain active.

The majority of nonmoving or slow-moving hosts (e.g. eggs, more or less sessile larvae, and pupae) and nonclustered hosts are marked both externally and internally. Strongly moving hosts that live in clusters are marked only internally. An external mark is likely to be useless, because moving hosts may frequently come into contact with each other and so may possibly transfer the mark; this would cause a large proportion of the unparasitized hosts to become marked and so escape parasitization.

Perception of the Marking Pheromone

The parasitoid can pick up traces of the marking pheromones with sensillae on antennae and ovipositor, and perhaps by sensillae on other parts of the body (tarsi?, see Salt 1937). The external marking pheromone of nonmoving hosts is discovered by the wasp through chemoreceptors on the antennae. Perception may result in immediate rejection: the parasitoid walks away from the host. Sometimes a parasitized host is provisionally accepted in spite of external examination, but the parasitoid may then reject the host after she has perceived the internal mark by sensillae on the ovipositor (see Figure 9.7). Some authors theorize that the second (internal) marking pheromone is necessary because the

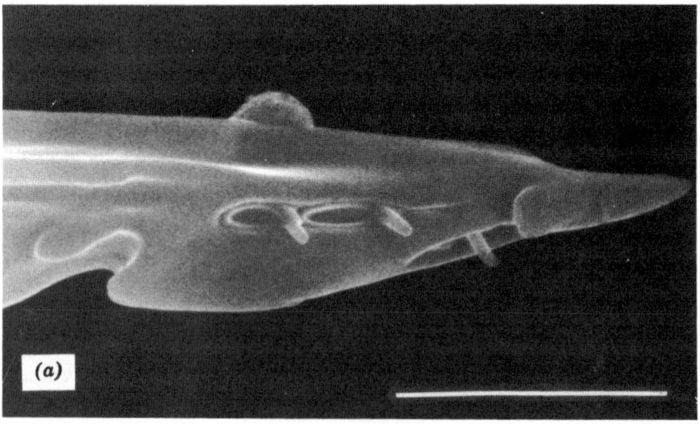

FIGURE 9.7. Scanning-electron micrographs of chemoreceptors on the ovipositors of (*a*) *Pseudeucoila bochei* (photograph by F. M. van der Wolk), (*b*) *Tetrastichus asparagi*, and (*c*) *Pachycrepoideus vindemiae* (photograph by the author). The length of the bar represents 5 μm.

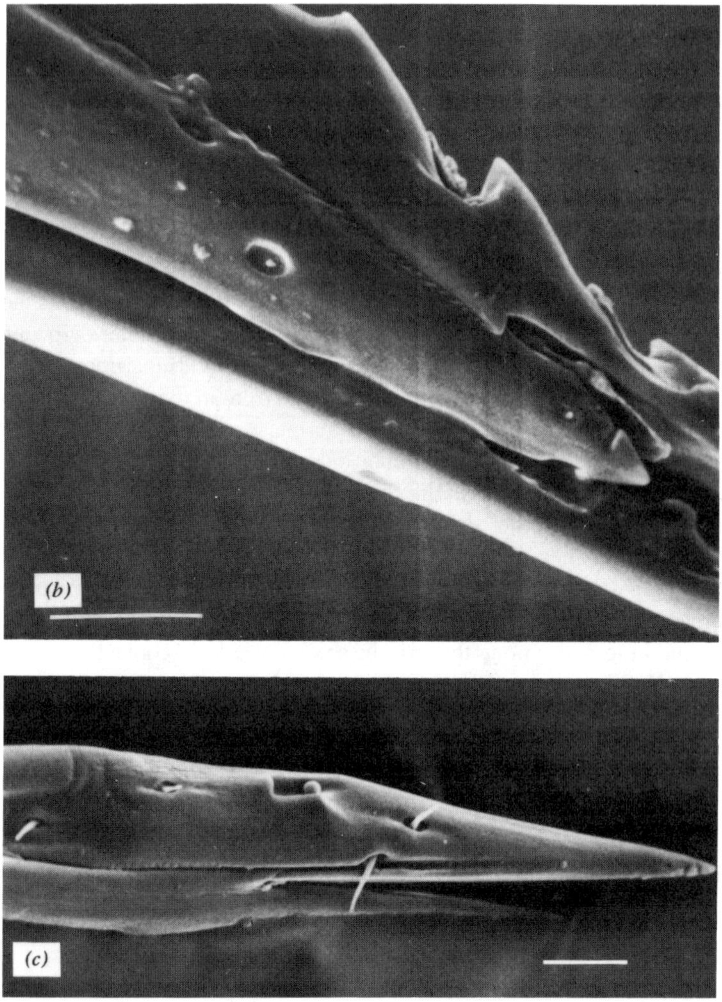

FIGURE 9.7. *(Continued)*

external marking pheromone might be washed off by rain, or it might diffuse and so become imperceptible after some days. The advantage of an external marking pheromone is, of course, that the parasitoid can discover it fast.

The perception of the internal marking pheromone of moving hosts is also done with sensillae on the ovipositor. A review of ways of marking the host and perception of the marking pheromone can be found in van Lenteren (1976).

Although a number of authors refer to antennae and ovipositors as playing a role in the perception of a marking pheromone (Ullyett 1936, Salt 1937, Simmonds 1956, Wilson 1961, Wilbert 1964, Wylie 1965, 1970, 1971, Jackson 1966, 1969, Hokyo et al. 1966, Fisher and Ganesalingam 1970, King and Rafai 1970, Rabb and Bradley 1970, Greany and Oatman 1972, Guillot and Vinson 1972, van Lenteren 1972, 1976, Rogers 1972, van Lenteren et al. 1976, 1978) and the nature of some marking pheromones is known (Guillot and Vinson 1972, Greany and Oatman 1972, Vinson and Guillot 1972), there are no antennograms or ovipositorgrams showing data of impulse frequencies or amplitudes of these organs after stimulation with material from unparasitized and parasitized hosts.

We did some work with the parasitoid *P. bochei,* which marks her host (*Drosophila* larvae) internally only and detects the mark with her ovipositor (van Lenteren 1972). Together with colleagues from the Department of Animal Physiology, I succeeded in recording impulses from an ovipositor with a simple setup. This work was continued by Lammers, who developed an elegant technique for standardizing the number of sensillae that were stimulated (Lammers and van der Starre 1978). His results showed that this parasitoid can detect with the ovipositor whether she is stinging agar, yeast (the medium in which the hosts live), or host hemolymph. We have not yet been able to detect differences between the patterns resulting from stimulation with hemolymph of parasitized and unparasitized hosts. Apparently, therefore, the parasitoid does not simply inject a marking pheromone that can be traced at a later contact.

Vinson (1977) mentions that external host-marking pheromones (after isolation, identification, and synthesis) may be used for biological pest control in two ways. First, in biological weed control, spraying of host-marking pheromone of the parasitoid of a phytophagous insect that is used for weed control might prevent parasitization of that phytophagous insect. Second, in biological insect control, spraying of the marking pheromone of a hyperparasitoid might prevent parasitization of the primary parasitoid. Application of these pheromones seems, however, to be a rather utopian goal at the moment, and application on a large scale would lead to selection of the parasitoids that are not inhibited from oviposition by the marking pheromone.

BIOLOGICAL SIGNIFICANCE OF HOST DISCRIMINATION

Several obvious advantages of host discrimination can be found in the literature. They are listed here, starting with those most commonly mentioned.

1 Discrimination may prevent wastage of parasitoid eggs. In solitary and gregarious parasitoids, ovipositing more than once in the same host results in too many parasitoids developing per host, which leads to elimination of supernumerary larvae or in development of very small adults (Salt 1961).

2 Discrimination may prevent wastage of hosts, because hosts that are stung and parasitized frequently show a considerably higher mortality than hosts parasitized once. So often both host and parasitoid are wasted.

3 Discrimination is a means to save time, provided laying an egg takes considerably more time than probing a host, which in fact is the case for many parasitoids.

4 Discrimination initiates migration after a number of probes at or in parasitized hosts. This advantage was discovered only recently, and bears important consequences for theories on optimal foraging (van Lenteren 1976, Hassell and Southwood 1978, van Lenteren and Bakker 1978, Waage 1979).

I think that being able to discover that hosts are parasitized *combined with* the initiation of migration after a number of contacts with parasitized hosts shows the importance of the phenomenon for population dynamics and biological pest control very clearly: the parasitoids move away from places with parasitized hosts, thereby saving time to spend in searching for and ovipositing in unparasitized sites.

IMPORTANCE OF HOST DISCRIMINATION FOR BIOLOGICAL PEST CONTROL AND POPULATION DYNAMICS

The handbooks on biological control and some monographs on parasitoids were checked on the point of whether host discrimination is mentioned as an important selection criterion for parasitoids (Clausen 1940, Flanders 1947, Doutt 1959, DeBach 1964, 1974, Askew 1971, Huffaker 1971, Huffaker and Messenger 1976, Ridgway and Vinson 1977).

Almost all these authors just cited mention the terms "superparasitism," "host discrimination," and "host marking," and distinguish between "discrimination" and "restraint." They all say that host discrimination occurs throughout most families of hymenopterous parasitoids, and they usually relate the occurrence of superparasitism to high parasitoid densities relative to host densities, thus do not consider superparasitism a cause for concern in the field. Only Flanders (1947) and Messenger et al. (1976) state explicitly that an effective parastioid should possess the ability to discriminate.

And here disillusion set in. When I began to work on biological control I thought that the presence or absence of host discrimination might be among the important selection criteria by which new useful parasitoids could be chosen. I found so many articles stating that the parasitoids studied lacked the ability to discriminate that it seemed obvious to select for control purposes only the parasitoids that do discriminate. In this chapter I presented arguments that cast many doubts on the alleged absence of host discrimination in so many species. Moreover, many species that are not efficient in controlling their host do clearly possess the ability to discriminate. All parasitoids tested by our research group(more than 15 species) showed the ability to discriminate, and I believe now that most hymenopterous parasitoids have this capability. In the chapter that follows, Prokopy presents a number of recent examples of the detection of marking pheromones in phytophagous insects. These insects also avoid ovipositing on or in marked objects, and it is interesting tonote the parallels in behavior among these insect groups.

Perhaps the selection criterion must be refined somewhat more to be useful: if several potential candidates for control applications are to be tested, the differences in the relation between the ability to refrain from oviposition in parasitized hosts and the initiation of migration may indicate the practical value of the species. An example: say we have three species, all able to discriminate. Parasitoids of the first species migrate after a few probes in parasitized hosts, and this means that many hosts will remain unparasitized at that patch. Parasitoids of the second species tend to migrate less soon; they refrain from oviposition for a long time and start subsequently to superparasitize. Most or all hosts will eventually be parasitized on that patch, but the wasps will not go to other places. Parasitoids of the third species migrate after a relatively high number (higher than for the first species) of successive negative contacts with parasitized hosts (e.g. after about 15 probes). This means that most hosts will be parasitized before the wasps leave the patch. Of course we would decide that concerning this aspect, the third species is the best.

In terms of optimal foraging theory, the parasitoids should not show a fixed-time, fixed-number, or fixed-searching-time behavior but a fixed-rate behavior. Migration from a depleted patch should be initiated by a rather high number of contacts with already parasitized hosts.

I realize that it takes some time to determine which parasitoid possesses the optimal strategy for biological control, but especially when easy control was not obtained, a detailed analysis of the causes of the failure revealed solutions in our biological control research (van Lenteren and Woets 1977). Therefore I still strongly support the opinion of Salt mentioned at the beginning of this chapter: "It is naive to suppose that natural enemies can be effectively used for control with inadequate knowledge of their biological nature."

ACKNOWLEDGMENTS

G. Salt unraveled the history of the research on host discrimination to me, and therefore, with great pleasure, I dedicate this chapter to him. K. Bakker and Nora Croin Michielsen read the manuscript critically and suggested improvements, Hetty Vogelaar typed it, and G. P. G. Hock drew the figures.

REFERENCES

Askew, R. R. 1971. Parasitic Insects. Heinemann, London.

Ayyar, P. N. K. 1941. Host-selection by *Spathius critolous* Nixon, an important parasite of *Pempherulus affinis* (Faust) in South India. Indian J. Entomol. 3:197–213.

Bakker, K., S. N. Bagchee, W. R. van Zwet, and E. Meelis. 1967. Host discrimination in *Pseudeucoila bochei* (Hymenoptera: Cynipidae). Entomol. Exp. Appl. 10:295–311.

Bakker, K., H. J. P. Eijsachers, J. C. van Lenteren, and E. Meelis. 1972. Some models describing the distribution of eggs of the parasite *Pseudeucoila bochei* (Hym., Cynip.) over its hosts, larvae of *Drosophila melanogaster*. Oecologia (Berlin) 10:29–57.

Beling, I. 1932. Zur Biologie von *Nemeritis canescens* Grav. (Hym., Ophion.). I. Züchtungserfahrungen und ökologische Beobachtungen. Z. angew. Entomol. 19:223–249.

Carl, K. P. 1976. The natural enemies of the pear-slug, *Caliroa cerasi* (L.) (Hymenoptera, Tenthredinidae), in Europe. Z. angew. Entomol. 80:138–161.

Caudri, L. W. D. 1941. The braconid *Alysia manducator* Panzer in its relation to the blow-fly *Calliphora erythrocephala* Meigen. D. Phil. thesis, Leiden University.

Chacko, M. J. 1969. Superparasitism in *Trichogramma evanescens minutum* Riley, an egg parasite of sugarcane and maize borers in India. II. Causes of superparasitism. Beitr. Entomol. 19:637–642.

Chamberlin, F. S., and J. N. Tenhet. 1926. *Cardiochiles nigriceps* Vier., an important parasite of the tobacco budworm, *Heliothis virescens* Fab. J. Agric. Res. 33:21–27.

Chrystal, R. N. 1930. Studies of the Sirex parasites. The biology and post-embryonic development of Ibalia leucospoides Hochenw. (Hymenoptera–Cynipoidea). Oxf. For. Mem. 11, Clarendon Press, Oxford.

Clausen, C. P. 1940. Entomophagous Insects. McGraw-Hill, New York.

DeBach, P. 1944. Environmental contamination by an insect parasite and the effect on host selection. Ann. Entomol. Soc. Am. 37:30–74.

DeBach, P. 1964. The scope of biological control. P. 3–20 , *In* P. DeBach (ed.), Biological Control of Insect Pests and Weeds. Chapman and Hall, London.

DeBach, P. 1974. Biological Control by Natural Enemies. Cambridge University Press, Cambridge.

Doutt, R. L. 1959. The biology of parasitic Hymenoptera. Annu. Rev. Entomol. 4:161–182.

Doutt, R. L., and P. DeBach. 1964. Some biological control concepts and questions. P. 118–142, *In* P. DeBach (ed.), Biological Control of Insect Pests and Weeds. Chapman and Hall, London.

Fisher, R. C. 1971. Aspects of the physiology of endoparasitic Hymenoptera. Biol. Rev. 46:243–278.

Fisher, R. C., and V. K. Ganesalingam. 1970. Changes in the composition of host haemolymph after attack by an insect parasitoid. Nature (London) 227:191–192.

Fiske, W. F. 1910. Superparasitism: an important factor in the natural control of insects. J. Econ. Entomol. 3:88–97.

Fiske, W. F., and W. R. Thompson. 1909. Notes on the parasites of the Saturniidae. J. Econ. Entomol. 2:450–460.

Flanders, S. E. 1947. Elements of host discovery exemplified by parasitic Hymenoptera. Ecology 28:299–309.

Force, D., and P. S. Messenger. 1965. Laboratory studies on competition among three parasites of the spotted alfalfa aphid *Therioaphis maculata* (Buckton). Ecology 46:853–859.

Fulton, P. B. 1933. Notes on *Habrocytus cerealellae*. Ann. Entomol. Soc. Am. 26:536–553.

Gerling, D., and A. Schwartz. 1974. Host selection by *Telenomus remus,* a parasite of *Spodoptera littoralis* eggs. Entomol. Exp. Appl. 17:391–396.

Greany, P. D., and E. R. Oatman. 1972. Analysis of host discrimination in the parasite *Orgilus lepidus* (Hymenoptera: Braconidae). Ann. Entomol. Soc. Am. 65:377–383.

Guillot, F. S., and S. B. Vinson. 1972. Sources of substances which elicit a behavioural response from the insect parasitoid, *Campoletis perdistinctus*. Nature (London) 235:169–170.

Hart, J. 't, J. de Jonge, C. Collé, M. Dicke, J. C. van Lenteren, and P. Ramakers. 1978. Host selection, host discrimination and functional response of *Aphidius matricariae* Haliday (Hymenoptera: Braconidae), a parasite of the green peach aphid, *Myzus persicae* (Sulz.). Proc. Int. Symp. Crop Protection. Meded. Fac. Landbouww. Rijksuniv. Gent 43:441–453.

Hase, A. 1925. Beiträge zur Lebensgeschichte der Schlupfwespe *Trichogramma evanescens* Westwood. Zur Kenntnis wirtschaftlich wichtiger Tierformen. 5. Arb. Biol. Reichsanst. 14:171–221.

Hassell, M. P. and T. R. E. Southwood. 1978. Foraging strategies of insects. Ann. Rev. Ecol. Syst. 1978:75–98.

Hays, D. B., and S. B. Vinson. 1971. Acceptance of *Heliothis virescens* (F.) (Lepidoptera, Noctuidae) as a host by the parasite *Cardiochiles nigriceps* Viereck (Hymenoptera, Braconidae). Anim. Behav. 19:344–352.

Hill, C. C. 1926. *Platygaster hiemalis* Forbes, a parasite of the hessian fly. J. Agric. Res. 32:261–275.

Hokyo, N., M. Shiga, and F. Nakasuji. 1966. The effect of intra- and interspecific conditioning of host eggs on the ovipositional behaviour of two scelionid egg parasites of the southern green stink bug, *Nezara viridula* L. Jap. J. Ecol. 16:67–71.

Howard, L. O. 1897. A study in insect parasitism: a consideration of the parasites of the white-marked Tussock Moth, with an account of their habits and interrelations, and with description of new species. U.S. Department of Agriculture Technical Series 5:5–57.

Huffaker, C. B. (ed.) 1971. Biological Control. Plenum Press, New York.

Huffaker, C. B., and P. S. Messenger (eds.). 1976. Theory and Practice of Biological Control. Academic Press, New York.

Jackson, D. J. 1966. Observations on the biology of *Caraphractus cinctus* Walker (Hymenoptera: Mymaridae), a parasitoid of the eggs of Dytiscidae (Coleoptera). III. The adult life and sex ratio. Trans. R. Entomol. Soc. London 118:23–49.

Jackson, D. J. 1969. Observations on the female reproductive organs and the poison apparatus of *Caraphractus cinctus* Walker (Hymenoptera: Mymaridae). Zool. J. Linn. Soc. 48:59–81.

Jenni, W. 1947. Beziehung zwischen Geslechtsverhältnis und Parasitierungsgrad einer in *Drosophila* larven schmarotzenden Gallwespe (*Eucoila* sp.). Rev. Suisse Zool. 54:252–258.

Jenni, W. 1951. Beitrag zur Morphologie und Biologie der Cynipide *Pseudeucoila bochei* Weld, eines Larvenparasiten von *Drosophila melanogaster* Meig. Acta Zool. 32:177–254.

Jørgensen, O. F. 1975. Competition among larvae of *Pimplopterus dubius* Hgn. (Hymenoptera: Ichneumonidae), a parasitoid of *Epinotia tedella* Cl. (Lepidoptera: Tortricidae). Z. angew. Entomol. 79:301–309.

King, P. E., and J. Rafai. 1970. Host discrimination in a gregarious parasitoid *Nasonia vitripennis* (Walker) (Hymenoptera: Pteromalidae). J. Exp. Biol. 53:245–254.

Klomp, H., B. J. Teerink, and W. C. Ma. 1980. Discrimination between parasitized and unparasitized hosts in the egg parasite *Trichogramma embryophagum* (Hym.: Trichogrammatidae): A matter of learning and forgetting. Neth. J. Zool. 30:254–277.

Lammers, H., and H. van der Starre. 1978. Electrophysiological responses of chemoreceptors on the ovipositor of the parasitic wasp *Pseudeucoila bochei*, involved in host discrimination. Proc. 3rd Congr. European Chemoreception Research Organization: 115.

Lawrence, P. O., P. D. Greany, J. L. Nation, and R. M. Baranowski. 1978. Oviposition behavior of *Biosteres longicaudatus*, a parasite of the Caribbean fruit fly, *Anastrepha suspensa*. Ann. Entomol. Soc. Am. 71:253–256.

Lloyd, D. C. 1935. Random distribution of parasite progeny. Nature (London) 135:472–473.

Lloyd, D. C. 1938. A study of some factors governing the choice of hosts and distribution of progeny by the chalcid *Ooencyrtus kuvanae* Howard. Proc. R. Soc. London, Phil. Trans., Ser. B 229:275–321.

Lloyd, D. C. 1940. Host selection by hymenopterous parasites of the moth *Plutella maculipennis* Curtis. Proc. R. Soc. London 128:451–484.

Lloyd, D. C. 1956. Studies of parasite oviposition behaviour. I. *Mastrus carpocapsae* Cushman (Hymenoptera: Ichneumonidae). Can. Entomol. 88:80–89.

Messenger, P. S., F. Wilson, and M. J. Whitten. 1976. Variation, fitness, and adaptability of natural enemies. P. 209–231. *In* C. B. Huffaker and P. S. Messenger (eds.), *Theory and Practice of Biological Control*. Academic Press, New York.

Nell, H. W., L. A. Sevenster-van der Lelie, J. C. van Lenteren, and J. Woets. 1976. The parasite-host relationship between *Encarsia formosa* (Hymenoptera: Aphelinidae) and *Trialeurodes vaporariorum* (Homoptera: Aleyrodidae). II. Selection of host stages for oviposition and feeding by the parasite. Z. angew. Entomol. 81:372–376.

Price, P. W. 1970. Trail odors: recognition by insects parasitic on cocoons. Science 170:546–547.

Rabb, R. L., and J. R. Bradley. 1970. Marking host eggs by *Telenomus sphingis*. Ann. Entomol. Soc. Am. 63:1053–1056.

Ridgway, R. L., and S. B. Vinson. (eds.) 1977. Biological Control by Augmentation of Natural Enemies. Insect and Mite Control with Parasites and Predators. Plenum Press, New York.

Rogers, D. 1972. The ichneumon wasp *Ventura canescens:* oviposition and avoidance of superparasitism. Entomol. Exp. Appl. 15:190–194.

Rogers, D. 1972. The ichneumon wasp *Ventura canescens*: oviposition and avoidance of superparasitism. Entomol. Exp. Appl. 15:190–194.

Rogers, D. J., and M. P. Hassell. 1974. General models for insect parasite and predator searching behaviour: interference. J. Anim. Ecol. 43:239–253.

Rosenberg, H. T. 1934. The biology and distribution in France of larval parasites of *Cydia pomonella* L. Bull. Entomol. Res. 25:201–256.

Salt, G. 1934. Experimental studies in insect parasitism. II. Superparasitism. Proc. R. Soc. London, Ser. B 114:455–476.

Salt, G. 1937. Experimental studies in insect parasitism. V. The sense used by *Trichogramma* to distinguish between parasitized and unparasitized hosts. Proc. R. Soc. London, Ser. B 122:57–75.

Salt, G. 1958. Parasite behaviour and the control of insect pests. Endeavour 17:145–148.

Salt, G. 1961. Competition among insect parasitoids. Symp. Soc. Exp. Biol. XV. Mechanisms in biological competition: 96–119.

Samson-Boshuizen, M., J. C. van Lenteren, and K. Bakker, 1974. Success of parasitization of *Pseudeucoila bochei* Weld (Hym., Cynip.): a matter of experience. Neth. J. Zool. 24:67–85.

Schröder, D. 1974. A study of the interactions between the internal larval parasites of *Rhyacionia buoliana* (Lepidoptera: Olethreutidae). Entomophaga 19:145–171.

Simmonds, F. J. 1956. Superparasitism by *Spalangia drosophilae* Ashm. Bull. Entomol. Res. 47:361–376.

Thompson, W. R. 1924. La théorie mathématique de l'action des parasites entomophages et le facteur du hasard. Ann. Fac. Sci. Marseille 2:69–89.

Thorpe, W. H. 1956. Learning and Instinct in Animals. Methuen, London.

Ullyett, G. C. 1936. Host selection by *Microplectron fuscipennis* Zett. (Chalcididae, Hymenoptera). Proc. R. Soc. London, Ser. B 120:253–291.

Ullyett, G. C. 1943. Some aspects of parasitism in field populations of *Plutella maculipennis* Curt. J. Entomol. Soc. S. Afr. 6:65–80.

Ullyett, G. C. 1945. Distribution of progeny by *Microbracon hebetor* Say. J. Entomol. Soc. S. Afr. 8:123–131.

Ullyett, G. C. 1949. Distribution of progeny by *Chelonus texanus* Cress. (Hymenoptera: Braconidae). Can. Entomol. 81:25–44.

van Lenteren, J. C. 1972. Contact chemoreceptors on the ovipositor of *Pseudeucoila bochei* Weld (Cynipidae). Neth. J. Zool. 22:347–350.

van Lenteren, J. C. 1976. The development of host discrimination and the prevention of superparasitism in the parasite *Pseudeucoila bochei* (Hym.: Cynipidae). Neth. J. Zool. 26:1–83.

van Lenteren, J. C. 1980. Evaluation of control capabilities of natural enemies: does art have to become science? Neth. J. Zool. 30:369–381.

van Lenteren, J. C., and K. Bakker. 1975. Discrimination between parasitized and unparasitized hosts in the parasitic wasp *Pseudeucoila bochei*: a matter of learning. Nature (London) 254:417–419.

van Lenteren, J. C., H. W. Nell, L. A. Sevenster-van der Lelie, and J. Woets. 1976. The parasite-host relationship between *Encarsia formosa* (Hymenoptera: Aphelinidae) and *Trialeurodes vaporariorum* (Homoptera: Aleyrodidae). III. Discrimination between parasitized and unparasitized hosts by the parasite. Z. angew. Entomol. 81:377–380.

van Lenteren, J. C., and J. Woets. 1977. Development and establishment of biological control of some glasshouse pests in the Netherlands. P. 81–87. *In* F. F. Smith and R. E. Webb (eds.), Pest Management in Protected Culture Crops. U.S. Department of Agriculture AS ARD-NE-85.

van Lenteren, J. C., K. Bakker, and J. J. M. van Alphen. 1978. How to analyze host discrimination. Ecol. Entomol. 3:71–75.

van Lenteren, J. C., and K. Bakker. 1978. Behavioural aspects of the functional response of a parasite (*Pseudeucoila bochei* Weld) to its host *(Drosophila melanogaster)*. Neth. J. Zool. 28:213–233.

Varley, G. C. 1941. On the search for hosts and egg distribution of some chalcid parasites of the knapweed gall-fly. Parasitology 33:47–66.

Venkatraman, T. V. 1964. Experimental studies in superparasitism and multiparasitism in *Horogenes cerophaga* (Grav.) and *Hymenobosmina rapi* (Cam.), the larval parasites of *Plutella maculipennis* (Curt.). Indian J. Entomol. 16:1–42.

Vinson, S. B. 1977. Behavioral chemicals in the augmentation of natural enemies. P. 237–279, *in* R. L. Ridgway and S. B. Vinson (eds.), Biological Control by Augmentation of Natural Enemies. Insect and Mite Control with Parasites and Predators. Plenum Press, New York.

Vinson, S. B., and F. S. Guillot. 1972. Host marking: Source of a substance that results in host discrimination in insect parasitoids. Entomophaga 17:241–245.

Waage, J. K. 1979. Foraging for patchily-distributed hosts by the parasitoid *Nemeritis canescens* (Grav.). J. Anim. Ecol. 48:353–371.

Walker, G. C. 1937. A mathematical analysis of superparasitism by *Collyria calcitrator* Grav. Parasitology 29:477–503.

Wilbert, H. 1964. Das Ausleseverhalten von *Aphelinus semiflavus* Howard und die Abwehrreaktionen seiner Wirte. Beitr. Entomol. 14:1–221.

Wilson, F. 1961. Adult reproductive behaviour in *Asolcus basalis* (Hymenoptera: Scelionidae). Aust. J. Zool. 9:739–751.

Wylie, H. G. 1958. Factors that affect host finding by *Nasonia vitripennis* (Walk.) (Hymenoptera: Pteromalidae). Can. Entomol. 90:597–608.

Wylie, H. G. 1965. Discrimination between parasitized and unparasitized housefly pupae by females of *Nasonia vitripennis* (Walk.)(Hym.: Pteromalidae). Can. Entomol. 97:279–286.

Wylie, H. G. 1970. Oviposition restraint of *Nasonia vitripennis* (Hymenoptera: Pteromalidae) on hosts parasitized by another hymenopterous species. Can Entomol. 102:886–894.

Wylie, H. G. 1971. Oviposition restraint of *Muscidifurax zaraptor* (Hymenoptera: Pteromalidae) on parasitized housefly pupae. Can. Entomol. 103:1537–1544.

CHAPTER TEN

EPIDEICTIC PHEROMONES THAT INFLUENCE SPACING PATTERNS OF PHYTOPHAGOUS INSECTS

RONALD J. PROKOPY

Department of Entomology
University of Massachusetts
Amherst, Massachusetts

As pointed out by Brown and Oriens (1970), "The dispersion of animals in space and time results, in a proximate sense, from the direct response of individuals to features of the environment and to the presence or absence of other individuals of the species."

This chapter focuses on the dispersion of phytophagous insects in space. The particular aim is to elucidate the role that pheromones may play in mediating a tendency toward uniform intraspecific distribution of a variety of phytophagous insects on exhaustible food resource units, such as individual plants or plant parts. Toward the end, examples are given of the use of such pheromones in effectively managing pest insects.

OPTIMAL INSECT DENSITY RANGE ON EXHAUSTIBLE FOOD RESOURCES

Consider the proposition of an optimal density range of individuals on an exhaustible food resource. The literature is replete with evidence that many animals, at one time or another, are involved in some form of competition for food. Accordingly, it is reasonable to suggest that if competition becomes severe, and individuals exceed the carrying capacity of a unit of resource, the effects on these individuals may be detrimental, taking the form of reduced fitness. On the other hand, it is also reasonable to suggest that in some species, too low a density of individuals on a unit of resource may give rise to equally deleterious consequences, with selection favoring some degree of aggregation of individuals. The concept of an optimal density range (Peters and Barbosa 1977) of individuals on a unit of resource would therefore seem to have validity. Because this concept constitutes the foundation of this chapter, let us illustrate it with a few examples.

Adult Douglas fir beetles *(Dendroctonus pseudotsugae)* attack living or recently fallen trees and initiate the formation of adult and larval galleries under the bark. McMullen and Atkins (1961) assessed various consequences of different adult density levels in logs. They found that at a density of 8 adults/1000 cm^2 of bark surface, the average length of adult galleries was greater, the number of eggs per centimeter of gallery was greater, the percentage of mortality of larvae was less, and the eventual number of adult progeny per female was greater than at higher adult densities. On the other hand, if the number of *D. pseudotsugae* adults colonizing a tree is fewer than 4–6/1000 cm^2 of bark surface, all beetles will die before the successful establishment of viable broods (Hedden and Pitman 1978). The precise reason for this is uncertain, but it probably involves the need for inoculation and exposure of a sufficient area of sapwood for establishment of pathogenic fungi, which are crucial to interruption of water conduction in the tree. This in turn, along with girdling of the tree by the beetles themselves, evidently prevents the tree from secreting oleoresin, which is toxic to the beetles (Smith 1963). Thus, a density range of six to eight colonizing beetles per 1000 cm^2 of bark surface appears to be optimal for maximum beetle fitness. Comparable situations exist among other bark beetles (Beaver 1974).

Mediterranean flour moths *(Ephestia kuehniella)* oviposit in flour, grain, and other foodstuffs, where the larvae feed and develop to maturity. Corbet (1971) reared *E. kuehniella* larvae at varying densities on food in 28-cm^3 containers and found that at the lowest density tested (one larva per container) as well as the highest (20 larvae per container), the number of days to pupation was greater, the generation time was longer, and the fecundity of females was lower than at an intermediate density (five larvae per container). At both the lowest and highest density levels, the larvae engaged in considerable wandering before they spun cocoons in which to pupate, apparently to increase their chances of reaching an area where larval density was, for reasons not yet determined, more "optimal." The greater wandering resulted in lighter weight and less fecund adults.

Black bean aphids *(Aphis fabae)* feed on phloem sap at the growing points of field beans and other legumes, where they are capable of rapid parthenogenetic reproduction. Way and Banks (1967) caged *A. fabae* nymphs in varying densities on individual plants, and found that the resulting virginoparous apterous adults reproduced more rapidly when the initial nymphal population was eight per plant than when it was two to four or 16–32 per plant. As subsequently shown by Dixon and Wratten (1971), isolated *A. fabae* reared on leaves also supporting small aggregations of other *A. fabae* were heavier at maturity and more fecund than isolated *A. fabae* reared on uninfested leaves. Apparently, through their combined feeding activities, a small aggregation of *A. fabae* is better able to divert nutrients (especially amino acids) to its leaf from other leaves than is an individual *A. fabae* (Way and Cammell 1970). Other benefits, such as enhanced protection from predators by aphid-tending ants, may also accrue to a greater degree among aggregated than dispersed *A. fabae* (Nault et al. 1976). On the other hand, too large an aggregation creates a local shortage of food (Way and Banks 1967).

Apple maggot flies *(Rhagoletis pomonella)* oviposit singly into hawthorn, apple, and other host fruits, where the larvae, being unable to move from one fruit to another, feed and develop to maturity in the flesh. Averill and Prokopy (unpublished data) allowed female *R. pomonella* to oviposit in the native host of this insect, hawthorn fruits of average size (12 mm diameter), and found that irrespective of the initial density of hatching eggs (one, two, or three per fruit), rarely did more than one larva per fruit develop to maturity. Thus the "optimal" density level of *R. pomonella* larvae is only about one healthy individual per hawthorn fruit.

These four examples (one each from four major orders of phytophagous pest insects) briefly illustrate the validity of the concept of an "optimal density range" of individuals on an exhaustible unit of food resource, among a number of insect species. The selective disadvantages of overcrowding may be manifest in increased mortality, reduced size, reduced development rate and fecundity, altered behavioral traits, and ultimately reduced fitness of individuals (see Peters and Barbosa 1977 for additional references). The selective disadvantages of not aggregating may be manifest in the inabilities to overcome host defense mechanisms, effectively mobilize or utilize host nutrients, effectively defend against or survive attack by natural enemies (Eisner and Kafatos 1962, Taylor 1976, Wood 1977, Kalin and Knerer 1977, Blum et al. 1978), or find a mate (Park 1933, Rowell 1978). Of course, for a large number of insects, aggregating on a food resource unit has no apparent adaptive value, and in fact, as in *R. pomonella,* the most selectively advantageous density level for these insects is only one healthy individual per resource unit.

Selection pressure toward early establishment of an optimal density range may be particularly strong in species where the stage or stages feeding on the resource have limited capability of successfully dispersing to a second unit of resource within the patch or to a new patch. Thus in the following cases there may be little selection pressure for early establishment of an optimal density range, with the ultimate density range being quite broad: (1) the feeding stage is sufficiently mobile to readily exploit other units or patches of resource, (2) resource patches are clustered or "coarse grain" in distribution, (3) it is advantageous to colonize a comparatively nonexhaustible resource that has proved to support development of conspecifics in the recent past, or (4) species are environmentally constrained to fully exploit highly ephermeral resources. In such cases, selection may favor a high degree of population clumping. Under conditions of very relaxed selection, population distribution may assume a pattern resembling randomness (Myers 1976, Myers and Campbell 1976, Wiens 1976, Jones 1977, Taylor and Taylor 1977, Hassell and Southwood 1978, Jaenike 1978).

ROLE OF EPIDEICTIC PHEROMONES IN PREVENTING OVERCROWDING

Now that we have briefly explored some of the selective advantages and disadvantages of various levels of phytophagous insect densities on single units of food resource, we can turn to the question: what sorts of behavioral mechanisms do insects utilize to achieve density levels within the optimal range? It turns out that a variety of mechanisms may be so employed, including physical combat (Pritchard 1969, Hubbell and Johnson 1977), and use of visual (Gilbert 1977, Rothschild and Schoonhoven 1977), acoustical (Russ 1969), and chemical stimuli. In the preceding chapter, van Lenteren explored the role of pheromones in host discrimination by insect parasitoids. In the following chapter, Roelofs illustrates how some phytophagous insects employ pheromones to achieve aggregation. Here, we explore the other side of that coin and illustrate how phytophagous insects of various sorts employ pheromones to prevent overcrowding.

For simplicity, such pheromones might be termed "dispersion pheromones" (Shorey 1976). However, this terminology implies a pheromonal role not only in influencing the spacing of individuals but also in alarming conspecifics to the approach of a natural enemy. "Epideictic pheromone" would seem to be a more appropriate term. Wynne-Edwards (1962) used "epideictic" as a term to describe specific sorts of animal display that have evolved principally to provide population density information to a population. Corbet (1971) used "epideictic pheromone" to describe a pheromone of *E. kuehniella* that "functions in the regulation of population density, controlling the dispersion of individuals." Epideictic pheromones that elicit dispersal from presently or potentially overcrowded food resources, and thereby act to partition intraspecific foraging activities, are the topic of discussion in this chapter. It is strongly advised that this term not be extended, as Wynne-Edwards did, to imply any influence on female fecundity or to imply group selection rooted in altruistic behavior, as opposed to individual or kin selection. Pheromones that regulate the density of males competing for females (in the sense of Hirai et al. 1978), or partition female and progeny food resources solely through male pheromonal defense of mating territory, are not discussed. Although the focus here will be on competition for food resources as regulated by epideictic pheromones, it is recognized that competition for space and refugia may also be regulated.

With this as background, let us proceed to examine some of the evidence for the existence of epideictic pheromones in phytophagous insects, including a number of pest species.

Coleoptera

Tenebrionidae

The first convincing evidence for the existence of an epideictic pheromone in a coleopterous insect, or in fact in any phytophagous insect, appears to be the work of Loconti and Roth (1953) on the red flour beetle *(Tribolium castaneum)*. The adults feed and oviposit in flour and a variety of grains. The larvae likewise feed and develop to maturity there.

Park (1934) demonstrated that *Tribolium* adults introduced into a batch of flour previously infested by conspecific adults produced fewer eggs than did adults in uninfested flour. Alexander and Barton (1943) then found that *T. castaneum* adults release considerable amounts of ethylquinone into flour, and suggested that this substance may somehow play a role in regulating *Tribolium* egg production. But it was Loconti and Roth (1953) who collected volatile secretions from more than 1.5 million *T. castaneum* beetles and identified three quinones. It was proved conclusively that two of these (2-ethyl-1,4-benzoquinone and 2-methyl-1,4-benzoquinone) effectively repel starved beetles from entering or feeding on batches of fresh flour. Subsequently, Naylor (1959, 1961) showed that volatile secretions from both sexes (but especially females) of *T. castaneum* as well as *Tribolium confusum* repelled conspecific females from feeding and ovipositing in fresh flour. The consequence was a uniform distribution of females among available feeding and oviposition sites, and a lessening of the deleterious effects of too large a population at any one site.

Recently, Faustini (unpub. data) found that, under conditions of low population density and ample food supply, production of benzoquinones by *T. castaneum* adults is low. At higher population densities, however, Faustini demonstrated that, not only does benzoquinone production increase, but, surprisingly, benzoquinones interact with droplets of aggregation pheromone secreted from male prothoracic glands causing chemical dissolution or breakdown of the aggregation pheromone.

Scolytidae

One of the most intensively studied groups of phytophagous insects known to emit epideictic pheromones is *Dendroctonus* bark beetle adults. Females are the first to attack a tree. Soon after arrival, the female bores a hole through the bark and releases a pheromone to which both males and other females are attracted. One male may join each female and contribute to production of attractant (Birch 1978).

In the southern pine beetle *(Dendroctonus frontalis),* females initiating the attack release the pheromones frontalin (Kinzer et al. 1969) and *trans*-verbenol (Renwick 1967) shortly after landing on the bark. Initial boring results in exudation of tree resins, including α-pinene. The combination of frontalin and α-pinene elicits aggregation of a large number of beetles, particularly males (Renwick and Vité 1969). The aggregating males in turn produce in the hindgut the epideictic pheromone verbenone (Renwick 1967), which, when released in very small amounts, apparently enhances the male-arresting effects of frontalin and α-pinene (Rudinsky 1973a). But when released in larger amounts, verbenone inhibits the landing of males on sites from which frontalin and α-pinene are being released, and in still larger amounts, it inhibits the landing of females as well (Renwick and Vité 1969)

Endobrevicomin, a second epideictic pheromone produced in smaller amounts by the males, likewise inhibits the landing of both sexes on sites treated with frontalin and α-pinene (Vité and Renwick 1971, Payne et al. 1977, 1978). The antennae of both sexes have numerous sensilla that are receptive to both these epideictic pheromones (Dickens and Payne 1977). Such inhibition to landing by high concentrations of these two epideictic pheromones is accompanied by cessation first of resin exudation from the tree, and subsequently of frontalin production by the now successfully established females (Renwick and Vité 1969). The result is that flying beetles shift their attack to less densely populated parts of the tree or to neighboring trees (Renwick and Vité 1970).

Similar phenomena occur in the Douglas fir beetle, the mountain pine beetle *(Dendroctonus ponderosae),* and the western pine beetle *(Dendroctonus brevicomis)* (Renwick and Vité 1970, Rudinsky et al. 1974a, Birch 1978). In *D. pseudotsugae,* the principal female-produced aggregating pheromone is a combination of frontalin and seudenol. On arrival, males stridulate outside the female's entrance hole. According to Rudinsky et al. (1976), sonic signals produced by males stimulate females to release the pheromone 3,2-methylcyclohexenone (3,2-MCH). This finding could not, however, be confirmed by Pitman and Vité (1974), possibly on account of genetic or physiologic differences between the beetle populations examined (Stock et al. 1979). The males produce substantial quantities of 3,2-MCH (Pitman and Vité 1974), as well as the isomer 3,3-MCH

(Libbey et al. 1976). Both compounds are reported to be released by males upon stimulation by female sound (Rudinsky et al. 1976).

When released in high concentration, both 3,2-MCH and 3,3-MCH appear to function epideictically, inhibiting in-flight approach landings of additional males and females in the area around the entrance hole, and thereby acting as a spacing mechanism to prevent overcrowding of beetles at that site. The combined effect of aggregating and epideictic pheromone use by *D. pseudotsugae* is regulation of the density of colonizing beetles within the optimal range of 6–8/1000 cm^2 of bark surface, noted earlier.

Dermestidae

The Khapra beetle (*Trogoderma granarium*), a notorius pest of stored grains, has life history characteristics similar to those of *Tribolium*. During the course of studies on the population dynamics of *T. granarium,* Finger et al. (1965) observed that the larvae preferred to aggregate and feed on grains of wheat manually cracked open or previously gnawed on by other conspecifics. Finger et al. proceeded to macerate 17 larvae (= largest number tested) and found that other larvae were positively attracted to the odor of the macerated ones. Subsequently, Tannert and Hein (1976) performed analogous tests and found that an extract of 100 *T. granarium* larvae was not attractive but instead was olfactorily repellent to other larvae, as well as to adults of both sexes; the repellent effect lasted about 6–8 hours. Differences in experimental procedure and source of larval population must be taken into account; these studies nevertheless suggest that at low concentration, the larval pheromone of *T. granarium* may function to aggregate conspecifics on a unit of resource that apparently requires simultaneous feeding by a multiple number of foragers to effect optimal feeding efficiency, while at high concentration, the pheromone functions to disperse conspecifics.

Bruchidae

The Azuki bean weevil (*Callosobruchus chinensis*) oviposits singly on beans in storehouses. The larva chews its way into the bean, and being unable to move to another bean, is constrained to complete development on the bean selected by its mother.

In a series of tests, Yoshida (1961), Umeya (1966), and Oshima et al. (1973) demonstrated that in association with the act of oviposition, the female *C. chinensis* deposits on the bean surface a marking pheromone that deters further egg laying on that bean. The oviposition-deterring substance is present in both sexes, but it is released in much greater amount by females. In an effort (not yet complete) to identify the pheromone of the female, Oshima et al. (1973) found a mixture of fatty acids (especially palmitic, stearic, oleic, linolic, and linolenic) and hydrocarbons (especially eicosane) to be active. The population consequence of marking pheromone deposition by *C. chinensis* is uniformity in egg distribution among available beans, and thereby a reduction in the number of occasions in which two or more larvae are forced to compete (with deleterious effects on fitness) for the limited resources of a single bean (Utida 1943, Yoshida 1961, Mitchell 1975). Similar biology and oviposition-deterring pheromone deposition behavior also are characteristic of the bean weevils *Callosobruchus maculatus, Callosobruchus rhodesianus,* and *Zabrotes subfasciatus* (Yoshida 1961, Umeya 1966).

The life history of the dry bean weevil (*Acanthoscelides obtectus*) is a bit different. In nature, females lay several eggs per clutch in holes that they gnaw along the suture of ripening bean pods. In granaries, at low weevil population densities, the eggs are scattered randomly in clutches under or next to stored beans. At high weevil population densities, the eggs are distributed more uniformly (Umeya and Kato 1970). Of all the larval stages, only the first instars have legs, an adaptation that enables them to disperse and locate suitable beans.

Szentési (unpublished data) showed that in association with exploring or ovipositing on a bean, female *A. obtectus* (and to a lesser extent, males as well) deposit on the bean surface a water-soluble marking pheromone that deters repeated oviposition. The pheromone also deters first instar larvae from entering marked beans. Although several larvae may develop in a single bean without any apparent reduction in fitness, too great a number would lead to rapid depletion of the food supply. Hence the epideictic pheromone is valuable in regulating egg and larval distribution at high weevil densities.

Chrysomelidae

The twelve-spotted asparagus beetle (*Crioceris duodecimpunctata*) oviposits mainly between the needles of asparagus plants. After hatching, the larva searches for a berry, enters it, and eats the flesh and young seeds. If the first berry entered is too small for completion of larval development, the second or third instar will move to a second berry and complete development there.

Van Alphen and Boer (1980, unpublished data) offered second and third instar larvae a choice between larval-infested and uninfested berries, and found that the larvae attempted to gnaw entry holes into the former much less often than into the latter. Subsequent tests suggested that an epideictic pheromone emitted by the original larval occupant was the principal factor deterring other larvae from entering an infested berry, thereby providing the mechanism mediating the tendency toward uniform distribution of larvae among berries on a plant (Fink 1913).

Curculionidae

Female cotton boll weevils (*Anthonomus grandis*) chew cavities in developing cotton blossom buds (squares) and deposit a single egg in each cavity. The larvae feed inside the bud or boll, with usually only one larva developing per bud (Hunter and Pierce 1912).

Cross (see Hedin et al. 1974) observed that following oviposition, female boll weevils secrete a frasslike substance (possibly containing a pheromone) into the oviposition cavity in the bud. He suggested that other females tended to avoid such a bud until the ratio of females to buds became "unfavorable." Though not supported by the data and conclusions of Jenkins et al. (1975) on the distribution of boll weevil ovipositions under field conditions, this suggestion by Cross of a possible oviposition-deterring pheromone in *A. grandis* deserves further evaluation.

Lepidoptera

Pyralidae

The first evidence for an epideictic pheromone in a lepidopteran appears to be the work of Corbet (1971) on *E. kuehniella*. As already mentioned, the adults oviposit in flour and the larvae develop to maturity there.

Corbet observed that when last instar flour moth larvae meet head to head, they deposit on the substrate droplets of secretion from their mandibular glands. Corbet made an ether extract of these glands, and went on to demonstrate that larvae moved less and pupated sooner in the presence of intermediate quantities of extract than in the presence of larger or smaller quantities. The larval response mediated by the epideictic mandibular gland pheromone ultimately had the effect of regulating larval density within an optimal range, and consequently toward greater fitness of subsequent adults, as discussed earlier.

Later, Corbet (1973) showed that the oviposition behavior of the adults was also influenced by this pheromone, such that more eggs were laid in the presence of an intermediate amount of pheromone than in the presence of a larger or smaller amount. The larger amount deterred egg laying, whereas the smaller amount elicited more egg laying than did a complete absence of pheromone, possibly because a small amount of pheromone indicated to the moths that that particular site was capable of supporting larval growth but was not overpopulated.

Mudd and Corbet (1973) indicated that the pheromone is a lipid (not yet identified), and suggested that the larvae of three additional stored products pests, the dried currant moth (*Ephestia cautella*), the Indian meal moth (*Plodia interpunctella*), and the cocoa moth (*Ephestia elutella*), likewise contain and release this lipid as an epideictic pheromone.

Tortricidae

Spruce budworm females (*Choristoneura fumiferana*) respond to sex pheromone produced by females of their own species by walking, antennal grooming, extension of the ovipositor, and oviposition (Palanaswamy and Seabrook 1978). The pheromone is perceived by antennal receptors. The observations and experiments of Palanaswamy and Seabrook led them to postulate that at high female population densities, thus at a high concentration of pheromone in the air, the sex pheromone may function epideictically to elicit dispersal of ovipositing females away from highly populated areas. This suggestion, if confirmed by future research, would have important implications with respect to the possibility that female sex pheromone components of other lepidopterans function in a multifaceted manner as epideictic pheromones in areas of high female population density.

Noctuidae

Female *Hadena bicruris* moths drink nectar from flowers of *Melandrium album*, after which they often deposit an egg into the flower. The larva feeds on the seed capsules. Brantjes (1976) offered the adults a choice of flowers, and found that those that already had received an egg had a lower probability of receiving a second or third one than did uninfested flowers. Further observations and tests led Brantjes to conclude that the female deposits a pheromone in association with the egg. The pheromone is apparently perceived via olfaction by incoming females prior to actual arrival, and is of comparatively short duration (less than one day). The oviposition-deterring effects of the pheromone are manifest in a lessening of competition among the larvae, in that only one larva per flower can survive.

The corn earworm (*Heliothis zea*) lays eggs singly on corn, cotton, legumes, and other hosts. On corn, the egg is laid on the silk of the developing ear. The larva feeds on the silk and kernels. It is infrequent that more than one or two larvae feed and survive on the same ear (Scott 1977). Gross and Jones (1977) made a water extract of frass deposited by last instar *H. zea* larvae and sprayed it on the silk of uninfested corn ears in the field. Over the next 3 days, they found that the ears treated with larval frass extract received only 17% as many eggs from native females as ears treated with water alone. In an unreplicated test, Nordlund and Lewis (unpublished data) made a hexane extract of the hair scales the females leave behind on the egg and plant surface after oviposition. They offered females in laboratory cages a choice between pieces of filter paper treated with such extract and paper treated with hexane only. Far fewer eggs were deposited on the extract-treated papers than on the others. Thus, in *H. zea,* the adults as well as the larvae appear to deposit oviposition-deterring pheromone.

Female cabbage loopers (*Trichoplusia ni*) deposit eggs singly on the lower surface of cabbage and a variety of other host plant leaves. The larvae feed on the leaf tissue. Recently, Renwick and Radke (1980) found that cabbage leaves treated with an aqueous solution of *T. ni* larval frass received considerably fewer eggs than did leaves treated with water alone. The evidence suggests the existence of an oviposition-deterring pheromone produced by *T. ni* larvae.

Data of Saad and Scott (1979) on *H. zea* and *Heliothis armigera,* and inferences by Palanaswamy and Seabrook (1978) on *T. ni,* indicate that in high concentration, sex pheromone produced by females of these species may elicit dispersal of conspecific virgin and/or ovipositing females away from highly populated areas, similar to the postulated situation in *C. fumiferana*.

Pieridae

Female cabbage butterflies, *Pieris brassicae* and *Pieris rapae,* deposit their eggs in clutches of 30 or so (*P. brassicae*) or singly (*P. rapae*) on leaves of cabbage and other host plants, where the larvae feed and develop to maturity.

Rothschild and Schoonhoven (1977) clearly demonstrated that *P. brassicae* females deposit on the egg and adjacent leaf surface an epideictic pheromone that deters repeated egg laying at or near that site. The pheromone, not yet identified, is water soluble, and is released from the tip of the abdomen during or just after oviposition. Subsequent behavioral and electrophysiological experiments by Behan and Schoonhoven (1978) revealed that the pheromone is produced in the female accessory glands as well as being present within the eggs, is perceived olfactorily from a distance via at least two types of sensilla on each segment of the antennae, and is perceived also on direct contact by receptors on the tarsi. The pheromone retains activity for the remarkably long period of at least 7 weeks (Schoonhoven et al. 1980). Though larval frass itself was not assayed for activity, the presence of feeding larvae on a leaf also was deterrent to oviposition (Rothschild and Schoonhoven 1977), as was an extract of male adult gut tissues (Schoonhoven, unpublished data).

The sort of selective advantage gained by *P. brassicae* in depositing more than one egg per clutch is uncertain, but it may be related to enhanced capability of grouped *P. brassicae* larvae to defend themselves against natural enemies

(Rothschild, personal communication). The oviposition-deterring effect of multiple *P. brassicae* egg clutches per plant would tend to inhibit overloading any one plant with eggs.

Rothschild and Schoonhoven (1977) presented evidence indicating that a European biotype of *P. rapae* likewise emits an oviposition-deterring pheromone from the abdomen during egg laying, but studies by Ives (1978) and Traynier (1979) suggest that this is probably not the case with Australian biotypes.

Nymphalidae

The monarch butterfly (*Danaus plexippus*) oviposits singly on milkweed plants. Studies by Rothschild and Schoonhoven (1977) suggested that like *P. brassicae, D. plexippus* females deposit on the egg surface an oviposition-deterring pheromone that has the effect of limiting the number of larvae on any one host plant. Later studies by Dixon et al. (1978) confirmed this suggestion, although the oviposition-deterring effect appears to be a rather mild one when compared with *P. brassicae*.

Diptera

Tephritidae

Within a single family, the greatest number of phytophagous insect species known to emit epideictic pheromones are frugivorous Tephritidae, possibly owing to a larger number of candidate species examined. The females oviposit in the flesh of growing fruits, and the larvae feed and develop there until maturity.

In 1971 it was discovered that immediately following oviposition into a host fruit, a *Rhagoletis pomonella* female circles the fruit, trailing her extended ovipositor on the surface, and in so doing deposits a marking pheromone that deters repeated egg-laying attempts (Prokopy 1972, 1981). A typical pheromone trail is about 53 mm long and 0.05 mm wide (Averill and Prokopy 1980a). Some or all of the pheromonal components appear to be produced and accumulated principally in the midgut and hindgut (Averill and Prokopy 1980a, Prokopy et al. 1979). The pheromone is perceived only upon direct contact by sensilla on the tarsi (particularly D-hairs on the ventral surface of the second, third, and fourth tarsomeres of the foretarsi) and the labellum (particularly the short marginal hairs) (Prokopy and Spatcher 1977, Crnjar et al. 1978). The pheromone is water soluble, but under dry conditions it retains substantial biological activity for at least 14 days (Averill and Prokopy, unpublished data). The female deposits a behaviorally and neurophysiologically active amount of the pheromone in its feces, the biological significance of which remains to be determined (Prokopy et al. 1979).

Naive *R. pomonella* females are behaviorally unresponsive to the pheromone, apparently requiring, as do some parasitoids that emit epideictic pheromones (van Lenteren, Chapter 9), at least one or two bouts of prior pheromonal contact or "learning" before exhibiting a normal response (Roitberg and Prokopy, 1981). Females deprived of oviposition sites for increasing lengths of time show proportionately greater tendency to bore into pheromone-marked fruit than do nondeprived females (Prokopy 1972). On the other hand, repeated contact with pheromone-marked fruit elicits a high degree of "nervous" or "flighty" activity in females. In fact, several successive arrivals on marked fruit in host trees soon elicits dispersal from the tree (Roitberg and Prokopy, unpublished data).

The marking pheromone appears to be the principal factor mediating the observed uniformity (Le Roux and Mukerji 1963, Cameron and Morrison 1974) in distribution of *R. pomonella* eggs among apples in nature. As already pointed out, usually only one larva is able to complete development in a hawthorn fruit of typical-size (~ 12 mm), compared with about 15 larvae completing development in an apple of typical size (~ 8 cm). The earliest-developing larva in a hawthorn fruit appears to have a competitive advantage over subsequent ones (Averill and Prokopy, unpublished data). The pheromone may need to deter oviposition only for the length of the egg incubation period (~ 3–4 days) to give the first larva this advantage. It would therefore be a competitive disadvantage for the original or another female to oviposit in a fruit that had already received the number of eggs equivalent to its larval carrying capacity. Similarly, distinct advantage accrues to a female that marks a fruit: rapid reinfestation is unlikely.

Not only does the marking pheromone substance deposited by an *R. pomonella* female function as an oviposition deterrent, it also functions to arrest males, apparently providing a signal to males that a female has recently been in the vicinity (Prokopy and Bush 1972).

Since 1971, oviposition-deterring pheromones have been discovered in 11 other species of tephritid flies, including walnut husk fly (*Rhagoletis completa*) (Cirio 1972), black cherry fruit fly (*R. fausta*) (Prokopy 1975), European cherry fruit fly (*R. cerasi*) (Katsoyannos 1975), eastern and western cherry fruit flies (*R. cingulata* and *R. indifferens*) (Prokopy et al. 1976), blueberry maggot fly (*R. mendax*) (Prokopy et al. 1976), two species of dogwood berry flies (*R. cornivora* and *R. tabellaria*) (Prokopy et al. 1976), rose hip fly (*R. basiola*) (Averill and Prokopy, 1980b), Caribbean fruit fly (*Anastrepha suspensa*) (Prokopy et al. 1977), and Mediterranean fruit fly (*Ceratitis capitata*) (Figure 10.1) (Prokopy et al. 1978). The evidence (Prokopy 1976) that species in several other tephritid genera (e.g. *Aciurina, Carpomyia, Dacus, Epochra, Eurosta,* and *Zonosemata*) tend to distribute eggs uniformly among available sites suggests that epideictic pheromones regulating oviposition may be utilized by species in these genera as well.

In tests that assayed individuals from different *Rhagoletis* species or popula-

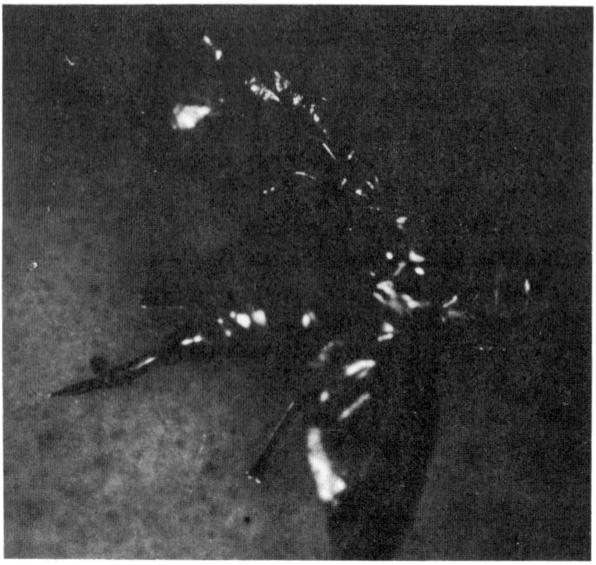

FIGURE 10.1. A *Ceratitis capitata* female dragging her ovipositor after ovipositing. (photo by Grant K. Uchida).

tions for cross-recognition of marking pheromones, Prokopy et al. (1976) found that (1) species belonging to different species groups did not recognize each other's pheromones, (2) species within the same species group varied in reaction from complete cross-recognition to none, and (3) wild populations of the same species always recognized each other's marking pheromone. Interestingly, a laboratory population (Geneva, New York, strain) of *R. pomonella* cultured on apples for about 15 generations was found to deposit pheromone having less deterrent effect on oviposition than pheromone deposited by wild flies (Prokopy et al. 1976). Furthermore, a laboratory population (Honolulu, Hawaii, strain) of *C. capitata* flies cultured on artificial media for more than 200 generations was almost completely unresponsive to an amount of pheromone that effectively deterred wild flies (Prokopy et al. 1978). Whether these differences between laboratory and wild flies are due to dietary factors, to selection having operated in different directions under different circumstances, or to genetic drift is uncertain. Nevertheless, these differences clearly demonstrate cause for concern about the usefulness of information obtained in future assays of laboratory-cultured individuals of any insect species on release of, or response to, epideictic pheromones.

No one has yet chemically identified the oviposition-deterring pheromone of any tephritid species. However, excellent progress toward this end has been made by Hurter et al. (1976) and Boller et al. (unpublished data) with *R. cerasi,* wherein, as in *R. pomonella,* the oviposition-deterring pheromone is perceived principally via D-hair receptors on the tarsi (Städler and Katsoyannos 1980) and functions also as a male arrestant (Katsoyannos 1975). The pheromone of *R. cerasi* is quite polar, has low volatility and a molecular weight of less than 10,000, and is stable in boiling water for at least 10 minutes (Hurter et al. 1976).

Anthomyiidae

Hylemya spp. deposits its eggs singly in developing flower buds of the perrenial herb *Polemonium foliosissimum.* The larvae feed on the seeds of the developing ovary and are apparently unable to successfully move from one ovary to another.

Zimmerman (1979) observed that females of this species spend a shorter time in preovipositional search behavior on uninfested buds than on infested ones. He went on to show that following oviposition in *P. foliosissimum,* and in association with the "glue" that cements the egg to the sepal, the female deposits an epideictic pheromone that deters repeated oviposition. As Zimmerman postulates, the pheromone is probably the principal factor mediating the observed highly significant tendency toward uniform distribution of larvae of *Hylemya* spp. in *P. foliosissimum.*

Subsequently, Zimmerman (1980) obtained the surprising result that *Hylemya* spp. females do not deposit oviposition deterring pheromone after egglaying in flowers of *Ipomopsis aggregata* even though the same females are capable of depositing and responding to such pheromone on *P. foliosissimum.*

Recently, Raina (unpublished data) demonstrated an oviposition deterring pheromone in another anthomyiid, *Atherigona soccata.* The pheromone is associated with water soluble cement that glues the eggs to host plant leaves, and apparently mediates the strong tendency toward uniform distribution of eggs on sorghum leaves by this species (Ogwaro 1978).

Homoptera

The evidence for existence of epideictic pheromones in Homoptera is circumstantial and indirect, but in certain aphids it is nevertheless enticing enough to merit consideration.

The release of alarm pheromone in response to predator attack has been demonstrated in a number of aphid species (reviewed by Nault and Montgomery 1977). Such pheromones are highly volatile, and their effects usually dissipate within a few minutes. However, Greenway et al. (1978) and Griffiths et al. (1978) found that in the green peach aphid (*Myzus persicae*) some less volatile extracts of aphid bodies, in particular certain fatty acids with chain lengths of C_8-C_{13} (e.g. dodecanoic acid), deter settling and larviposition by apterous adults when applied to an artificial membrane (Parafilm^R) mimicking a leaf or when incorporated into an artificial diet sandwiched between two such membranes. Dodecanoic acid remained substantially effective in deterring adult (and larval) settling on treated membranes for at least 4 days. Apparently the aphids sense this deterrent via the mouthparts rather than via the antennae or tarsi.

These authors did not suggest whether or how the aphids might release these deterrent fatty acids. Conceivably, some of them may be released in aphid cornicle droplets upon contact stimulation by neighboring aphids (Phelan and Miller 1979). Also, some may be present in aphid salivary secretion and injected via the stylets into the phloem during feeding. If so, the situation would be analogous to that in the lime aphid (*Eucallipterus tiliae*) wherein the larvae have been shown by Kidd (1977) to receive information on the state of aphid crowding through substances in the leaves, possibly salivary substances injected by conspecifics. Kidd showed that after molting to alate adults, the lime aphids that as larvae had been exposed to a high concentration of aphid-secreted substance in the plant sap, flew from plants in greater numbers than did alates that had not been so exposed as larvae.

Though not conclusive evidence for the existence of epideictic pheromones in aphids, these suggestions, if confirmed by future research, offer a unique way of transmitting an epideictic pheromonal message in phytophagous insects with sucking mouthparts: injection of pheromone into the plant sap and reception via sensilla associated with the mouthparts. Such pheromones could well be one of the mechanisms by which, as pointed out earlier, aphids are able to maintain population density levels on individual plants within an optimal range.

Orthoptera

The existence of aggregation pheromones in several orthopterans is well established (reviewed in Shorey 1973), particularly among the Blattidae and Acrididae. In at least one orthopteran, the European house cricket (*Acheta domestica*), which is an omnivorous forager, there is also convincing evidence for the existence of an epideictic pheromone.

Sexton and Hess (1968) confined adult house crickets in varying densities on wooden blocks and then presented these blocks to other adults in terraria. During the first 24 hours of exposure, many more females as well as males were occupying control blocks than were occupying blocks that had been visited previously by either sex. After 30 hours, however, the deterring effects of previsited blocks were no longer detectable, but instead, males and females began aggregating on these blocks. Quite possibly, the pheromone that regulates such aggregation and dispersal in *A. domestica* is a single substance, eliciting aggregation of adults at favorable resource sites when at low concentration, and dispersal of adults from high population density areas when at high concentration.

Hymenoptera

The social Hymenoptera employ pheromones in a variety of situations to regulate a multitude of behavioral responses (Blum and Brand 1972, Blum 1977, Hölldobler 1978). In his outstanding treatise on the biology of social insects, Wilson (1971) suggests that many if not most species of social wasps, bees, and ants are able to recognize members of their own colony by colony-specific odors, which he suggests are probably comprised in most cases of a combination of pheromonal and environmentally acquired components. In some species, the odors are deposited to recruit other workers to food resources (e.g. Blum et al. 1970, Tumlinson et al. 1971), whereas in other species, including subsocial ones, pheromones are employed by males to mark and advertise male territory and/or to attract females (e.g. Alcock 1975, Raw 1975, Gwynne 1978, Evans and O'Neill 1978). Although the adaptive functions of territorial defense by male animals are still incompletely understood (Verner 1977), there is evidence (Wilson 1975, Davies 1978, Walker 1981) to suggest that in some cases, one function is protection of a sufficient food supply for the female or eventual offspring. Though we do not deal here with any specific examples of this sort, we briefly examine two cases of pheromone deposition in social Hymenoptera by females having the effect of partitioning female food-foraging activities. Such pheromones can therefore be considered as having epideictic effects.

Formicidae

The African weaver ant (*Oecophylla longinoda*) is a dominant ant species in African forest canopies, where the workers forage for insect honeydews, plant exudates, insect prey, and a variety of other foodstuffs. Although plant tissue per se plays little role in the diet of weaver ants, the colonies of 500,000 or more workers exert a definite influence on the structure of trees through the folding or weaving together of thousands of leaves to form nests (Hölldobler and Wilson 1977).

Hölldobler and Wilson (1978) observed that *O. longinoda* colonies are more or less uniformly distributed within a habitat, and that each colony occupies one to several trees over which it maintains exclusive possession. In a series of tests, these researchers demonstrated that *O. longinoda* workers mark newly acquired home range territory with randomly placed droplets of pheromone emitted principally from the rectal vesicle. The pheromone, which remains active for at least 12 days, enables workers to distinguish their own domain from that of alien conspecifics. Aliens respond to the pheromone with aversion and aggressive displays. Hölldobler and Wilson conclude that this territorial pheromone aids in effecting a partitioning of food resources among neighboring colonies.

Evidence indicates that worker ants of *Myrmica rubra, Myrmica scabrinodis,* and *Atta cephalotes* likewise deposit territorial marking pheromones (Cammaerts et al. 1977, 1978, Jaffé et al. 1979).

Anthophoridae

In Texas, female *Xylocopa virginica texana* bees forage for nectar from *Passiflora incarnata* flowers. Frankie and Vinson (1977) noticed that in conjunction with such nectar-foraging visits, the females seemed as though they might be depositing some sort of chemical marker on the flowers. When they applied a hexane extract of the contents of female DuFour's glands to several flowers, they found that although such treated flowers received about the same number of female approaches as flowers treated with hexane alone, in nearly every case of a female approaching the former type, she abruptly turned away just before landing and never attempted to take any nectar. The DuFour's gland repellent pheromone was behaviorally effective for about 10 minutes, and Vinson et al. (1978) found that it was in part a mixture of the methyl esters of myristic and palmitic acids. Although the females may utilize information from the pheromone-marked flowers to ''trapline'' within and between flower patches, Frankie and Vinson suggested that the 10-minute duration of pheromone effectiveness may be selectively adjusted to the time needed by the flower to replenish its nectar supply and once again render it productive to foragers. Thus the pheromone apparently acts to enhance overall foraging efficiency of females within and between *Passiflora* flower patches.

Ferguson and Free (1979) suggest that deposition of ''alarm'' pheromone at food resources by foraging workers may achieve a similar enhancement of foraging efficiency in *Apis mellifera*.

Tenthredinidae

European apple sawfly adults (*Hoplocampa testudinea*) oviposit singly into developing apple fruit during bloom. After consuming the limited supply of flesh in the original fruit, the larva moves to a second fruit, where it completes development.

Experimental findings suggest that *H. testudinea* adults deposit an oviposition-deterring pheromone in association with the egg, and thereby are able to recognize an occupied fruit (Roitberg and Prokopy, unpub. data). It appears that the pheromone may be emitted during the 2-3 second period in which the female touches her mouthparts to the oviposition puncture site immediately after egglaying. In addition, when larvae were given a choice of entering uninfested second fruits compared with larval-infested ones or ones treated with a water extract of larval frass, they showed a strong preference for the uninfested type (Roitberg and Prokopy, unpublished data). This suggests the existence in the larval frass of an epideictic pheromone that regulates larval entry of second fruits.

KAIROMONAL AND ALLOMONAL EFFECTS OF EPIDEICTIC PHEROMONES

As pointed out by Blum (1977), there are a number of cases of pheromones emitted by insects that have "backfired" in the sense that they have become utilized by natural enemies as host-finding cues. This is true not only for sex and aggregating pheromones but apparently for epideictic pheromones as well. The following are examples of parasitoids or predators that make use of chemical information from substances having epideictic pheromonal effects as an aid in locating potential prey. (In none of these cases is it known for certain whether the kairomone has the identical chemical structure or structures as the epideictic pheromone).

Corbet (1971) found that the epideictic pheromonal secretion produced by the mandibular glands of *E. kuehniella* elicits ovipositional thrusts by the larval ichneumonid parasitoid *Venturia canescens*. Components of *H. zea* larval frass, which has epideictic pheromonal effects, were shown by Lewis et al. (1976) to elicit host-seeking flight activity of *Microplitis croceipes*, a braconid wasp parasitoid of *H. zea* larvae. In addition, the epideictic phenomenal substance deposited by *H. zea* females on the egg and plant surface after oviposition stimulates host-searching activity of adults of at least two egg parasitoids, *Trichogramma evanescens* and *Trichogramma pretiosum*, as well as larvae of an egg predator, *Chrysopa carnea* (Lewis et al. 1972, 1977, Nordlund et al. 1977). The oviposition-deterring pheromonal substance of *R. pomonella* acts to retain and elicit antennal tapping of female *Opius lectus* parasitoids on marked fruit (Prokopy and Webster 1978). *O. lectus* females oviposit in the developing eggs and young larvae of *R. pomonella* in the fruit flesh. A number of clerid and trogositid predators of bark beetles utilize pheromones of their hosts as kairomones (Borden 1977). The data to date suggest, however, that at least in the cases of *D. pseudotsugae* and *D. frontalis*, epideictic pheromones are not among those that influence the prey-seeking behavior of such predators (Vité and Williamson 1970, Furniss et al. 1974, Richerson and Payne 1979).

There is at least one case of an epideictic pheromone for which an allomonal effect on a competing species has been demonstrated. Verbenone, an epideictic pheromone secreted by *D. brevicomis* adults, inhibits tree colonization by adults of the competing bark beetle *Ips paraconfusus* (Byers and Wood 1980). Similarly, pheromone emitted by *D. frontalis* adults inhibits nearby landing of *Ips grandicollis* adults, but it is not yet known whether it is sex pheromone or epideictic pheromone components (or both) that are responsible for this inhibition (Birch et al. 1980).

It would not be surprising if additional examples of epideictic pheromones having allomonal effects on competing species were to turn up, especially in cases of interacting species that have been competing for similar resources in the same habitat over a long period of evolutionary time.

USE OF EPIDEICTIC PHEROMONES IN PEST CONTROL

The application of epideictic pheromones to important crops or forests, or their products, would seem to offer excellent potential as a highly selective, behavior-manipulating method of protecting such resources against pest attack. To date, this approach to management has been tried on two groups of pest insects: *Dendroctonus* bark beetles and *Rhagoletis* fruit flies. Probable constraints on successful use of epideictic pheromones as a unilateral management technique are considered in the section on *Rhagoletis* and in the conclusions.

Dendroctonus

The first substantive attempt to achieve management of a *Dendroctonus* species solely through application of synthetic epideictic pheromone was by Furniss et al. (1972, 1974). From April to August, these workers allowed the landing-inhibiting pheromone 3,2-MCH to diffuse from liquid dispensers at three different rates and at three different spacings around freshly felled Douglas fir trees in plots measuring about 350 m^2 in an area populated by *D. pseudotsugae*. At the optimal MCH concentration (1−2 mg/day) and spacing (~ 1 g/acre/day), adult attacks and numbers of progeny were, respectively, 96 and 91% fewer than in untreated control plots. Subsequently, Furniss et al. (1977) found that small wax-coated beads impregnated with 3,2-MCH and placed in cans at about 1 m height and 3 m above ground were even more effective, reducing both adult attacks and progeny numbers on felled trees by 98% compared with untreated controls. Similarly, Kline et al. (1974) and Rudinsky et al. (1974b) showed that when applied in granular or liquid formulation, 3,2-MCH, a component of the landing-inhibiting epideictic pheromone of the spruce beetle (*Dendroctonus rufipennis*), very effectively protected felled Sitka spruce trees against attack by *D. rufipennis* for at least 35 and 60 days, respectively.

Hedden and Pitman (1978) hypothesized that in large plantations, use of 3,2-MCH in conjunction with Douglure® (a combination of the tree attractants

α-pinene and camphene, and the aggregating pheromones frontalin and seudenol) might be a more effective method of managing *D. pseudotsugae* than use of 3,2-MCH alone. They placed slow-release polyethylene-enclosed formulations of varying ratios of Douglure and 3,2-MCH on or near live host trees in plots 40 m in diameter. They found that in certain combinations, MCH and Douglure effectively reduced attacks below the critical level of four to six beetles per 1000 cm^2 of bark surface necessary for successful colony establishment. Their findings led them to propose a new approach to future management of *D. pseudotsugae:* use of Douglure to aggregate scattered populations of beetles in stands of uninfested trees, coupled with employment of MCH to shut off the attack at a density below the critical threshold for colonization and tree mortality. In effect, the proposed technique would modify the pests' behavior in such a manner that the host tree would become a lethal trap, or pest population "sink."

Recently Richerson and Payne (1979) evaluated the potential of the epideictic pheromones verbenone and endobrevicomin (which was mixed with exobrevicomin) for controlling *D. frontalis*. They affixed polyethylene caps containing various combinations of these pheromones to loblolly pine trees and found that when together, the pheromones effected a 92% reduction in gallery formation and a 94% reduction in egg laying, compared with adjacent untreated control trees. Despite this, however, the pheromone-treated trees died, very possibly as the result of a much greater number of *Ips avulsus* beetles landing and colonizing them compared with the controls. This prompted Richerson and Payne to conclude that use of landing-inhibiting epideictic pheromones alone will probably not protect host trees against *D. frontalis* attack.

Rhagoletis

Two field tests have been carried out on the application of oviposition-deterring pheromones for control of *Rhagoletis,* both with *R. cerasi* in Switzerland. In the first, Katsoyannos and Boller (1976) made five spray applications (to runoff) to two entire cherry trees and to individual branchlets of 18 other trees of a 0.2% aqueous solution of partially purified pheromone (gathered after ~ 200,000 ovipositions) together with a wetting agent. Despite frequent and heavy rainfalls during the test, there was a 69% reduction in number of oviposition punctures and a 77% reduction in number of larva-infested fruit compared with untreated controls. In their second test, conducted the following year when there was only 13 mm of rainfall during the test period, Katsoyannos and Boller (1980) sprayed the same concentration of pheromone solution on four entire cherry trees and achieved with one treatment an 85% reduction in larval infestation and with two treatments a 90% reduction compared with adjacent untreated trees.

These highly encouraging results with *R. cerasi* notwithstanding, it is doubtful that application of oviposition-deterring pheromone in and of itself would lead to truly effective long-term control of *R. cerasi* or any other pest in a large agroecosystem. In the absence of acceptable untreated egg-laying sites, the threshold level for oviposition at treated sites might become so low that deterred females would, over time, eventually oviposit in spite of the presence of deterrent. In addition, there is the possibility that initially deterred females might eventually undergo physiological habituation or adaptation to deterrent pheromone applications, especially under conditions of large-area, high-concentration treatments over long time intervals. For long-term success, therefore, application of synthetic oviposition-deterrent pheromone will in most cases probably require integration with use of attractive olfactory-visual stimuli to trap or capture out deterred females before (1) their oviposition ''drive'' has reached supernormal levels, (2) they have become physiologically habituated or adapted to pheromone treatments, and (3) the fly population has undergone significant selection for reduced responsiveness to pheromone.

CONCLUSION

In conclusion, I shall summarize some of the major points and offer some speculations on the future of basic and applied research on epideictic pheromones in phytophagous insects.

First, the evidence is rather convincing that for many phytophagous insects, there is a range within which the density of individuals on an exhaustible unit of food resource is optimal, with selection favoring a high degree of individual fitness and full exploitation of the benefits of the resource, but not overcrowding. In some species, such as *Dendroctonus* bark beetles, *Ephestia* flour moths, aphids, and *P. brassicae,* the optimal density range may be rather high, with aggregations of individuals being selectively advantageous in overcoming host defense mechanisms, mobilizing or utilizing host nutrients, defending against or avoiding attack by natural enemies, and so on. On the other hand, in other species such as *Callosobruchus* bean weevils, *H. zea, Rhagoletis* fruit flies, and *H. testudinea,* the optimal density level may be as low as one individual per unit of food resource.

Second, direct or circumstantial evidence suggests that a considerable number of phytophagous insects—spanning at least six orders (Coleoptera, Lepidoptera, Diptera, Homoptera, Hymenoptera, and Orthoptera), 16 families, and 33 pest species—release epideictic pheromones that influence host exploitation processes. Just as attractant pheromones elicit selectively advantageous levels of aggregation, epideictic pheromones elicit the dispersal of arriving conspecifics away from food resources already occupied at densities near or above the upper end of the optimal range. By acting to discourage or prevent overcrowding of any one unit of resource, epideictic pheromones mediate a tendency toward uniformity in spacing among available resources within the patch or habitat. The evidence reveals that epideictic pheromones may be released from larvae, eggs and/or adults and may elicit responses from larvae and/or adults. The pheromones may be perceived olfactorily at a distance and/or upon direct tactile contact, and may deter individuals from landing, feeding, and/or ovipositing.

As the host selection process of phytophagous insects becomes better understood in detail, it is likely that numerous additional taxa will be found to use epideictic pheromones to prevent overcrowding. This result would appear to be most likely among species wherein the stage(s) feeding on an exhaustible resource unit had limited capability of successfully dispersing to a second unit, and/or among species that tend to distribute themselves uniformly among available units. Observed low or high population density per se, on a unit of resource, should not be taken as suggestive of the existence or nonexistence of an epideictic pheromone. Periodic scarcity in resource abundance can result in periodic overcrowding, even among species which employ epideictic pheromones. Furthermore, a density that we might initially interpret as being one of apparent overcrowding, hence suggestive of lack of epideictic pheromone utilization, may actually lie well within the optimal range for that species.

It is also important to recognize that laboratory-cultured individuals involved in epideictic pheromone assay programs could produce an unnatural quantity or quality of pheromone, as in *R. pomonella,* or exhibit an unnatural response to pheromone, as in *C. capitata.*

Third, the evidence suggests that at least some epideictic pheromones have multiple functions, hence are illustrative of the phenomenon of "pheromonal parsimony" (Blum 1970). For example, *D. frontalis, T. granarium, E. keuhniella,* and *A. domestica* all release pheromones that act to attract or arrest conspecifics when at low or moderate concentration but act epideictically at high concentration. There is an increasing number of cases of insect allomones that have been found to function also as alarm pheromones (Blum 1977). It is tempting to speculate, therefore, that some substances whose presently described function is that of an aggregating pheromone, alarm pheromone, allomone, or even a component of a sex pheromone, may, at certain concentrations and in certain contexts, turn out to function as epideictic pheromones as well. Modified components of ingested food, excretory products, substances used to cement or glue eggs to plant surfaces, and chemical adherents on moth scales might be particularly likely candidates. So also might female-released sex pheromones that are dabbed on the plant surface, such as that of the pink bollworm (*Pectinophora gossypiella*) (Colwell et al. 1978), especially if pheromone dabbing and mating initiation are restricted to larval host plants.

Several examples of epideictic pheromones having kairomonal effects on natural enemies were given, as was an example of an epideictic pheromone having an allomonal effect on a competing species. The probability is great that at least some of the insect-secreted substances whose only described biological significance to date is that of being a kairomone to a parasitoid or predator are in fact epideictic pheromones released by the host.

Fourth, examples were presented of how epideictic pheromones have been evaluated for direct control of pests such as *Dendroctonus* and *Rhagoletis*. As the number of key phytophagous pests found to secrete and respond to epideictic pheromones increases, and as the chemical identity of the active pheromonal components becomes better known, it would seem that use of synthetic epideictic pheromones to disrupt landing, feeding, or oviposition behavior would take its place alongside pheromones that disrupt mating behavior as a selective, effective technique of phytophagous insect pest control.

Of course, as pointed out, there may be a number of stumbling blocks along the road to successful implementation of this technique. For example, consider (1) eventual physiological habituation or adaptation to the presence of epideictic pheromone, (2) in the face of continuous deprivation of available resource sites, a lowering of the threshold level for oviposition such that the pest would attack the plant despite the presence of epideictic pheromone, (3) eventual decrease in the concentration of pheromone, which in some species would then elicit aggregation rather than deterrence, and (4) selection for individuals exhibiting reduced responsiveness to pheromone. Hence it will probably be necessary in many cases to employ optimally distributed synthetic epideictic pheromones in conjunction with optimally distributed attractants to trap out or otherwise eliminate deterred individuals.

Fifth, as pointed out by Gross in Chapter 8, a number of potential advantages might be gained from the application of synthetic kairomones in managing parasitoids and predators in agricultural croplands. Special attention should be given to the cases of kairomones that are actually epideictic pheromones produced by the host. What might be devised as an ideal physical pattern of distributing kairomone particles for eliciting maximum host-searching efficiency by a natural enemy could, on the one hand, turn out to have added benefit by deterring the pest itself from landing, feeding, or ovipositing. On the other hand, the "ideal" pattern could backfire if it called for application of a low concentration of kairomone, which elicited pest aggregation rather than having epideictic effects.

Finally, apart from (1) discovering the chemical identity and characterizing the extrinsic and intrinsic origin of the active components of epideictic pheromones, (2) determining the energy cost to the individual of pheromone production, and (3) assaying for possible differences in pheromone quality or quantity between conspecifics from different populations, there are many intriguing questions that, if answered, would provide much better understanding of how epideictic pheromones function, thus enhancing their potential use for pest management.

For example, how, at the levels of the sensory receptor and central nervous systems, do epideictic pheromones influence or inhibit appetitive feeding or oviposition behavior, and for how long do these influences persist (see Barton Browne 1977)? Do all epideictic pheromones function solely by eliciting orientation away from the site of pheromone release, or do some function partly or wholly by inhibiting any further movement toward food resource units (see Borden 1977)? Answers to these two questions would have considerable bearing on choice of pattern for synthetic epideictic pheromone and attractant distribution.

What is the degree of variability among individuals in response to epideictic pheromones? Do most phytophagous insects require a period of "learning" or prior experience with an epideictic pheromone before exhibiting a response, as does *R. pomonella,* and does responsiveness fade toward the end of life?

As addressed by van Lenteren and Bakker (1978) in regard to parasitoid responses to hosts, what is the nature of the host-searching pattern, and what is the nature of the functional and numerical response of phytophagous insects to varying densities of plant hosts or parts of hosts, marked or unmarked with epideictic pheromone? How strongly does the overall spatial structure of the habitat (Hassell and Southwood 1978) influence the response to marked plants or plant parts? Is the optimal foraging strategy (Parker and Stuart 1976, Pyke et al. 1977, Hassell and Southwood 1978, Hubbard and Cook 1978, Waage 1979) of the phytophagous insects that utilize epideictic pheromones a strategy of adjusting the length of host patch visit to (1) the rate of encounter with unmarked hosts in the patch, (2) the rate of encounter with marked hosts, or (3) a combination of these? Is there an "evolutionarily stable strategy" (Maynard Smith 1974, Parker and Stuart 1976) in which the fitness payoff of an individual searching strategy in a habitat of mixed unmarked and marked host plants is dependent on the frequency of other searching strategies in the population, leading to a mixture of strategies maintained in stable frequencies? Indeed, to what extent might there be species polymorphism (Southwood 1977) with respect to varying levels of response to epideictic pheromones and overcrowding? If an effective parasitoid or predator were to make increasingly efficient use of an epideictic pheromone as a kairomonal cue in host location, would selection eventually favor diminished production of the pheromone by the surviving host?

These and many other unanswered questions pose a wealth of exciting research opportunities for those motivated to pursue basic and applied investigations on epideictic pheromones influencing the spacing patterns of phytophagous insects.

ACKNOWLEDGMENTS

Part of the research on *Rhagoletis pomonella* reported here was supported by Massachusetts Agricultural Experiment Station Project No. 380 and the Science and Education Administration of the U.S. Department of Agriculture under Grant No. 7800168 from the Competitive Research Grants Office and Cooperative Agreement No. 12-14-1001-1205. I appreciate the constructive suggestions of the following persons on an earlier draft of this manuscript: A. L. Averill, B. D. Roitberg, M. C. Birch, E. F. Boller, J. S. Elkinton, J. A. Renwick, L. M. Schoonhoven, J. C. van Lenteren, and J. K. Waage.

REFERENCES

Alexander, P., and D. H. R. Barton. 1943. The excretion of ethylquinone by the flour beetle. Biochem. J. 37:463–465.

Alcock, J. 1975. Territorial behavior by males of *Philanthus multimaculatus* (Hymenoptera: Sphecidae) with a review of territoriality in male sphecids. Anim. Behav. 23:889–895.

Averill, A. L., and R. J. Prokopy. 1980a. Release of oviposition-deterring pheromone by apple maggot flies, *Rhagoletis pomonella*. J. N.Y. Entomol. Soc. 88:34.

Averill, A. L. and R. J. Prokopy. 1980b. Oviposition deterring fruit marking pheromone in *Rhagoletis basiola*. J. Fla. Entomol. Soc. (In Press).

Barton Browne, L. 1977. Host-related responses and their suppression: Some behavioral considerations. P. 117–127. *In* H. H. Shorey and J. J. McKelvey, Jr. (eds.), Chemical Control of Insect Behavior. Wiley, New York.

Beaver, R. A. 1974. Intraspecific competition among bark beetle larvae (Coleoptera: Scolytidae). J. Anim. Ecol. 43:455–467.

Behan, M., and L. M. Schoonhoven. 1978. Chemoreception of an oviposition deterrent associated with eggs in *Pieris brassicae*. Entomol. Exp. Appl. 24:163–179.

Birch, M. C. 1978. Chemical communication in pine bark beetles. Am. Sci. 66:409–419.

Birch, M. C., P. Svihra, T. D. Paine, and J. C. Miller. 1980. Influence of chemically mediated behavior on host tree colonization by four cohabiting species of bark beetles. J. Chem. Ecol. 6:395–414.

Blum, M. S. 1970. The chemical basis of insect sociality. P. 61–94. *In* M. Beroza (ed.), Chemicals Controlling Insect Behavior. Academic Press, New York.

Blum, M. S. 1977. Behavioral responses of Hymenoptera to pheromones and allomones. P. 149–167. *In* H. H. Shorey and J. J. McKelvey, Jr. (eds.), Chemical Control of Insect Behavior. Wiley, New York.

Blum, M. S., and J. M. Brand. 1972. Social insect pheromones: Their chemistry and function. Am. Zool. 12:553–576.

Blum, M. S., R. M. Crewe, W. E. Kerr, L. H. Keith, A. W. Garrison, and M. M. Walker. 1970. Citral in stingless bees: Isolation and functions in trail-laying and robbing. J. Insect Physiol. 16:1637–1648.

Blum, M. S., J. B. Wallace, R. M. Duffield, J. M. Brand, H. M. Fales, and E. A. Sokoloski. 1978. Chrysomelidial in the defensive secretion of the leaf beetle *Gastrophysa cyanea*. J. Chem. Ecol. 4:47–53.

Borden, J. H. 1977. Behavioral responses of Coleoptera to pheromones, allomones, and kairomones. P. 169–198. *In* H. H. Shorey and J. J. McKelvey, Jr. (eds.), Chemical Control of Insect Behavior. Wiley, New York.

Brantjes, N. B. M. 1976. Prevention of superparasitation of *Melandrium* flowers by *Hadena*. Oecologia 24:1–6.

Brown, J. L., and G. H. Orians. 1970. Spacing patterns in mobile animals. Annu. Rev. Ecol. Syst. 1:239–262.

Byers, J. A., and D. L. Wood. 1980. Interspecific inhibition of the response of the bark beetles, *Dendroctonus brevicomis* and *Ips paraconfusus* to their pheromones in the field. J. Chem. Ecol. 6:149–164.

Cameron, P. J., and F. O. Morrison. 1974. Sampling methods for estimating the abundance and distribution of all life stages of the apple maggot, *Rhagoletis pomonella*. Can. Entomol. 106:1025–1034.

Cammaerts, M. C., E. D. Morgan, and R. C. Tyler. 1977. Territorial marking in the ant *Myrmica rubra* L. (Formicidae). Biol. Behav. 2:263–272.

Cammaerts, M. C., M. R. Inwood, E. D. Morgan, K. Parry, and R. C. Tyler. 1978. Comparative study of the pheromones emitted by workers of the ants *Myrmica rubra* and *Myrmica scabrinodis*. J. Insect Physiol. 24:207–214.

Cirio, U. 1972. Osservazioni sul comportamento di ovideposizione della *Rhagoletis completa* in laboratorio. Proc. 9th Congr. Ital. Entomol., Siena. 99–117.

Colwell, A. E., H. H. Shorey, P. Baumer, and S. E. Vanvorhiskey. 1978. Sex pheromone scent marking by females of *Pectinophora gossypiella*. J. Chem. Ecol. 4:717–721.

Corbet, S. A. 1971. Mandibular gland secretion of larvae of the flour moth, *Anagasta kuehniella*, contains an epideictic pheromone and elicits oviposition movements in a hymenopteran parasite. Nature (London) 232:481–484.

Corbet, S. A. 1973. Oviposition pheromone in larval mandibular glands of *Ephestia kuehniella*. Nature (London) 243:537–538.

Crnjar, R. M., R. J. Prokopy, and V. G. Dethier. 1978. Electrophysiological identification of oviposition-deterring pheromone receptors in *Rhagoletis pomonella*. J. N.Y. Entomol. Soc. 86:283–284.

Davies, N. B. 1978. Ecological questions about territorial behavior. P. 317–350. *In* J. R. Krebs and N. B. Davies (eds.), Behavioral Ecology, An Evolutionary Approach. Sinauer Associates, Sunderland, MA.

Dickens, J. C., and T. L. Payne. 1977. Bark beetle olfaction: Pheromone receptor system in *Dendroctonus frontalis*. J. Insect Physiol. 23:481–489.

Dixon, A. F. G., and S. D. Wratten. 1971. Laboratory studies on aggregation, size, and fecundity in the black bean aphid, *Aphis fabae*. Bull. Entomol. Res. 61:97–111.

Dixon, C. A., J. M. Erickson, D. N. Kellert, and M. Rothschild. 1978. Some adaptations between *Danaus plexippus* and its food plant, with notes on *Danaus chrysippus* and *Euploea core*. J. Zool. London 185:437–467.

Eisner, T. E., and F. C. Kafatos. 1962. Defense mechanisms of arthropods. X. A pheromone promoting aggregation in an aposematic distasteful insect. Psyche 69:53–61.

Evans, H. E., and K. M. O'Neill. 1978. Alternative mating strategies in the digger wasp *Philanthus zebratus* Cresson. Proc. Natl. Acad. Sci. U.S.A. 75:1901–1903.

Ferguson, A. W. and J. B. Free. 1979. Production of a forage-marking pheromone by the honeybee. J. Apic. Res. 18:128–135.

Fink, D. E. 1913. The asparagus miner and the twelve-spotted asparagus beetle. Cornell University Agriculture Experiment Station Bulletin No. 331:422–435.

Finger, A., D. Heller, and A. Shulov. 1965. Olfactory response of the Khapra beetle (*Trogoderma granarium*) larva to factors from larvae of the same species. Ecology 46:542–543.

Frankie, G. W., and S. B. Vinson. 1977. Scent marking of passion flowers in Texas by females of *Xylocopa virginica texana* J. Kansas Entomol. Soc. 50:613–625.

Furniss, M. M., L. N. Kline, R. F. Schmitz, and J. A. Rudinsky. 1972. Tests of three pheromones to induce or disrupt aggregation of Douglas fir beetles on live trees. Ann. Entomol. Soc. Am. 65:1227–1232.

Furniss, M. M., G. E. Daterman, L. N. Kline, M. D. McGregor, G. C. Trostle, L. F. Pettinger, and J. A. Rudinsky. 1974. Effectiveness of the Douglas fir beetle anti-aggregation pheromone methylcyclohexenone at three concentrations and spacings around felled host trees. Can. Entomol. 106:381–392.

Furniss, M. M., J. W. Young, M. D. McGregor, R. L. Livingston, and D. R. Hamel. 1977. Effectiveness of controlled release formulations of MCH for preventing Douglas fir beetle infestation in felled trees. Can. Entomol. 109:1063–1069.

Gilbert, L. E. 1977. Insect-plant coevolution and ecosystem substructure. Colloq. In. C.N.R.S. 265:399–413.

Greenway, A. R., D. C. Griffiths, and S. L. Lloyd. 1978. Response of *Myzus persicae* to components of aphid extracts and to carboxylic acids. Entomol. Exp. Appl. 24:169–174.

Griffiths, D. C., A. R. Greenway, and S. L. Lloyd. 1978. The influence of repellent materials and aphid extracts on settling behavior and larviposition of *Myzus persicae*. Bull. Entomol. Res. 68:613–619.

Gross, H. R., and R. L. Jones. 1977. Evaluation of oviposition deterrents for *Heliothis zea* on corn. Paper presented at the National Meeting of the Entomological Society of America, Washington, D.C.

Gwynne, D. T. 1978. Male territoriality in the bumblebee wolf, *Philanthus bicinctus:* Observations on the behavior of individual males. Z. Tierpsychol. 47:89–103.

Hassel, M. P., and T. R. E. Southwood. 1978. Foraging strategies of insects. Annu. Rev. Ecol. Syst. 9:75–98.

Hedden, R. L., and G. B. Pitman. 1978. Attack density regulation: A new approach to the use of pheromones in Douglas fir beetle population management. J. Econ. Entomol. 71:633–637.

Hedin, P. A., R. C. Gueldner, R. D. Henson, and A. C. Thompson. 1974. Volatile constituents of male and female boll weevils and their frass. J. Insect Physiol. 20:2135–2142.

Hirai, K., H. H. Shorey, and L. K. Gaston. 1978. Competition among courting male moths: Male-to-male inhibitory pheromones. Science 202:644–645.

Hölldobler, B. 1978. Ethological aspects of chemical communication in ants. Adv. Study Behav. 8:75–115.

Hölldobler, B. K., and E. O. Wilson. 1977. Weaver ants. Sci. Am. 237:146–154.

Hölldobler, B. K., and E. O. Wilson. 1978. The mutliple recruitment systems of the African weaver ant *Oecophylla longinoda* Behav. Ecol. Sociobiol. 3:19–60.

Hubbard, S. F., and R. M. Cook. 1978. Optimal foraging by parasitoid wasps. J. Anim. Ecol. 47:593–604.

Hubbell, S. P., and L. K. Johnson. 1977. Competition and nest spacing in a tropical stingless bee community. Ecology 58:949–963.

Hunter, W. D., and W. D. Pierce. 1912. The Mexican boll weevil: A summary of the investigations of this insect up to Dec. 31, 1911. U.S. Senate Document No. 305.

Hurter, J., B. Katsoyannos, E. F. Boller, and P. Wirz. 1976. Beitrag zur Anreicherung und Teilweisen Reinigung des eiablageverhindernden Pheromons der Kirschenfliege, *Rhagoletis cerasi* L. (Dipt., Trypetidae). Z. Angew. Entomol. 80:50–56.

Ives, P. M. 1978. How discriminating are cabbage butterflies? Aust. J. Ecol. 3:261–276.

Jaenike, J. 1978. An optimal oviposition behavior in phytophagous insects. Theor. Pop. Biol. 14:350–356.

Jaffé, K., M. Bazire-Benazet, and P. E. Howse. 1979. An integumentary pheromone-secreting gland in *Atta* sp: territorial marking with a colony-specific pheromone in *Atta cephalotes*. J. Insect Physiol. 25:833–839.

Jenkins, J. N., W. L. Parrott, and J. W. Jones. 1975. Boll weevil oviposition behavior: Multiple punctured squares. Environ. Entomol. 4:861–867.

Jones, R. E. 1977. Movement patterns and egg distribution in cabbage butterflies. J. Anim. Ecol. 46:195–212.

Kalin, M., and G. Knerer. 1977. Group and mass effects in diprionid sawflies. Nature (London) 267:427–429.

Katsoyannos, B. I. 1975. Oviposition-deterring, male-arresting fruit marking pheromone in *Rhagoletis cerasi*. Environ. Entomol. 4:801–807.

Katsoyannos, B. I., and E. F. Boller. 1976. First field application of oviposition-deterring pheromone of European cherry fruit fly, *Rhagoletis cerasi*. Environ. Entomol. 5:151–152.

Katsoyannos, B. I., and E. F. Boller. 1980. Second field application of oviposition-deterring pheromone of the European cherry fruit fly, *Rhagoletis cerasi*. Z. Angew. Entomol. 89:278–281.

Kidd, N. A. C. 1977. The influence of population density on the flight behavior of the lime aphid, *Eucallipterus tiliae*. Entomol. Exp. Appl. 22:251–261.

Kinzer, G. W., A. F. Fentiman, T. F. Page, R. L. Foltz, J. P. Vité, and G. B. Pitman. 1969. Bark beetle attractants: Identification, synthesis, and field bioassay of a new compound isolated from *Dendroctonus*. Nature (London) 221:477–478.

Kline, L. N., R. F. Schmitz, J. A. Rudinsky, and M. M. Furniss. 1974. Repression of spruce beetle attraction by methylcyclohexenone in Idaho. Can. Entomol. 106:485–491.

LeRoux, E. J., and M. K. Mukerji. 1963. Notes on the distribution of immature stages of the apple maggot, *Rhagoletis pomonella*, on apples in Quebec. Ann. Entomol. Soc. Quebec 8:60–70.

Lewis, W. J., R. L. Jones, and A. N. Sparks. 1972. A host-seeking stimulant for the egg parasite *Trichogramma evanescens:* Its source and a demonstration of its laboratory and field activity. Ann. Entomol. Soc. Am. 65:1087–1089.

Lewis, W. J., R. L. Jones, H. R. Gross, and D. A. Nordlund. 1976. The role of kairomones and other behavioral chemicals in host finding by parasitic insects. Behav. Biol. 16:267–289.

Lewis, W. J., D. A. Nordlund, H. R. Gross, R. L. Jones, and S. L. Jones. 1977. Kairomones and their use for management of entomophagous insects. V. Moth scales as a stimulus for predation of *Heliothis zea* eggs by *Chrysopa carnea* larvae. J. Chem. Ecol. 3:483–487.

Libbey, L. M., M. E. Morgan, T. B. Putman, and J. A. Rudinsky. 1976. Isomer of antiaggregative pheromone identified from male Douglas fir beetle. J. Insect Physiol. 22:871–873.

Loconti, J. D., and L. M. Roth. 1953. Composition of the odorous secretion of *Tribolium castaneum*. Ann. Entomol. Soc. Am. 46:281–289.

Maynard Smith, J. 1974. The theory of games and evolution of animal conflicts. J. Theor. Biol. 47:209–222.

McMullen, L. H., and M. D. Atkins. 1961. Intraspecific competition as a factor in the natural control of the Douglas fir beetle. Forest Sci. 7:197–203.

Mitchell, R. 1975. The evolution of oviposition tactics in the bean weevil, *Callosobruchus maculatus*. Ecology 56:696–702.

Mudd, A., and S. A. Corbet. 1973. Mandibular gland secretion of larvae of the stored products pests *Anagasta kuehniella, Ephestia cautella, Plodia interpunctella,* and *Ephestia elutella*. Entomol. Exp. Appl. 16:291–293.

Myers, J. H. 1976. Distribution and dispersal in populations capable of resource depletion: A simulation model. Oecologia 23:255–269.

Myers, J. H., and B. J. Campbell. 1976. Distribution and dispersal in populations capable of resource depletion: A field study on cinnabar moth. Oecologia 24:7–20.

Nault, L. R., and M. E. Montgomery. 1977. Aphid pheromones. P. 527–545. *In* K. F. Harris and K. Maramorosch (eds.), Aphids as Virus Vectors. Academic Press, New York.

Nault, L. R., M. E. Montgomery, and W. S. Bowers. 1976. Ant-aphid association: Role of aphid alarm pheromone. Science 192:1349–1351.

Naylor, A. F. 1959. An experimental analysis of dispersal in the flour beetle *Tribolium confusum*. Ecology 40:453–465.

Naylor, A. F. 1961. Dispersal in the red flour beetle *Tribolium castaneum*. Ecology 42:231–237.

Nordlund, D. A., W. J. Lewis, J. W. Todd, and R. B. Chalfant. 1977. Kairomones and their use for management of entomophagous insects. VII. The involvement of various stimuli in the differential response of *Trichogramma pretiosum* to two suitable hosts. J. Chem. Ecol. 3:513–518.

Ogwaro, K. 1978. Ovipositional behavior and host plant preference of the sorghum shoot fly, *Atherigona soccata*. Entomol. Exp. Appl. 23:189–199.

Oshima, K., H. Honda, and I. Yamamoto. 1973. Isolation of an oviposition marker from Azuki bean weevil, *Callosobruchus chinensis*. Agric. Biol. Chem. 37:2679–2680.

Palanaswamy, P., and W. D. Seabrook. 1978. Behavioral responses of the female Eastern spruce budworm, *Choristoneura fumiferana*, to the sex pheromone of her own species. J. Chem. Ecol. 4:649–655.

Park, T. 1933. Studies in population physiology. II. Factors regulating initial growth of *Tribolium confusum* population. J. Exp. Zool. 68:17–41.

Park, T. 1934. Studies in population physiology. III. The effect of conditioned flour upon the productivity and population decline of *Tribolium confusum* J. Exp. Zool. 68:167–182.

Parker, G. A., and R. A. Stuart. 1976. Animal behavior as a strategy optimizer: Evolution of resource assessment strategies and optimal emigration thresholds. Am. Nat. 110:1055–1076.

Payne, T. L., J. E. Coster, and P. C. Johnson. 1977. Effects of slow release formulations of synthetic endo- and exo-brevicomin on southern pine beetle flight and landing behavior. J. Chem. Ecol. 3:133–141.

Payne, T. L., J. E. Coster, J. V. Richerson, L. J. Edson, and E. R. Hart. 1978. Field response of the southern pine beetle to behavioral chemicals. Environ. Entomol. 7:578–582.

Peters, T. M., and P. Barbosa. 1977. Influence of population density on size, fecundity, and developmental rate of insects in culture. Annu. Rev. Entomol. 22:431–450.

Phelan, P. L., and J. R. Miller. 1979. Efficacy of chemical repellents against alighting aphids. Paper presented at the National Meeting of the Entomological Society of America, Denver, CO.

Pitman, G. B., and J. P. Vité. 1974. Biosynthesis of methylcyclo-hexenone by male Douglas fir beetle. Environ. Entomol. 3:886–887.

Pritchard, G. 1969. The ecology of a natural population of Queensland fruit fly, *Dacus tryoni*. II. The distribution of eggs and its relation to behavior. Aust. J. Zool. 17:293–311.

Prokopy, R. J. 1972. Evidence for a marking pheromone deterring repeated oviposition in apple maggot flies. Environ. Entomol. 1:326–332.

Prokopy, R. J. 1975. Oviposition-deterring fruit marking pheromone in *Rhagoletis fausta*. Environ. Entomol. 4:298–300.

Prokopy, R. J. 1976. Significance of fly marking of oviposition site (in Tephritidae). P. 23–27. *In* V. Deluchhi (ed.), Studies in Biological Control. Cambridge University Press, Cambridge.

Prokopy, R. J. 1981. Oviposition deterring pheromone system of apple maggot flies. *In* E. R. Mitchell (ed.) *Management of Insect Pests with Semiochemicals*. Plenum Publ. Corp. New York (in press).

Prokopy, R. J., and G. L. Bush. 1972. Mating behavior in *Rhagoletis pomonella*. III. Male aggregation in response to an arrestant. Can. Entomol. 104:275–283.

Prokopy, R. J., and P. J. Spatcher. 1977. Location of receptors for oviposition deterring pheromone in *Rhagoletis pomonella*. Ann. Entomol. Soc. Am. 70:960–962.

Prokopy, R. J., and R. P. Webster. 1978. Oviposition deterring pheromone of *Rhagoletis pomonella:* A kairomone for its parasitoid *Opius lectus*. J. Chem. Ecol. 4:481–494.

Prokopy, R. J., W. H. Reissig, and V. Moericke. 1976. Marking pheromones deterring repeated oviposition in *Rhagoletis* flies. Entomol. Exp. Appl. 20:170–178.

Prokopy, R. J., P. D. Greany, and D. L. Chambers. 1977. Oviposition deterring pheromone in *Anastrepha suspensa*. Environ. Entomol. 6:463–465.

Prokopy, R. J., J. R. Ziegler, and T. T. Y. Wong. 1978. Deterrence of repeated oviposition by fruit-making pheromone in *Ceratitis capitata* (Diptera: Tephritidae). J. Chem. Ecol. 4:55–63.

Prokopy, R. J., A. L. Averill, C. M. Bardinelli, R. M. Crnjar, and P. J. Spatcher. 1979. Physiology of oviposition deterring pheromone production in *Rhagoletis pomonella*. Paper presented at the National Meeting of the Entomological Society of America, Denver, CO.

Pyke, G. H., H. R. Pulliam, and E. L. Charnov. 1977. Optimal foraging: A selective review of theory and tests. Q. Rev. Biol. 52:137–154.

Raw, A. 1975. Territoriality and scent marking by *Centris* males in Jamaica. Behavior 54:311–321.

Renwick, J. A. A. 1967. Identification of two oxygenated terpenes from the bark beetles *Dendroctonus frontalis* and *Dendroctonus brevicomis*. Contrib. Boyce Thompson Inst. 23:355–360.

Renwick, J. A. A., and J. P. Vité. 1969. Bark beetle attractants: Mechanism of colonization by *Dendroctonus frontalis*. Nature (London) 244:1222–1223.

Renwick, J. A. A., and J. P. Vité. 1970. Systems of chemical communication in *Dendroctonus*. Contrib. Boyce Thompson Inst. 24:283–292.

Renwick, J. A. A., and C. D. Radke. 1980. An oviposition deterrent associated with frass from feeding larvae of the cabbage looper, *Trichoplusia ni*. Envir. Entomol. 9:318–320.

Richerson, J. V., and T. L. Payne. 1979. Effects of bark beetle inhibitors on landing and attack behavior of the southern pine beetle and beetle associates. Environ. Entomol. 8:360–364.

Roitberg, B. D. and R. J. Prokopy. 1981. Experience required for pheromone recognition by the apple maggot fly. Nature (in press).

Rothschild, M., and L. M. Schoonhoven. 1977. Assessment of egg load by *Pieris brassicae*. Nature (London) 266:352–355.

Rowell, H. F. 1978. Food plant specificity in neotropical rain-forest acridids. Entomol. Exp. Appl. 24:451–462.

Rudinsky, J. A. 1973a. Multiple functions of the southern pine beetle pheromone verbenone. Environ. Entomol. 2:511–514.

Rudinsky, J. A. 1973b. Multiple functions of the Douglas fir beetle pheromone 3-methyl-2-cyclohexen-1-one. Environ. Entomol. 2:579–585.

Rudinsky, J. A., M. Morgan, L. M. Libbey, and R. R. Michael. 1973. Sound production in Scolytidae: 3-Methyl-2-cyclohexen-1-one released by the female Douglas fir beetle in response to male sonic signal. Environ. Entomol. 2:505–509.

Rudinsky, J. A., M. E. Morgan, L. M. Libbey, and T. B. Putnam. 1974a. Antiaggregative rivalry pheromone of the mountain pine beetle and a new arrestant of the southern pine beetle. Environ. Entomol. 3:90–98.

Rudinsky, J. A., C. Sartwell, T. M. Graves, and M. E. Morgan. 1974b. Granular formulation of methylcyclohexenone: An antiaggregative pheromone of the Douglas fir and spruce bark beetles. Z. Angew. Entomol. 75:254–263.

Rudinsky, J. A., L. C. Ryker, R. R. Michael, L. M. Libbey, and M. E. Morgan. 1976. Sound production in Scolytidae: Female sonic stimulus of male pheromone release in two *Dendroctonus* beetles. J. Insect Physiol. 22:1675–1681.

Russ, K. 1969. Beitrage zum Territorialverhalten der Raupen des Springwurmwicklers, *Sparganothis pilleriana* (Lepidoptera: Tortricidae). Pflanzenschutzberichte 40:1–9.

Saad, A. D., and D. R. Scott. 1979. *Heliothis armigera* and *H. zea:* Virgin females repelled by other females. Paper presented at the National Meeting of the Entomological Society of America, Denver, CO.

Schoonhoven, L. M., T. Sparnay, W. van Wissen, and J. Meerman. 1980. Seven weeks' persistence of an oviposition-deterrent pheromone. J. Chem. Ecol. (in press).

Scott, D. R. 1977. The corn earworm in Southwestern Idaho: Infestation levels and damage to processing corn and sweet corn seed. J. Econ. Entomol. 70:709–713.

Sexton, O. J., and E. H. Hess. 1968. A pheromone-like dispersant affecting the local distribution of the European house cricket, *Acheta domestica*. Biol. Bull. 134:490–502.

Shorey, H. H. 1973. Behavioral responses to insect pheromones. Annu. Rev. Entomol. 18:349–380.

Shorey, H. H. 1976. *Animal Communication by Pheromones*. Academic Press, New York.

Smith, R. H. 1963. Toxicity of pine resin vapors to three species of *Dendroctonus* bark beetles. J. Econ. Entomol. 56:827–831.

Southwood, T. R. E. 1977. Habitat, the templet for ecological strategies? J. Anim. Ecol. 46:337–365.

Städler, E., and B. I. Katsoyannos. 1980. Das Makierungspheromon der Kirschenfliege: Erste electrophysiologische Untersuchung der Tarsalen Sinnesorgan der Weibchen. Mitt. Schweiz. Entomol. Ges. (in press).

Stock, M. W., G. B. Pitman, and J. D. Guenther. 1979. Genetic differences between Douglas-fir beetles (*Dendroctonus pseudotsugae*) from Idaho and coastal Oregon. Ann. Entomol. Soc. Am. 72:394–397.

Tannert, W., and B. C. Hein. 1976. Nachweis und Funktion eines repellent Pheromones der Larven von *Trogoderma granarium*. Zool. Jahrb. Physiol. 80:69–81.

Taylor, L. R., and R. J. Taylor. 1977. Aggregation, migration, and population mechanics. Nature (London) 265:413–421.

Taylor, R. J. 1976. The value of clumping to prey. Oecologia 30:285–294.

Traynier, R. M. M. 1979. Long-term changes in the oviposition behavior of the cabbage butterfly, *Pieris rapae,* induced by contact with plants. Physiol. Entomol. 4:87–96.

Tumlinson. J. H., R. M. Silverstein, J. C. Moser, R. G. Brownlee, and J. M. Ruth. 1971. Identification of the trail pheromone of a leaf-cutting ant, *Atta texana*. Nature (London) 234:348–349.

Umeya, K. 1966. Studies on the comparative ecology of bean weevils. I. On the egg distribution and the oviposition behaviors of three species of bean weevils infesting Azuki bean. Res. Bull. Plant Prot. Jpn. 3:1–11.

Umeya, K., and T. Kato. 1970. Studies on the comparative ecology of bean weevils. V. Distribution of eggs and larvae of *Acanthoscelides obtectus* in relation to its oviposition and boring behavior. Res. Popul. Ecol. 12:35–50.

Utida, S. 1943. Studies on experimental populations of the Azuki bean weevil, *Callosobruchus chinensis*. VIII. Statistical analysis of the frequency distribution of the emerging weevils on bean. Mem. Coll. Agric. Kyoto Imp. Univ. 54:1–22.

Van Alphen, J. J. M., and H. Boer. 1980. Avoidance of scramble competition between larvae of the spotted asparagus beetle, *Crioceris duodecimpunctata* by discrimination between unoccupied and occupied asparagus berries. Neth. J. Zool. 30:136–143.

Van Lenteren, J. C., and K. Bakker. 1978. Behavioral aspects of the functional responses of a parasite (*Pseudeucoila bochei*) to its host (*Drosophila melanogaster*). Neth. J. Zool. 28:213–233.

Verner, J. 1977. On the adaptive significance of territoriality. Am. Nat. 111:769–775.

Vinson, S. B., G. W. Frankie, M. S. Blum, and J. W. Wheeler. 1978. Isolation, identification, and function of the Dufour gland secretion of *Xylocopa virginica texana*. J. Chem. Ecol. 4:315–323.

Vité, J. P., and J. A. A. Renwick. 1971. Inhibition of *Dendroctonus frontalis* response to frontalin by isomers of brevicomin. Naturwissenschaften 8:418–419.

Vité, J. P., and D. L. Williamson. 1970. *Thanasimus dubius:* Prey perception. J. Insect Physiol. 16:233–239.

Waage, J. K. 1979. Foraging for patchily-distributed hosts by the parasitoid, *Nemeritis canescens*. J. Anim. Ecol. 48:353–371.

Walker, T. J. 1981. Do populations self-regulate? *In* C. B. Huffaker and R. L. Rabb (eds.), Environmental Entomology. John Wiley and Sons, New York. (In press).

Way, M. J., and C. J. Banks. 1967. Intra-specific mechanisms in relation to the natural regulation of numbers of *Aphis fabae*. Ann. Appl. Biol. 59:189–205.

Way, M. J., and M. E. Cammell. 1970. Self-regulation in aphid populations. pp. 232–242. *In* P. J. den Boer and G. R. Gradwell (eds.), Dynamics of Populations. Proc. Adv. Study Inst. Dynamics Numbers Popul. (Oosterbeek).

Wiens, J. A. 1976. Population responses to patchy environments. Annu. Rev. Ecol. Syst. 7:81–120.

Wood, T. K. 1977. Defense in *Umbonia crassicornis:* Role of pronotum and adult aggregation. Ann. Entomol. Soc. Am. 70:524–528.

Wilson, E. O. 1971. The Insect Societies. Harvard University Press, Cambridge, MA.

Wilson, E. O. 1975. Sociobiology. Harvard University Press, Cambridge, MA.

Wynne-Edwards, V. C. 1962. Animal Dispersion in Relation to Social Behavior. Hafner, New York.

Yoshida, T. 1961. Oviposition behavior of two species of bean weevils and interspecific competition between them. Mem. Fac. Lib. Arts Educ., Miyazaki Univ. 11:41–65.

Zimmerman, M. 1979. Oviposition behavior and the existence of an oviposition deterring pheromone in *Hylemya*. Environ. Entomol. 8:277–279.

Zimmerman, M. 1980. Selective deposition of an oviposition-deterring pheromone by *Hylemya*. Envir. Entomol. 9:321–324.

CHAPTER ELEVEN

ATTRACTIVE AND AGGREGATING PHEROMONES

WENDELL L. ROELOFS

New York State Agricultural Experiment Station
Geneva, New York

Sex pheromones have become the most publicized of all insect semiochemicals in many ways, particularly in reference to semiochemicals used for pest control. Sex pheromones are a specific class of semiochemicals restricted to intraspecific communication used in the mating process. For centuries naturalists have noted the amazing phenomenon of a single female moth luring hundreds of male moths to her vicinity. It long has been recognized that this reliance on a mating communication system could represent a vulnerable area in the biology of pest species on which to base a very selective insect control strategy. Since the announcement that the first sex pheromone had been decoded by Professor Butenandt and co-workers (1959) after a 30-year effort on the domestic silkworm moth (*Bombyx mori*), much research time and money have been devoted to insect sex pheromones. Opinions vary on the value of these efforts, but some conclusions are obvious. From a purely academic standpoint, pheromone research has greatly expanded our knowledge of many aspects of chemical communication. These include integrated studies involving chemistry, physiology, behavior, biology, neurophysiology, and biochemistry. In addition to basic chemical and behavioral information on pheromone emission and perception, this research has been particularly important in developing an understanding of the basic mechanisms involved in the poorly understood area of olfactory reception. It also has made surprising contributions to aspects of insect taxonomy and evolution.

Most publicity, however, is centered on the practical application of insect pheromones in pest control. Many people express a negative impression concerning sex pheromones in insect control, whereas others are cautiously optimistic. Some people have been disappointed to find out that this approach is not a panacea for all their pest problems, while others impatiently expected a new approach to be developed and made commercially available as soon as the messages were decoded. In fact, such an approach cannot be used for all pest species, and the development of the commerical usage of sex pheromones has taken a few years. However, the field has progressed steadily, and sex pheromones now play an important role in pest management programs. This chapter describes some of the basic steps taken to reach the present stage of development. Detailed information and references can be obtained from the following review sources: Shorey and McKelvey (1977), Roelofs and Cardé (1977), Roelofs (1978), Ritter (1979), and the report "Establishing Efficacy of Six Pheromones in Insect Control" sponsored by the American Institute of Biological Sciences, submitted to the U.S. Environmental Protection Agency (EPA) and published by the Entomological Society of America (Buckholder, Cardé, Chambers, Roelofs, Shorey, and Wood).

CHARACTERIZATION OF SEX PHEROMONE COMPONENTS

In accordance to the original definition of a sex pheromone (Karlson and Butenandt 1959, Karlson and Lusher 1959) it is important to rigorously identify the chemical structures of the natural components obtained either from the organism or from its effluvium. It also is important to prove that each chemical elicits behavioral activity that is definable as part of a natural sequence at the concentration used in the natural communication system. Problems have been encountered in both parts of this definition.

Identification of Active Components

Pheromone components can be obtained from the insects by a variety of techniques. The pheromone can be extracted from whole insects or from segments containing the pheromone gland. In one study, pheromone of the Indian meal moth (*Plodia interpunctella*) was identified from whole body extracts of 670,000 males and females, although other researchers simultaneously identified the same pheromone by extracting filter papers exposed to 19,000 virgin female moths in glass jars. For many moth species it has been found that the best collection method is to strip out the glands themselves and to use pure gland extract, thus reducing the number of moths needed.

A study leading to the identification of the four pheromone components of the cotton boll weevil (*Anthonomus grandis*) involved the extraction of more than 4 million weevils and 54.7 kg of fecal material. Later, airborne collections with Porapak Q were found to be the best approach for some insects. Two beetle-produced pheromones (I, II of Figure 11.1) and a third host-produced

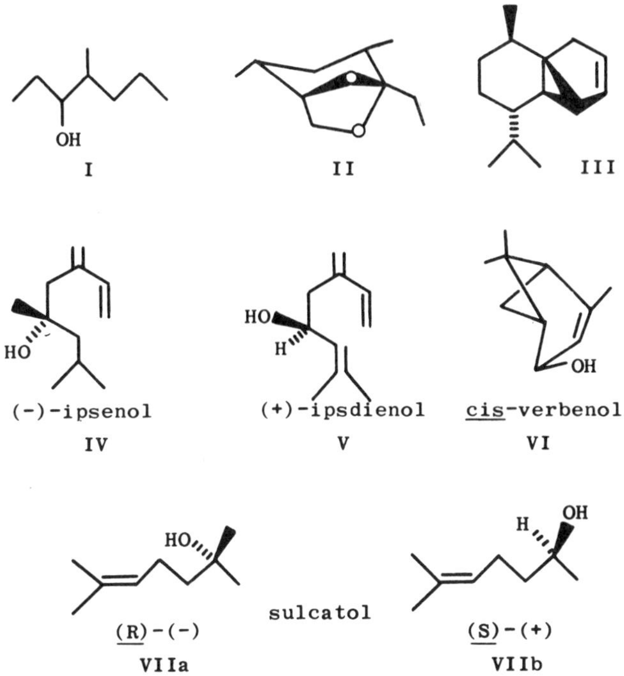

FIGURE 11.1. The structure of compounds identified as pheromone components of various insect species.

component (III) were identified by this method from volatiles in the air around virgin female smaller European elm bark beetles (*Scolytus multistriatus*) as they tunneled in elm logs. In this study batches of 4000–7000 virgin females were aerated continuously for 7 days and the volatiles trapped on Porapak Q. Airborne collections of pheromone-emitting insects have been extremely useful in obtaining active components that were not found in pheromone gland extracts.

The isolation of active components is dependent on an appropriate bioassay technique. As discussed in the next section, it is important to use bioassay techniques that allow one to assess activity for the entire blend of pheromone components. With many species, the electroantennogram (EAG) (Roelofs 1979a) is extremely useful for detecting various active components in fractionated extract or effluvium. The EAG technique utilizes the insect's highly selective and sensitive antenna as a detecting instrument. Since the majority of olfactory sensilla, particularly with moth antennae, are specifically attuned to the pheromone component, fractions containing pheromone components elicit greater antennal response than other fractions. The individual compounds elicit EAG responses, whereas the confirmatory behavioral tests usually require a

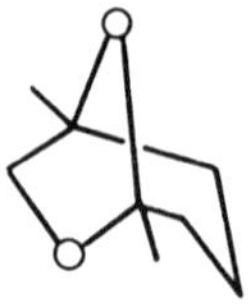

(−)-frontalin
VIII

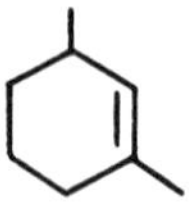

seudenol
IX

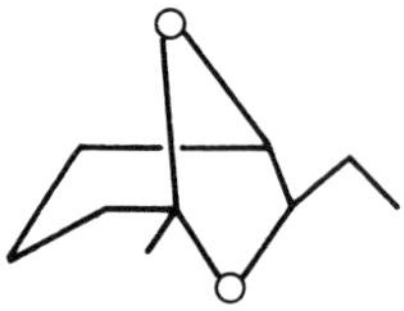

(+)-exo-brevicomin
X

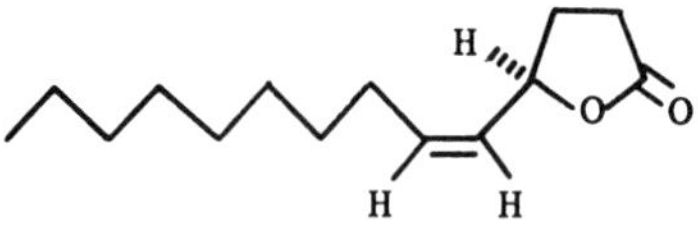

XI

(7R,8S)-(+)-disparlure
XII

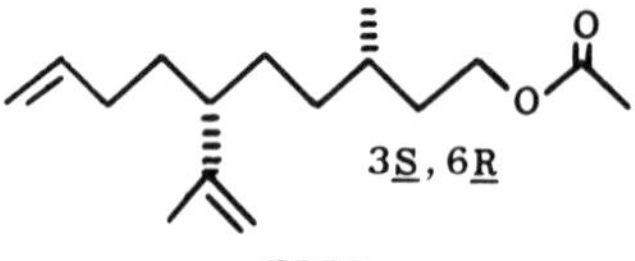

XIII

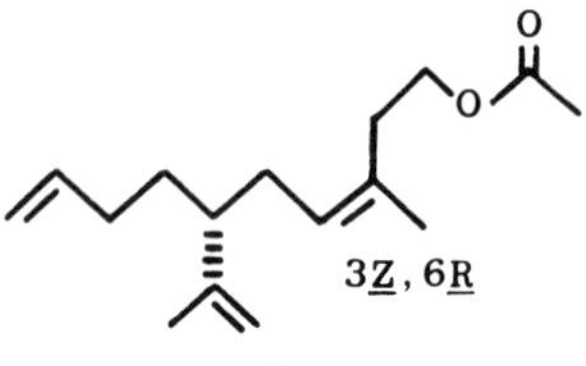

XIV

FIGURE 11.1. *(Continued)*

precise blending of several components. This specific monitoring of pheromone components allows the researcher to isolate the main pheromone component quickly from a small number of insects.

The sensory cell specificity for pheromone component molecular characteristics also can be used in predicting double bond positions and configurations. EAG analyses with a complete library of monounsaturated acetates, aldehydes, and alcohols can indicate the carbon chain length and sites of unsaturation for which the antenna is specifically attuned. The predicted structure then can be proved by chemical and physical analyses of isolated components.

Any ideas that a single unique compound acted as a pheromone for each species were quickly dispelled in 1966 when Silverstein et al. (1966) reported that the pheromone produced by the male bark beetle (*Ips paraconfusus*), which attracts both male and female beetles, was a mixture of three compounds. The compounds (−)-ipsenol, (+)-ipsdienol, and (+)-*cis*-verbenol (IV−VI in Figure 11.1) were not active by themselves but only in mixture. Since then, many pheromones have been found to be blends of several components. Many species of Lepidoptera utilize precise blends of geometric isomers. Pheromone blends of several leafroller species are given in Table 11.1 to illustrate the type of specificity that can be obtained by utilizing several compounds in various ratios. Males of these species respond very precisely to their respective blends at the natural release rates; thus the pheromones represent a major reproductive isolating mechanism for many of these species. Blends composed of various mixtures of unsaturated long-chain acetates, alcohols, and aldehydes have been found quite commonly as attractants for lepidopterous species; this is best exemplified by the attractants of more than 80 species in the Noctuidae (Roelofs 1979b) and of more than 120 species of Tortricidae.

It has become important to identify not only all the minor components of a pheromone blend, but also to define the isomeric purity of all components.

In many cases, the optical purity cannot be assessed directly with natural components containing asymmetric centers, but bioassays of all possible synthetic enantiomers have helped to define the most active isomers. With *I. paraconfusus* the aggregation pheromone is a mixture of 100% (−)-ipsenol, 90%(+)/10%(−)-ipsidienol, and (+)-*cis*-verbenol, whereas the major aggregation pheromone component for *Ips pini* (pine engraver beetle) (Light and Birch 1979) in California is the other isomer, 100% (−)-ipsdienol, which greatly reduces aggregation of *I. paraconfusus* beetles. Optical isomer blends have been determined for a number of other beetle species, including a 65(+)/35(−)-

TABLE 11.1. Tortricid Moth Pheromone Blends Utilizing Specific Blends of Geometric Isomers

	Isomer Blends (%)		
Species	*Z*11−14:Ac	*E*11−14:Ac	Other Components
Choristoneura rosaceana	97	3	*Z*11−4:OH
Argyrotaenia velutinana	92	8	12:Ac
Adoxophyes fasciata	90	10	*Z*9−14:Ac, 10-Me-12:Ac
Archips mortuanus	90	10	*Z*9−14:Ac, 12:Ac
Archips argyrospilus	60	40	*Z*9−14:Ac, 12:Ac
Archips podana	50	50	
Archips semiferanus	30	70	
Sparganothis directana	24	76	*E*9, 11-12:Ac
Archips cerasivoranus	15	85	
Platynota stultana	12	88	*E*11-12:OH

sulcatol (VII) mixture in ambrosia beetles (*Gnathotrichus sulcatus*), an 85(−)/15(+) frontalin (VIII) mixture in southern pine beetles (*Dendroctonus frontalis*), and a 50(+)/50(−)-seudenol (IX) mixture in Douglas fir beetles (*Dendroctonus pseudotsugae*), whereas the five-spined engraver beetle (*Ips grandicollis*) aggregates best to only (−)-ipsenol (IV), and the western pine beetle (*Dendroctonus brevicomis*) uses a mixture of pure isomers of (+)-exobrevicomin (X), (−)-frontalin (VIII), and myrcene. The pheromone for the Japanese beetle (*Popillia japonica*) was found to be the (−)-isomer of the lactone (XI). With the European elm bark beetle, one of the pheromone components, multistriatin (II), was found to be a single enantiomer (1*S*, 2*R*, 4*S*, 5*R*) (−)-α-multistriatin. In this case the four diastereomers, α, β, γ, δ, were synthesized and the specific rotations compared to that of the natural material.

Optical isomers have not been found as frequently in Lepidoptera species to date, but the gypsy moth (*Lymantria dispar*) pheromone, disparlure (XII), does contain an epoxide optical center and the (+)-isomer was found to be the only active isomer. There are four possible geometric and optical isomers of each of the two components of the California red scale (*Aonidiella aurantii*) pheromone (Roelofs et al. 1978), but synthesis and bioassays of all possible isomers showed that only one isomer of each component (XIII-3*S*, 6*R*, and XIV-3*Z*, 6*R*) was significantly more active than the others.

Another interesting case of specificity was reported with the sex pheromones of diprionid sawflies. At least 12 species from two genera respond to the acetate or propionate of 3,7-dimethylpentadecan-2-ol. This structure contains three asymmetric centers, and in the case of the white pine sawfly (*Neodiprion pinetum*) (Kraemer et al. 1979), the males responded quite specifically to the (−)-2*S*,3*S*,7*S* isomer. Specificity thus can be effected among the species by utilizing different functional groups and different optical isomers.

Behavior

Identification of chemical constituents in female gland extracts or effluvia is only part of the process in determining the pheromone blend. It must be shown that each chemical plays some role in the precopulatory behavioral sequence. This is no easy task for many of the minor components. In some instances the components can be used at different times in the mating sequence. For example, with the western pine beetle, the females initiate an attack by finding a suitable host tree and boring through the outer bark into the phloem tissue. The components exobrevicomin and myrcene are released in this process and attract predominantly male beetles to the host tree. The males then release frontalin, and the combination of pheromone attracts beetles in a 1−1 ratio throughout the mass attack.

In other instances, minor components are present in the emitted pheromone but function predominantly in close-range courtship behavior. With Lepidoptera a classification system has been reported (Roelofs and Cardé 1977) to separate the various pheromone components. *Primary pheromone components* are those involved in long-distance (>1 m), positive anemotaxis, whereas *secondary pheromone components* are chemicals that are not essential for long-distance responses but are involved in mediating close-range responses, such as landing, wing fanning, hair pencilling, and other courtship responses. Detailed analyses of these behaviors is extremely difficult. In an in-depth study of the Oriental fruit moth *Grapholitha molesta* (Cardé et al. 1975, Baker and Cardé 1979), the chemical-mediated precopulatory behavioral sequence was analyzed. Chemicals that are present in the emitted pheromone, but with observable activity only in close-range behavior, can have other less obvious effects throughout the entire sequence. Sustained-flight tunnel (Miller and Roelofs 1978) studies with various mixtures and a range of concentrations indicate that some secondary pheromone components can lower the response thresholds for behaviors associated with primary pheromone components. It became obvious that even the natural blend of pheromone components elicits reduced responses if used at too low or too high concentrations (Roelofs 1978). In a very strict sense, the chemicals are pheromone components for that species only when used in the correct ratio and at concentrations that mediate the natural sequence of precopulatory behavioral responses.

PHEROMONE USAGE IN PEST CONTROL

The preceding section briefly discusses some the research involved in defining a sex pheromone for various pest species. It does not do justice, however, to the time and effort required in many cases to reach that stage of development. What is simply given as a pheromone blend of several components most likely represents years taken to establish a viable laboratory culture, develop appropriate techniques for isolating the pheromone components, identify the chemicals, synthesize all possible geometric and optical isomers, and conduct extensive laboratory and field behavioral studies. Although pheromones have been researched extensively for a decade and a half, it only has been in the last 5 years that the accumulated knowledge of pheromone chemistry and behavior has allowed intensive studies on the application of pheromones in pest control. Again, it required years of research to develop technology on use of the chemicals and to accumulate knowledge of when to use the technique in the overall pest control program.

The first stages of development have progressed extremely well. Pheromone components have been identified for a huge number of pest species. Pest species for which attractants are known and field studies for insect control that have been initiated are listed in Table 11.2. There are many more pest species for which attractants are known and could be the subject of insect control research programs. This developmental phase of the program also takes much time, and to date there are complete pheromone efficacy tests with only a few pest species. The section that follows briefly reviews the steps being taken to increase pheromone usage in pest control.

Monitoring Pest Populations

The most immediate commercial usage of pheromone-baited traps is to monitor the presence and density of pest populations in pest management programs. This application has been on a commercial level for a decade and is common throughout the world for many pest species. An example of this can be taken from the New York State Apple Pest Management project, established in 1973 to integrate pest management techniques into a control program that would reduce pesticide usage without reducing quality and quantity of fruit. The program was based on a continuous monitoring of orchards for pertinent weather data, pests, chemicals used, and beneficial organisms. Pheromone traps were used for the five major lepidopterous pest species and sprays were applied only when necessary. Growers in the program were able to decrease insecticide and miticide usage up to 50%. Over the past four years the growers in this program have spent about $60/ha less than other growers in the area without any decrease in quality and quantity of fruit.

The pheromone traps can be combined with other information to help decide whether pesticides should be applied, and if so, when. The first moth appearance can be used as a biological fix point for the accumulation of thermal units, whereas moth catches over a certain interval can help to predict population densities and time of egg hatch. Predicting egg hatch is important for some pest species because insecticide spray can be most effective on the vulnerable, newly hatched larvae as they disperse to their feeding sites. In the Netherlands, pheromone traps are dispensed to apple growers by the Horticultural Advisory Service and used to determine the initiation of moth flight of the summer fruit tortrix moth (*Adoxophyes orana*) in each orchard. A precise timing of spray applications has reduced the conventional five to seven preventive sprays to three or four precisely timed and more effective sprays.

Alford et al. (1979) found that pheromone traps were useful for monitoring codling moth (*Laspeyresia pomonella*) and *Archips podana* populations in individual orchards. Insecticide sprays were not used for densities below five moths/trap/week for codling moth and 20 moths/trap/week for *A. podana*.

In a different type of study, sex pheromone traps were used to make inferences

TABLE 11.2. Examples of Pest Species for Which Attractants Are Known and Field Studies for Insect Control Have Been Initiated

FRUIT PESTS
Adoxophyes orana, summer fruit tortrix moth
Aonidiella aurantii, California red scale
Archips argyrospilus, fruit tree leafroller moth
Archips podana, fruit tree tortrix moth
Argyrotaenia velutinana, redbanded leafroller moth
Ceratitis capitata, Mediterranean fruit fly
Choristoneura rosaceana, obliquebanded leafroller moth
Dacus cucurbitae, melon fly
Dacus dorsalis, Oriental fruit fly
Endopiza viteana, grape berry moth
Euopecilia ambiguella, grape moth
Grapholitha molesta, Oriental fruit moth
Grapholitha prunivora, lesser appleworm moth
Laspeyresia pomonella, codling moth
Lobesia botrana, grapevine moth
Pandemis limitata, threelined leafroller moth
Phylonorycter blancardella, spotted tentiform leafminer
Prays citri, citrus flower moth
Synanthedon exitiosa, peachtree borer
Synanthedon pictipes, lesser peachtree borer
FOREST PESTS
Choristoneura fumiferana, eastern spruce budworm
Dendroctonus brevicomis, western pine beetle
Dendroctonus frontalis, southern pine beetle
Dendroctonus ponderosae, mountain pine beetle
Dendroctonus pseudotsugae, Douglas fir beetle
Dendroctonus rufipennis, spruce beetle
Eucosma sonomana, western pine shoot borer
Gnathotrichus sulcatus, ambrosia beetle
Ips pini, pine engraver beetle
Ips typographus, bark beetle
Lymantria dispar, gypsy moth
Orgyia pseudotsugata, Douglas fir tussock moth
Rhyacionia frustrana, Nantucket pine tip moth
Scolytus multistriatus, smaller European elm brak beetle
Zeiraphera diniana, larch bud moth

TABLE 11.2 *(Continued)*

FIELD PESTS
Adoxophyes fasciata, smaller tea tortrix
Agrotis ipsilon, black cutworm
Anthonomus grandis, cotton boll weevil
Chilo suppressalis, striped stem borer
Diparopsis castanea, red bollworm
Heliothis armigera, Old World bollworm
Heliothis virescens, tobacco bollworm
Heliothis zea, corn earworm
Pectinophora gossypiella, pink bollworm
Phthorimaea operculella, potato tuberworm
Popillia japonica, Japanese beetle
Spodoptera littoralis, Egyptian cotton leafworm
Spodoptera litura, tobacco cutworm
Trichoplusia ni, cabbage looper
STORED PRODUCT PESTS
Acanthoscelides obtectus, dried bean beetle
Ephestia cautella, almond moth
Ephestia elutella, tobacco moth (warehouse moth)
Ephestia kuehniella, Mediterranean flour moth
Plodia interpunctella, Indian meal moth
Trogoderma glabrum, dermestid beetle
Trogoderma granarium, Khapra beetle
HOUSEHOLD PESTS
Attagenus megatoma, black carpet beetle
Blatta orientalis, oriental cockroach
Blatella germanica, German cockroach
Musca domestica, housefly
Periplaneta americana, American cockroach
Reticulitermes virginicus, subterranean termite

about insect movements on Cyprus (Campion et al. 1977). It was suspected that populations of the agricultural pest, the Egyptian cotton leafworm (*Spodoptera littoralis*), were due in part to immigrants. However a network of 51 virgin female traps throughout the island shows that the spread from small population patches in June to distribution over the whole island by October could be explained by population growth within the island, together with localized movements away from areas of high populations.

Although the traps provide a specific, inexpensive, and lightweight monitoring tool, the interpretation of catch is usually very complex. Trap catches are not always directly proportional to pest populations in the immediate area and so the interpretation of trap catch should be done by experienced advisory agents. Pheromone traps used for monitoring purposes are not subject to insecticide registration requirements; therefore, they have been marketed commercially throughout the world for a large number of pest species.

In addition to pheromone traps for insect monitoring, potent attractants are used in the detection and control of several tephritid fruit fly pests. The substances, which attract males of these species, were discovered by assaying many chemicals. Methyl eugenol is used for Oriental fruit fly (*Dacus dorsalis*), and mango fruit fly (*Dacus zonatus*); cue-lure is used for melon fly (*Dacus cucurbitae*) and Queensland fruit fly (*Dacus tryoni*); and trimedlure is used for Mediterranean frut fly *(Ceratitis capitata)* and Natal fruit fly *(Ceratitis rosa)*. The attractants also are combined with pesticides to remove males from the fly population. Thixotropic liquid formulations that can be applied from the ground or air without leaving solid residues were developed. This technique is used in areas were fly population reduction is necessary, but difficult by other methods.

Pest Control by Trapping

Fruit Pests

A number of pheromone mass trapping tests have been carried out on lepidopterous fruit pests. A four year project with the redbanded leafroller moth (*Argyrotaenia velutinana*) in commercial apple orchards in New York state showed that mass trapping can be effective in controlling this pest, but the technique was not applicable in this situation because of the necessity to control a number of other pests at the same time. Also the cost of the high number of traps that would have been needed for control was prohibitive. The pheromone traps compete with female redbanded leafroller moths for males; thus the traps must be very efficient to prevent males from mating before eventually being captured by a trap. It has been calculated that the trap must capture at least 95% of the male population to effect a subsequent population decrease. If it is assumed that (1) the males are polygamous but do not mate or attempt to mate more than once each 24 hours, (2) the females are monogamous, (3) the overall sex ratio is 1–1, (4) the moths have an 80-day life span, and (5) the traps are equal to two females, it can be calculated that the trap-female ratio needed for 95% control is 2.5–1. It is obvious that a huge number of traps would be needed for high population densities. The optimum trap spacing and density, however, must be determined for each insect and trapping system, because some systems can be made more attractive than live females by releasing higher rates of pheromone and by trapping throughout the males' entire activity period each day or night. In most instances, the trap-out technique is best used when there is a single major pest that must be maintained at a low population density.

In 1969, in a heavily infested 8-ha commercial apple orchard, 2400 pheromone traps captured more than 17,000 redbanded leafroller males in the summer flight. This ratio was well below that needed for control, as indicated by the resulting 32% fruit damage. In another 6-ha orchard with a low pest population level, 1100 pheromone traps captured 723 males in the spring flight and 76 males in the summer flight. There was only 0.1% fruit damage in that year. The test area was enlarged to 16 ha in 1970 and the trap density reduced to 1 trap/tree. The experiment ran for 4 years, and fruit damage was only 0.09% in the test orchard when the test was terminated. A nearby 6-ha check orchard was set up in 1971, and the fruit damage due to redbanded leafroller was up to 12% in the first year.

The codling moth, a worldwide pest of apple, has been the subject of a number of mass trapping experiments. Madsen and Carty (1979) used 10 and 36 traps/ha and concluded that pheromone trapping for this pest is effective only under special circumstances. The codling moth populations must be very low initially, and the orchard must be isolated from outside sources of the pest. It was suggested that the male removal technique could be used with supplemental control measures. Other investigators also obtained insect suppression with mass trapping for codling moth, but not enough to achieve commercial control. MacLellan (1976) used trap densities of 5–12.4 traps/ha to significantly reduce first-generation damage, but not below the 1% commercial economic level. Hagley (1978) showed that a trap density of 34/ha did not control the spring generation adequately and suggested that the technique be combined with insecticide control of first-generation larvae.

Trap-out studies with the grape berry moth (*Endopiza viteana*) also showed that the technique suppresses the pest population, but not to commercially acceptable levels.

Field Pests

In Israel, the Agricultural Extension Service has initiated a large program for commerical control of the citrus flower moth (*Prays citri*). The program involves more than 20,000 pheromone traps in about 250 ha of lemon groves to eliminate insecticide applications from March to May.

Much effort has been made to utilize the mass trapping technique for control of the Egyptian cotton leafworm. An example of one of the large tests on this pest is that reported by Teich et al. (1979). The test plot increased from 1600 ha in 1975 to 1800 ha in 1977. Although a total of 4,171,000 males were captured in 1660 traps in 1977, there was a reported 40% reduction in egg clusters and a 26%

reduction in insecticide treatments. It is difficult to completely assess the efficacy of the mass trapping technique in this case, however, since insecticide treatments were used throughout the test and there was a large increase in the pest population at the end of the summer.

The related pest, *Spodoptera litura*, has been the subject of mass trapping programs in Japan for several years. A trap density of 1–2/ha in more than 2000 ha in 1978 resulted in approximately 40% reduction of insecticide spray applications.

The cotton boll weevil also has been the subject of much pheromone research. It is a major pest, and the use of insecticide for its control has created many other serious pest problems on cotton. One technique studied was to use pheromone traps to aggregate weevils in certain strips of cotton for insecticide treatments. This technique is used in conjunction with other management practices. Another technique is to use pheromone traps in the fields to suppress the weevil population. Combining this technique with a few insecticide applications gave excellent control of this pest. Traps used at a rate of 100/ha captured 76% of the overwintered weevils, estimated to be at a density of 60/ha.

Forest Pests

Much of the research on using attraction for insect control of forest pests has been carried out on various beetle species of Scolytidae. Pheromones and inhibitors have been identified for more than 17 species of these beetles, and several large federally funded projects have been initiated to investigate the potential of these agents in forest management strategies.

The mass trapping experiments with the western pine beetle illustrate the type of effort needed to set up and evaluate a project with a forest pest. The western pine beetle is responsible for an annual loss estimated at 1 billion board feet of ponderosa pine in the western United States. If the beetles do not aggregate successfully above a certain threshold attack density, they do not reproduce. The "trap-out" strategy was tested to determine whether it could be used to suppress mass attacks below the threshold attack density that kills the trees. One study was conducted in 1970 in a 65-km^2 test area at Bass Lake, California. Large vane traps, each consisting of four sticky-coated screen panels (0.76×2.0 m each), were erected 2.3 m above the ground to trap aggregating beetles. The test area was monitored for beetles with traps releasing 2 mg pheromone/ha placed in a 0.8-km grid from May through October. Two suppression plots and two check plots, each 2.56 km^2, were set up within the test area. The suppression traps released 20 mg pheromone/ha, and there were 60 traps/1.2 km^2. In the spring flight, the suppression traps caught an estimated 405,000 beetles and the survey traps caught an estimated 189,000 beetles. Although tree mortality due to the western pine beetle declined from 283 to 91 in the test area, there was an increase of beetle-infested trees in the suppression areas where a high pheromone concentration was present. A complex computer program was developed to analyze the relationship between beetle emergence and attack densities, trap catch, and tree mortality. Another test at McCloud Flats, California, utilized a much larger area and resulted in beetle captures of 2.6 million in 1971 and 4.3 million in 1972. The beetle population was too high for the number of traps in the area, but there was optimism that a better trap-beetle ratio could suppress mass attack densities below some threshold that would kill the trees.

Different trapping techniques were attempted with the European elm bark beetle, which is the principal vector for *Ceratocystis ulmi*, for Dutch elm disease pathogen. The beetle behavior is similar to that of the western pine beetle in that there is a dispersal flight of emerged beetles that can cover several kilometers. The virgin females find elm trees for breeding sites and tunnel into the inner bark while releasing an aggregating pheromone that attracts both sexes to the breeding site. Setting up traps in a grid with 30–50 m intervals in a 520-ha area in Detroit resulted in captures of 4 million beetles, but the Dutch elm disease increased in the test plot from 4.3 to 7.4%.

Other tests utilized the encirclement technique, in which the area is surrounded with pheromone traps to prevent immigration of the beetles. In Evanston, Illinois, 750 traps were used to encircle the city. A total of 3.5 million beetles were trapped and the beetle population within the city was reduced, but there was no corresponding decrease in Dutch elm disease. Other tests are being conducted in California, Colorado, Delaware, Illinois, New York, North Carolina, and Washington, D.C.

Another application of attractants for insect suppression had for its target the ambrosia beetle. These beetles bore pinholes into the sapwood of many coniferous tree species, especially western hemlock and Douglas fir. A test to reduce beetle attack on freshly sawn lumber was conducted in 1974 (McLean and Borden 1977). Two traps baited with sulcatol were placed on each load of freshly sawn lumber and appeared to suppress the number of beetles attacking the lumber.

Attractants have been tested for the control of many other beetle species (see Table 11.2), but the most ambitious program is the recently initiated mass trapping project in Sweden and Norway for *Ips typographus* (O'Sullivan 1979). The program costs for the first year are estimated at $18 million for Norway and $5 million for Sweden, to pay for more than 100 kg of pheromone and almost 1 million traps. The beetle killed an estimated 5 million trees in Norway and 1 million trees in Sweden in 1978. The beetle population density for these countries was projected to be 80 billion for 1979. Since other control measures have not been effective against this pest, it is hoped that the mass trapping effort will provide a means of checking the devastating beetle outburst.

Stored Product Pests

A recent review (Ritter 1979) of sex and food attractants of storage insects describes how mass trapping has become a valuable tool for the protection of stored products. Some of the notorious storage pests are dermestid beetles in the genus *Trogoderma*. Pheromones for many of these species have been identified and found to be various geometric and optical isomers of 14-methyl-8-hexadecanal (trogodermal). Traps baited with trogodermal were successful in capturing male Khapra beetles *(Trogoderma granarium)* in infested grain stores. These traps can be used for detection and surveillance of the pests, as well as for population suppression. The pheromone also can be used in conjunction with insecticides or pathogens. The mortality rate for some dermestid species expose to the spores (6.25×10^7 spores/g of medium) of *Mattesia trogodermae* is 100%. This pathogenic protozoan can be combined with pheromone in a powder to contaminate male beetles. Spores are spread by venereal transmission, by contaminating eggs during oviposition, by cannibalism of infested or dead larvae, and by elimination of spores in feces of infected larvae followed by ingestion of these spores by healthy larvae.

Pheromones are also known for a number of phycitid moth species in the genera *Ephestia* and *Plodia*. Pheromone trapping of these moths in a flour mill from June through October 1976 was successful in capturing 70–97% of the males of the three species present.

Household Pests

Pheromones have been identified and used in traps of various types for the German cockroach (*Blatella germanica*), the American cockroach (*Periplaneta americana*), several subterranean termite species in the genus *Reticulitermes*, and the housefly (*Musca domestica*).

Mating Disruption by Air Permeation

In many cases it is not economically feasible to use mass trapping techniques for the control of one or more insect pests. In some instances it is better to permeate the atmosphere with a low level of the attractant chemicals, to disrupt the natural mating communication system. The disruptant chemicals can confuse the males, or alter pheromone perception by effecting quantitative and qualitative changes in the chemical stimuli.

The technique had been suggested for many years, but it was not until 1967 (Gaston et al. 1967) that a preliminary field test showed that the technique had potential for insect control. The cabbage looper pheromone was evaporated from planchets located 1 m above the ground at 3-m spacing to generate a release rate of 100 mg/hour/ha. This pheromone concentration was sufficiently high to keep males from being lured to female moths in the test plot. Other experiments with the cabbage looper showed that the concentration of pheromone in the air was the critical factor for this species. Evaporators releasing 0.2 μg/hour and spaced 1 m apart were as disruptive as evaporators releasing 4×10^4 μg/hour at 400-m spacing. A release rate of 0.007 mg/hour/ha was found to be too low to disrupt the male cabbage looper moths.

Mating disruption tests subsequently were conducted on many pest species. Much of the effort was devoted to developing a suitable formulation or dispenser. The best controlled-release system would release active material at a constant rate for a specified period of time, with total discharge of all active material. The available dispensers all fall short of this ideal system, but some acceptable dispensers have been developed. Research with the dispensers generally involves testing various release rates per unit area for the effect on the responding insects. These tests are followed by larger area studies in which mating disruption can be assessed by comparing the percentage of mated females in test and check plots and determining the effects on subsequent populations by larval sampling before and after the adult flight in test and check plots and by crop damage determinations in both areas. A reduction in mating in the test plot does not necessarily correspond to population control, because many pest species disperse long distances and could arrive in a test area after mating.

Fruit Pests

Disruption tests have been conducted on a number of pest species, including the redbanded leafroller moth, the plum fruit moth (*Grapholitha funebrana*), the Oriental fruit moth, the lesser appleworm moth (*Grapholitha prunivora*), the codling moth, the grape berry moth, the peachtree borer (*Synanthedon exitiosa*), the European grapevine moth (*Lobesia botrana*), and the summer fruit tortrix moth. One of the first large plot tests was conducted in Australia with the Oriental fruit moth in apple orchards. Pheromone was released from closed polyethylene microcapillary tubes at a rate of 5–12 mg/hour/ha. It was found that the disruption method significantly reduced fruit damage and shoot damage compared to an orchard that received six insecticide applications.

The codling moth is another pest that has received much attention with respect to the mating disruption technique. It was found that high concentrations of pheromone decreased the number of matings with confined or tethered insects in the laboratory and in the field. Small plot tests with pheromone released from hollow fibers at a rate of 2 mg/hour/ha was effective in preventing males from being lured into female-baited traps and from mating with tethered virgin female moths. A larger plot test utilizing 87,500 chopped hollow polymeric fibers per hectare, with a pheromone release rate of 13 mg/hour/ha, showed that the disruption technique could provide codling moth control equal to that of a standard insecticide control program.

Vegetable and Field Pests

Disruption tests for the control of vegetable and field pests have been conducted on many species, including the cotton boll weevil, the pink bollworm (*Pectinophora gossypiella*), the red bollworm (*Diparopsis castanea*), the black cutworm (*Agrotis ipsilon*), the cabbage looper (*T. ni*), the European corn borer (*Ostrinia nubilalis*), the corn earworm (*Heliothis zea*), the tobacco budworm (*Heliothis virescens*), the fall armyworm (*Spodoptera frugiperda*), Japanese beetle, and the Egyptian cotton leafworm. The tests are similar to those of the fruit pests, with a variety of microcaps, hollow fibers, plastic laminates, and polyethylene capillary tubes used to release pheromone at certain rates. The smaller test plot experiments have generally been successful in showing that pheromone components can be released at rates (~2–20 mg/hour/ha) that disrupt mating. Efficacy tests have been the most complete with the pink bollworm. This pest has received considerable attention because the extensive use of chemical insecticides kills predators and parasitoids and usually leads to outbreaks of other pests, such as *Heliothis* species. It is a serious pest in most of the cotton-producing countries in the world, including the desert Southwest in the United States.

Pink bollworm disruption was tested in 1600 ha of cotton in the Coachella Valley of California in 1974. The results were encouraging, and a follow-up experiment in 1976 gave excellent results. Pheromone was released from hollow fiber hoops distributed on a 1 × 1 m grid throughout three fields of cotton (23 ha). A total of 230,000 hoops were used for five applications between mid-May and early September. A total of 33 g of pheromone/ha provided pink bollworm control equal to that of neighboring fields receiving an average of two to six insecticide applications per hectare.

A commercial company (Conrel Co.) developed the technology to mechanize the technique by dispensing sticky-coated chopped fibers from airplanes. In 1977 they utilized their pheromone disruption system on 9300 ha of cotton in the southwestern United States. Their data showed that growers were able to reduce conventional insecticide usage in these fields by 50–80% and that the attending relief in chemical stress on the plant and ecosystem resulted in lint yield improvements of as much as 20–50% in conditions of intense pink bollworm pressure coupled with outbreaks of *Heliothis* spp.

In 1978 efficacy was successfully shown by obtaining data on disruption of male orientation to pheromone traps, data on differences in levels of larval infestation between treated and untreated cotton fields, and data on mating disruption from mating table experiments. In 1978 a total of 51 g of pheromone/

ha was used for pink bollworm control from May to September. In February 1978 the EPA granted Conrel Company registration of this pheromone as formulated in a hollow fiber for protection of a field crop. In 1979 the pheromone disruption system also was used successfully for pink bollworm control in several South American countries.

Forest Pests

Disruption tests have been conducted on various forest pest species, including the gypsy moth, the eastern spruce budworm (*Choristoneura fumiferana*), the Douglas fir tussock moth (*Orgyia pseudotsugata*), the western pine shoot moth (*Eucosma sonomana*), the Nantucket pine tip moth (*Rhyacionia frustrana*), the southern pine beetle, the Douglas fir beetle, and the pine engraver beetle.

Research with the lepidopteran pests was similar to that reported above, but the determination of efficacy was plagued by the problems discussed in connection with mass trapping for the control of forest pests. Much effort has been put into the use of pheromones for the disruption of the gypsy moth and the eastern spruce budworm, but conclusive proof of efficacy has been lacking. A variety of release systems (e.g. microcaps, plastic laminates, and hollow fibers) have been used in these studies. Efficacy has been assessed by reductions in mating of populations of females that have been simulated by artificial placement or by reductions in egg densities or egg masses in moderate to sparse natural populations. Several tests have shown that the disruption technique can successfully suppress mating in densities that simulate a sparse population, but it is difficult to conclude what overall effect this control strategy might have on the natural population dynamics of these pests.

A number of tests have been conducted for mating disruption of beetle species in the forest, but, again, it is difficult to prove efficacy. Experiments have included dispensing frontalure-soaked rice seeds by aircraft over a pine forest for suppression of the southern pine beetle, releasing ipsenol from capillary tubes for suppression of the pine engraver beetle, and various tests in which chemicals are emitted near traps to suppress trap catch.

With some insect species, certain chemicals have been found that decrease the responding insects' responses to the sex pheromone. With beetle species, these antiaggregative chemicals have been used in disruption tests to reduce mass attack levels of certain species. An example is the use of 3-methyl-2-cyclohexen-1-one (MCH) as an antiaggregation compound for the Douglas fir beetle. This beetle is a major pest of the Douglas fir throughout western United States and British Columbia. The beetles breed in epidemic numbers in damaged trees, and the progeny attack en masse and kill selected trees. Once again the female beetles are the pioneers that make an initial attack on a new host. They undergo a dispersal flight until a lowered lipid content results in a lowered threshold level for host volatile responses. The beetles then are attracted to volatiles from uninfested Douglas fir trees and initiate gallery construction while releasing aggregating pheromone. A combination of the pheromones frontalin and seudenol and host volatiles, such as camphene, α-pinene, and ethanol, aggregates beetles to the host tree. The antiaggregating chemical MCH is eventually released to prevent more beetles from attacking the tree and causing an oversaturation of that host tree.

Experiments were conducted to determine the optimum release rate and spacing of evaporators of MCH to prevent an increase in Douglas fir beetle populations in attractive, susceptible host trees. An optimum rate of 50–100 mg/hour/ha was found in experiments with freshly felled Douglas fir trees. The use of an antiaggregating chemical has the advantage over attractant pheromones of not aggregating beetles in trees within or surrounding the test area.

Stored Product Pests

Mating disruption tests have been conducted for a number of *Ephestia* and *Plodia* species, particularly the almond moth (*Cadra cautella*) and the Indian meal moth. It was found that air permeation with (*Z,E*)-9,12-tetradecadienyl does reduce the mating frequency and rate of population growth with these species. The success of this technique was dependent on the population density, and, apparently, did not reduce mating in rooms with high numbers of insects.

A study with the black carpet beetle (*Attagenus megatoma*) showed the potential of the pheromone megatomic acid for disrupting mating of this pest in warehouses.

Pheromone Usage in the Future

It is obvious that pheromones are available for hundreds of pest species and that much research has been initiated for their use in pest management systems. Success of the huge government programs in various countries could open the way for additional programs on an international scale. Usage on a commercial basis depends on at least two major factors: the determination of efficacy and the registration of the pheromone as an insecticide. Proof of efficacy of pheromones can be difficult because appropriate checks are not always possible. It is important to realize, however, that the pheromone treatment does not necessarily have to result in 100% control of an insect pest, but it can have an important role in pest management systems in delaying pest population buildups or in maintaining the population density at some tolerable level. Pheromones would be used in conjunction with selected insecticides or other insect control agents in strategies designed to maintain natural parasitoids and predators as much as possible.

Registration of pheromones has been a frustrating exercise in the past few years, but changes are being made. Pheromones differ from insecticides in many ways and logically the registration procedures established for them should reflect these differences. One of the conclusions of a NATO-sponsored Advanced Research Institute on Chemical Ecology in 1978 (Ritter 1979) was that ''such chemicals which must provide insectistasis without insecticidal action should be clearly distinguished from insecticides'' by the regulatory agencies. In May 1979 a workshop in Sweden, ''Environmental Protection and Biological Control of Pest Organisms,'' involved regulatory agency representatives from nine countries. It was agreed that guidelines for testing biological control agents should be developed and that biological control agents present a lower risk potential than chemical pesticides.

On May 14, 1979, the *Federal Register* summarized the U.S. Environmental Protection Agency's proposed approach to registration of biorational pesticides. In the proposed policy, it was stated that the EPA would recognize biorational pesticides as different from conventional pesticides and that the agency would take into account the different modes of action and lower risks of adverse effects. It was further stated that the EPA would facilitate the registration of the biorational pesticides by assuring that requirements for the registration of these substances are appropriate to their nature and are not unduly burdensome. If the requirements take into account the low usage rate, the volatility, and the low toxicity of pheromone compounds, registration should be expedited.

In summary, research efforts involving pheromone chemistry, in-depth behavioral studies, and efficacy tests in the field now underway on a number of pest species are providing the background data necessary for the development of a new insect control technique. Commercial companies and federal agencies already are using this technique for several pest species, and new pheromone registration guidelines should open the door for commercial usage of a number of other pheromones in the next few years.

REFERENCES

Alford, D. V., P. W. Carden, E. B. Dennis, H. J. Gould, and J. D. R. Vernon. 1979. Monitoring codling and tortrix moths in United Kingdom apple orchards using pheromone traps. Ann. Appl. Biol. 91:165–178.

Baker, T. C., and R. T. Cardé. 1979. Courtship behavior of the Oriental fruit moth (*Grapholitha molesta*): Experimental analysis and consideration of the role of sexual selection in the evolution of courtship pheromones in the Lepidoptera. Ann. Entomol. Soc. Am. 72:173–188.

Butenandt, A., R. Beckmann, D. Stamm, and E. Hecker. 1959. Über den Sexuallockstoff des Seidenspinners *Bombyx mori*. Reindarstellung und Konstitution. 2. Naturforschung 146:283–284.

Campion, D. G., B. W. Bettany, J. B. McGinnigle, and L. R. Taylor. 1977. The distribution and migration of *Spodoptera littoralis* (Boisduval) (Lepidoptera: Noctuidae), in relation to meterology on Cyprus, interpreted from maps of pheromone trap samples. Bull. Entomol. Res. 67:501–522.

Cardé, R. T., T. C. Baker, and W. L. Roelofs. 1975. Behavioral role of individual components of a multichemical attractant system in the Oriental fruit moth. Nature (London) 253:348–349.

Gaston, L. K., H. H. Shorey, and C. A. Saario. 1967. Insect population control by the use of sex pheromones to inhibit orientation between the sexes. Nature (London) 213:1155.

Hagley, E. A. C. 1978. Sex pheromones and suppression of the codling moth (Lepidoptera: Olethreutidae). Can. Entomol. 110:781–783.

Karlson, P., and A. Butenandt. 1959. Pheromones (ectohormones) in insects. Annu. Rev. Entomol. 4:39.

Karlson, P., and M. Luscher. 1959. "Pheromones": A new term for a class of biologically active substances. Nature (London) 183:55.

Kraemer, M., H. C. Coppel, F. Matsumura, T. Kikukawa, and K. Mori. 1979. Field responses of the white pine sawfly, *Neodiprion pinetum,* to optical isomers of sawfly sex pheromones. Environ. Entomol. 8:519–520.

Light, D. M., and M. C. Birch. 1979. Inhibition of the attractive pheromone response in *Ips paraconfuses* by (*R*)-(−)-ipsdienol. Naturwissenschaften 66:159–160.

MacLellan, C. R. 1976. Suppression of codling moth (Lepidoptera: Olethreutidae) by sex pheromone trapping of males. Can. Entomol. 108:1037–1040.

Madsen, H. F., and B. E. Carty. 1979. Codling moth (Lepidoptera: Olethreutidae) suppression by male removal with sex pheromone traps in three British Columbia orchards. Can. Entomol. 111:627–630.

McLean, J. A., and J. H. Borden. 1977. Attack by *Gnathotrichus sulcatus* (Coleoptera: Scolytidae) on stumps and felled trees baited with sulcatol and ethanol. Can. Entomol. 109:675–689.

Miller, J. R., and W. L. Roelofs. 1978. Sustained-flight tunnel for measuring insect responses to wind-borne sex pheromones. J. Chem. Ecol. 4:187–198.

O'Sullivan, D. A. 1979. Pheromone lures help control bark beetles. C&E News, July 30. 10–13.

Ritter, F. J. 1979. Chemical Ecology: Odour Communication in Animals. Elsevier/North Holland, New York.

Roelofs, W. L. 1978. Threshold hypothesis for pheromone perception. J. Chem. Ecol. 4:685–699.

Roelofs, W. L. 1979a. Electroantennograms. Chemtech 9:222–227.

Roelofs, W. L. 1979b. Pheromone trap specificity and potency. Proc. Conf. Movement of Selected Species of Lepidoptera in the Southeastern U.S.A., Raleigh, NC, April 9–11.

Roelofs, W. L., and R. T. Carde. 1977. Responses of Lepidoptera to synthetic sex pheromone chemicals and their analogs. Annu. Rev. Entomol. 22:377–399.

Roelofs, W. L., M. Gieselmann, A. Cardé, H. Tashiro, D. Moreno, C. Henrick, and R. Anderson. 1978. Identification of the California red scale sex pheromone. J. Chem. Ecol. 4:211–224.

Shorey, H. H., and J. J. McKelvey, Jr. (eds.) 1977. Chemical Control of Insect Behavior: Theory and Application. Wiley, New York.

Silverstein, R. M., J. V. Rodin, and D. L. Wood. 1966. Sex attractants in the frass produced by male *Ips confusus* in ponderosa pine. Science 154:509–510.

Teich, I., S. Neumark, M. Jacobson, J. Klug, A. Shani, and R. M. Waters. 1979. Mass trapping of males of Egyptian cotton leafworm (*Spodoptera littoralis*) and large-scale synthesis of prodlure. P. 343–352. *In* F. J. Ritter (ed.), Chemical Ecology: Odour Communication in Animals. Elsevier/North Holland, New York.

SECTION IV

CHEMISTRY AND EVOLUTION OF SEMIOCHEMICALS

CHAPTER TWELVE

CHEMISTRY OF SEMIOCHEMICALS INVOLVED IN PARASITOID–HOST AND PREDATOR–PREY RELATIONSHIPS

RICHARD L. JONES

Department of Entomology, Fisheries, and Wildlife
University of Minnesota
St. Paul, Minnesota

In their search for suitable hosts or prey, entomophagous insects take advantage of a wide variety of semiochemicals produced by or otherwise associated with those hosts or prey. Chemically, these products range from long-chain, nonpolar, hydrocarbons to short-chain, polar, carboxylic acids. Functionally, they serve a wide range of roles in the hosts or prey, including pheromones, intermediary metabolites, components of the epicuticle, and excretory products.

Allelochemics affecting the behavior of several parasitoids and predators have been identified. In this chapter I discuss these chemicals in the contexts of identification, dose-activity and structure-activity relationships, type of activity effected, sources, and primary functions. In keeping with a biological theme, I discuss the chemicals in a format that follows a general pattern of host-seeking behavior for a parsitoid or predator.

Research on the identity and activity of allelochemics that mediate host or prey selection by entomophagous insects has been relatively scant, but interest has increased in recent years. There were a variety of reasons for this lack of research effort. One of these was our failure to recognize the importance of allelochemics in the behavior of entomophages. Closely related was our failure to fully appreciate intertrophic effects as discussed by Price (Chapter 13). In addition, an intense interest in insect pheromones has occupied most of the research efforts of those scientists and laboratories with the training and equipment to conduct chemical identification research.

HABITAT LOCATION

In seeking an environment that is likely to contain hosts or prey, an entomophage might use a variety of physical or chemical stimuli. This subject is discussed at length by Vinson (Chapter 4). The specific roles of chemicals in these processes have not been fully determined. All other factors being equal, we would expect semiochemicals to play a larger role in habitat location for a parasitoid with a monophagous host than for a parasitoid with a polyphagous host. The qualification is very important, however, since in most cases all factors are not equal. For example, Arthur (1962) showed that *Itoplectis conquisitor* preferred the odor of Scots pine to that of red pine, both of which are host trees for the European pine shoot moth (*Rhyacionia buoliana*), the parasitoid's host. Further investigation revealed that, because of bud structure and size, and the resultant host protection by red pine, the success rate of the parasitoid was higher on Scots pine. A preference for Scots pine naturally followed. Thus, although the pine shoot moth is polyphagous, its parasitoid exhibits a strong habitat preference.

A semiochemical can function in either of two ways to effect habitat location by entomophages. It may serve as an attractant to an individual located outside the habitat, or as an arrestant to a wandering individual that passes through the habitat. In addition to those produced by plants, a number of semiochemicals produced by an entomophage's hosts or prey might also function in either of these categories.

To date, two identified chemicals have been implicated in habitat location by parasitoids. Greany et al. (1977) isolated and identified ethanol from rotting peaches and demonstrated its role as an attractant for *Biosteres longicaudatus*, a braconid solitary endoparasitoid of peach-infesting tephritid fruit fly larvae. The bioassay measured the relative attraction of the females to a filter paper substrate containing the test chemical and to a filter paper control. These greenhouse bioassays also demonstrated that the parasitoids responded only under the high-intensity light and ample room for flight provided in the greenhouse. Laboratory bioassays were negative.

Subsequent tests showed that acetaldehyde, a common oxidation product of ethanol, was about 50 times as active as ethanol. Acetic acid, although active, was much less active than ethanol. The optimum response was observed with 100 μg of acetaldehyde, to which 96% of the responding parasitoids were attracted. Variation from this amount of chemical in either direction decreased attraction. An optimum response of 90% of the responding parasitoids was observed with 50 and 100 mg of ethanol, whereas 10 mg of acetic acid attracted only 73% of the responding parasitoids. (A 50% response represented no effect in the bioassays.) Other one-, two-, and three-carbon alcohols, aldehydes, and acids were not active. Since the purity of the chemicals was not stated, it is possible that the attraction to ethanol was due to impurities of acetaldehyde. The chemicals were not tested against each other to measure relative attraction, and no synergism tests were reported.

Read et al. (1970) determined that allyl isothiocyanate, a volatile constituent of crucifers, was very attractive to *Diaeretiella rapae*, a braconid parasitoid of crucifer-feeding aphids. In laboratory tests in a Y-tube olfactometer, females showed an optimum response to an unspecified quantity of a $10^{-5}\%$ aqueous solution of the chemical. Dose-response tests produced a bell-shaped curve with no responses to 10^{-6} and $10^{-2}\%$ solutions. A $10^{-1}\%$ solution repelled the parasitoids.

In regard to mode of action, it is evident that allyl isothiocyanate is a true attractant for *D. rapae* and that it plays a significant role in habitat selection by this parasitoid. In the case of *B. longicaudatus,* the case is not as clear. The bioassay used did not completely eliminate the possibility of aggregation via an arrestant. However, the fact that the parasitoid must be in flight to respond plus the author's observations of its behavior strongly indicate that the primary mode of action of acetaldehyde and ethanol is attraction.

In the generalized parasitoid search behavior sequences shown in Figure 12.1, both these groups of chemicals would function as cues (S_2) to elicit habitat-scanning activity.

SEARCH STIMULATION

Following the habitat selection process, a parasitoid is behaviorally responsive to stimuli (S_3, Figure 12.1) that indicate the presence of a host in the environment. Although semiochemicals that function as cues later in the sequence might also act at this point, at least one group of chemicals has been identified that appears to act exclusively as S_3-type stimuli (Lewis et al. 1975, 1979). These chemicals (docosane, tricosane, tetracosane, and pentacosane) were isolated and identified by Jones et al. (1973) from the scales of adult *Heliothis zea*. The chemicals elicited an increased rate of parasitism by the egg parasitoid *Trichogramma evanescens*, when applied to the host egg substratum. The best evidence indicates that these chemicals, of which tricosane is the most active, act via positive orthokinesis. Since *Trichogramma* spp. locates its host through vision, the result is an increased rate of host location and, consequently, parasitism. Bioassays in the laboratory and greenhouse demonstrated that tricosane applied to petri dish substrates at a rate of 6 pg/cm^2 or to pea seedling substrates at a rate of 250 pg/cm^2 increased parasitism up to twofold. No decreased activity with increased doses of these chemicals has been observed.

The source of these chemicals is probably the host epicuticle, although the

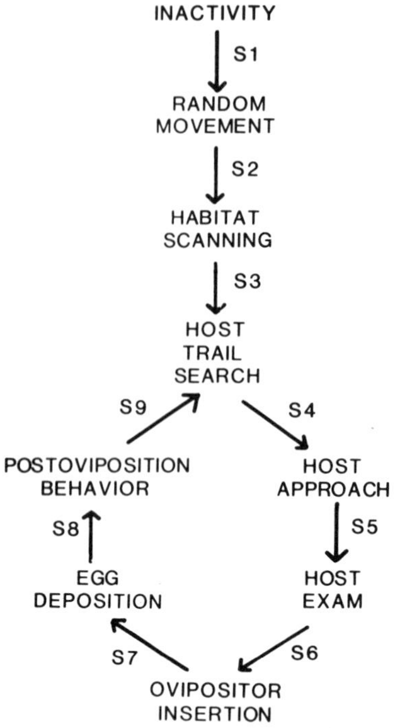

FIGURE 12.1 A general scheme of parasitization behavior with stimuli (S_1-S_9).

possibility that a specialized glandular secretion exists cannot be eliminated. If there is such a secretion, the chemicals probably function as some type of pheromone for the host.

HOST LOCATION

A number of chemicals have been identified that act as cues for parasitoids in locating hosts. When the host insect is mobile, (e.g. leaf-feeding immatures), certain chemicals associated with the host act as trail-following kairomones. Such cues would function as S_4 stimulants (Figure 12.1). These chemicals have been identified for two-host parasitoid systems.

Jones et al. (1971) identified 13-methylhentriacontane as the major chemical component of *H. zea* frass that elicits intense antennation and trail-following behavior in the parasitoid *Microplitis croceipes*. Bioassays were conducted by allowing the parasitoid to walk across filter paper to which the chemical had been applied in a 1-cm-diameter circle. Antennation of this spot by the parasitoid denoted a positive response. In bioassays conducted with 150 ng of chemical, 13-methylhentriacontane elicited a response of 2.3 (based on a system of 3.0—strongest and 0.0—no response), and scores of 0.0, 0.2, 0.5, and 1.7 were obtained for the 9-, 11-, 12-, and 15-methyl analogues, respectively, all of which were present in host frass. These results indicated a strong degree of specificity with regard to the position of the methyl group. Branched hydrocarbons of other chain lengths, although present in the host frass, elicited no response in the bioassay.

Tests among the methyl branched hentriacontanes revealed no synergism. Instead, mixtures of 50 ng of 13- and 15-methylhentriacontanes produced a score that was the average of the chemicals tested alone (0.9, 1.2, and 0.6). Thus the chemicals were diluting one another.

A series of closely related chemicals indentified by Vinson et al. (1975) exhibited a similar mode of action, as described for *M. croceipes*, in the parasitoid-host association of *Cardiochiles nigriceps* and *Heliothis virescens*. These chemicals, a series of methyl branched hen-, do-, and tritriacontanes, were isolated from the host's mandibular glands. Bioassays were conducted in the same fashion as described above for *M. croceipes* and the results are presented in Figure 12.2. The most active analogues, when tested at the 5-μg level, are 11-methylhentriacontane (C_{31}), 12-methyldotricontane (C_{32}), and 16-methyltritriacontane (C_{33}). These tests indicate that *C. nigriceps* does not exhibit the high degree of specificity for the methyl group position shown by *M. croceipes*. It is also interesting to note that excluding the 16-methyl C_{32} analogue, the most active methyl positions are 11-, 12-, and 13- for the C_{31}, C_{32}, and C_{33} analogues, respectively. In these cases the methyl groups are all equidistant from the other end of the molecule, perhaps implying that this is the critical part of the molecule for the receptor mechanism. The most active chemicals were subjected to a dose-response test, and these results are presented in Figure 12.3. A bell-shaped curve was found for all compounds, 5 μg most often being the optimum dose. However, the optimum dose for the 13-methyltritriacontane was 500 ng and, as illustrated in the figure, it is by far the most active chemical.

These compounds demonstrated no synergism when the methyl position was varied within the same chain. However a high degree of synergism was evident among the compounds with different chain lengths. The most active combinations are shown in Figure 12.3. In all cases the optimum dose was 500 ng (total for the three chemicals) with the 12-methyl C_{31}−12-methyl C_{32}−13-methyl C_{33} mixture being the most active.

In both the cases of *M. croceipes* and *C. nigriceps*, these hydrocarbons act as S_4-type cues (Figure 12.1). These chemicals also play an undefined role as S_5 and S_6 cues. For example, *M. croceipes* will insert the ovipositor into an unnatural host such as the European corn borer (*Ostrinia nubilalis*) only if

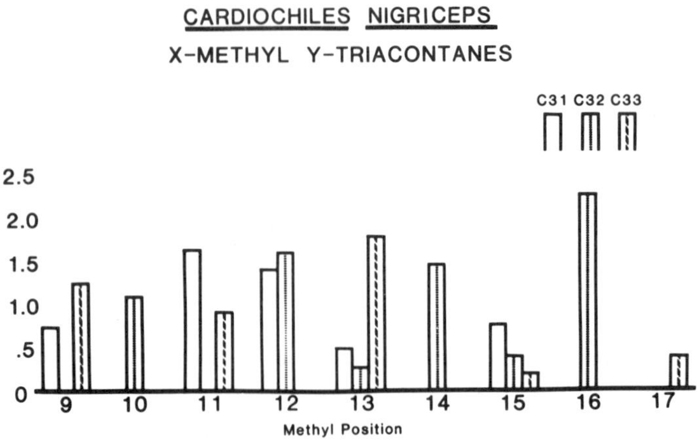

FIGURE 12.2. Response (Y) of *Cardiochiles nigriceps* to the methyl branched hydrocarbons C31, C32, and C33 (X).

13-methylhentriacontane has been applied to the corn borer cuticle. Also, *C. nigriceps* will sometimes unsheath the ovipositor and probe in the presence of its hydrocarbon cues. The effect of steroisomerism in these chemicals has not been studied.

The methyl branched hydrocarbons are present in the insect cuticle as well as in the mandibular glands. Their presence in frass is due to mandibular gland secretion of cast exocuticle consumption. In the cuticle they function as components of the epicuticular barrier to water movement. In the mandibular gland, they might serve to aid food mastication. In addition, they may well function as dispersal or ''epideictic'' pheromones as demonstrated by Corbet (1971) for *Anagasta kuehniella* larvae (Chapter 10).

The subject of long-chain hydrocarbons and insects has recently been reviewed by Nelson (1978). These compounds are widespread in the natural environment, occurring in a wide variety of insects as well as in plants. Thus it is evident that the context in which a parasitoid detects these chemicals is very important in host location.

Tucker and Leonard (1977) reported the use of long-chain hydrocarbons as S_5- and S_6-type cues for *Brachymeria intermedia*, a solitary chalcid endoparasitoid of lepidopteran pupae. These chemicals are present in the cuticle of the gypsy moth, *Lymantria dispar*. Among these hydrocarbons were heptacosane, nonacosane, and several dimethyl compounds in the pentatriacontane–tetracontane range. Of these, the dimethylpentatricontane analogue elicited the most activity in a bioassay that measured drumming activity by the parasitoid. Dose-activity or synergism tests with these chemicals have not been reported. As

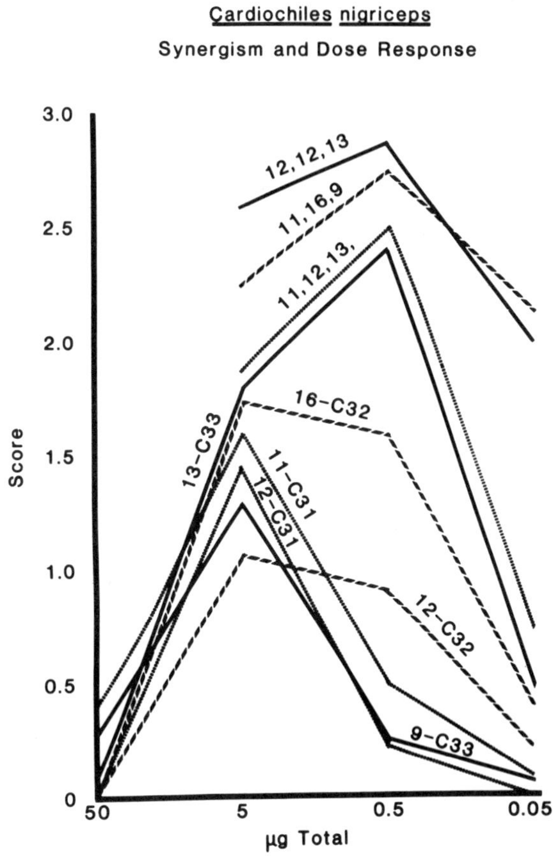

FIGURE 12.3. Synergism and dose response (Y) of *Cardiochiles nigriceps* to methyl branched hydrocarbons C31, C32, and C33 (X).

in the cases previously discussed, these chemicals serve a water barrier role for the pupae.

Bracon mellitor, a braconid ectoparasitoid of the boll weevil (*Anthonomus grandis*) uses several cholesterol esters produced by its host to determine host location. When exposed to the chemical, the parasitoid antennates the area and, if sufficiently stimulated, it will probe with the ovipositor (S_5 and S_6 cues). Henson et al. (1977) isolated and identified these cholesterol esters from the frass of the host larvae. They found ester of palmitic, stearic, oleic, linoleic, and linolenic acids; linoleate, the preponderant ester, accounted for 46% of the total. Bioassay of synthetic cholesterol linoleate elicited probing activity and thus confirmed the identification. Since this was the only synthetic tested, no structure-activity data are available. Additionally, there are no published data concerning dose responses to these chemicals.

As pointed out by the authors, these esters are common insect metabolic products. Obviously, selectivity is dependent on the exact ratios or quantities of these esters or their context, or on other mechanisms.

Another parasitoid that must probe for a hidden host, for which a kairomone has been identified, is *Orgilus lepidus*, a solitary braconid endoparasitoid of the potato tuberworm (*Phthorimaea operculella*). The host frass contains a chemical that elicits antennation and ovipositor probing by the parasitoid (S_5 and S_6, Figure 12.1). Hendry et al. (1973) isolated and identified the active component as heptanoic acid. This chemical elicited a bell-shaped dose-response curve from the parasitoid with an optimum response to 100 ng and no response to the 1-ng and 100-μg levels (Figure 12.4). A comparison of analogues showed that hexanoic acid elicited moderate activity, whereas valeric and octanoic acids showed a minimum activity (Figure 12.4). The authors report that the mandibular gland of the host is the probable source of the chemical. Its function there is unknown.

There are a number of reports of predators using host pheromones as cues for host location. These include the attraction of predacious clerids to bark beetles via bark beetle aggregation pheromones. Such cases have been reported for

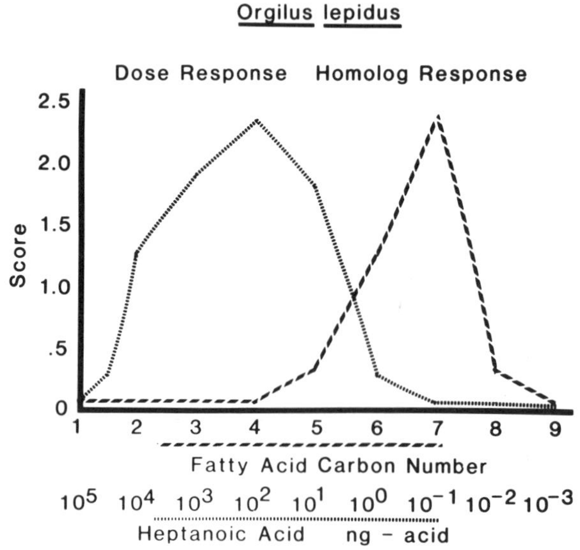

FIGURE 12.4. Response (Y) of *Orgilus lepidus* to dose and analogues of heptanoic acid (X).

Enoclerus lecontei, and the pheromone blend of *Ips paraconfusus* (Wood et al. 1968), as well as other bark beetle pheromones (Lanier et al. 1972). In the Wood study, *E. leconti* responded to ipsenol, ipsdienol, and mixtures of these two alone and in combination with *cis*-verbenol.

Thanasimus dubius and *Thanasimus undatulus* were reported to be attracted to frontalin, a pheromone of *Dendroctonus* species (Vité and Williamson 1970, Pitman 1973, Kline et al. 1974, Dyer et al. 1975). Bakke and Kvamme (1978) reported the attraction of *Thanasimus formicarius* and *Thanasimus rufipes* to a mixture of 3-methyl-3-buten-2-ol, *cis*-verbenol, and ipsdienol, the pheromone components of *Ips typographus*.

Other kairomones, which appear to be proteinaceous, have been isolated (Weseloh and Bartlett 1971, Weseloh 1977). In the first study, a proteinlike chemical (m.w. >150,000) isolated from the brown soft scale (*Coccus hesperidum*) elicited tapping, drilling, and probing behavior by an encyrtid polyphagous hyperparasitoid, *Cheiloneurus noxius*. This chemical is thus acting as an S_6-S_7 type of cue. In the latter study (Weseloh 1977), a proteinlike molecule isolated from the silk gland and silk of the gypsy moth was shown to elicit antennation and probing response by the braconid larval parasitoid *Apanteles melanoscelus*. This chemical acts as an S_6 cue, but in addition, it is probably active as an S_4 and S_5 cue. Since the source of this water-soluble protein is the silk gland, it is a sericin- or fibrinogenlike protein.

EGG DEPOSITION

Arthur et al. (1972) and Hegdekar and Arthur (1973) reported that certain components of the hemolymph of *Galleria mellonella* are essential to induce egg deposition following ovipositor insertion by the pupal parasitoid *I. conquisitor*. These correspond to S_7-type cues in Figure 12.1. The bioassays were conducted with Parafilm® tubes in the shape of pupa, which contained sample solutions. The parasitoid was allowed to oviposit in the tubes and the eggs were counted. Their results indicated that the quality and quantity of salts and amino acids, along with the proper pH, are the primary inducers of egg deposition by this parasitoid. Arginine, isoleucine, methionine, lysine, and leucine (0.5 M) in combination with serine (0.5 M) were the most active amino acids. Magnesium in the form of magnesium chloride at levels of 0.025–0.1 M increased deposition by 300%. Other ions had no activity or acted as inhibitors. A test of pH response showed an optimum response at pH 7 with a bell-shaped response curve. Oviposition was inhibited below pH 4 and above pH 9.

Another S_7-type of cue has been identified from leaves of the oak (*Quercus robur*). Hassell (1968) reported that the tachinid *Cyzenis albicans*, a parasitoid of the winter moth (*Operophtera brumata*) oviposits in the presence of sugars exuded by damaged leaves of that tree. Although no sugar was isolated with a bioassay, damaged leaves were shown to contain sucrose and fructose in the sap, and sucrose and honey were shown to simulate damaged leaves in eliciting egg deposition. Such a semiochemical would be called a synomone according to Nordlund and Lewis (1976) and Chapter 2. The primary function of simple sugars in the leaf is, of course, energy transport and storage.

A number of other semiochemicals have been characterized but not identified as cues for parasitoid oviposition. Nettles and Burks (1975) isolated and described a protein (m.w. ~ 30,000) from the frass of *H. virescens* larvae that acts as a cue for the larviposition of the tachinid *Archytas marmoratus*. This chemical serves as an S_7-type cue for this parasitoid. The authors reasoned that a nonvolatile S_7 cue ensured that larviposition took place in close proximity of the host, thus favoring survival. The authors did not speculate on the origin of this protein. They did determine that it was not of plant (host diet) origin. It seems likely that it could be an enzyme secreted into the gut, or perhaps it is a constituent of the peritrophic membrane. It is also possible that the protein is produced by a gut symbiont.

CONCLUSION

It is apparent from this chapter and others in this book that any habitat is saturated with chemicals from a wide variety of sources. Any insect species within a given habitat has evolved the facilities to use the chemicals there to its advantage. The result of this adaptation is that a given chemical can elicit quite different behavioral responses from a wide variety of species. This is exemplified by bark beetle pheromones and predators as discussed above, as well as by the combination kairomone–epideictic pheromone isolated by Mudd and Corbet (1973). Less specific examples have also been reported. In addition, some semiochemicals are closely related chemically to other semiochemicals. For example, Z-9-tricosene, a pheromone of the housefly (*Musca domestica*) is closely related to tricosane, a kairomone for *Trichogramma* spp. In fact, the monounsaturated C_{20}–C_{25} hydrocarbons have some kairomone activity for *Trichogramma* spp. (Jones 1975, unpublished).

These facts have important implications in the use of chemicals as behavior modifiers in insect management programs. For example, the use of an insect pheromone disruption or trapping technique could have a significant, and perhaps more intense, impact on any number of parasitoids and/or predators that might be equally involved in other plant-herbivore systems. Or the use of a semiochemical analogue might introduce a legitimate semiochemical for other species. Thus, evaluation and efficacy studies of semiochemicals must include studies of the total environmental impact or we may find that 10–15 years hence, we have more problems than solutions.

REFERENCES

Arthur, A. P. 1962. Influence of host tree on abundance of *Itoplectis conquisitor* (Say) (Hymenoptera: Icheumonidae), a polyphagous parasitoid of the European pine shoot moth, *Rhyacionia buoliana* (Schiff.) (Lepidoptera: Olethreutidae). Can. Entomol. 94:337–347.

Arthur, A. P., B. M. Hegdekar, and W. W. Batsch. 1972. A chemically defined, synthetic medium that induces oviposition in the parasite *Itoplectis conquisitor* (Hymenoptera: Ichneumonidae). Can. Entomol. 104:1251–1258.

Bakke, A., and T. Kvamme. 1978. Kairomone response by the predators *Thanasimus formicarius* and *Thanasimus rufipes* to the synthetic pheromone of *Ips typographus*. Norw. J. Entomol. 25:41–43.

Corbet, S. A. 1971. Mandibular gland secretion of larvae of the flour moth, *Anagasta kühniella,* contains an epideictic pheromone and elicits oviposition movements in a hymenopteran parasite. Nature (London) 232:481–484.

Dyer, E. D. A., P. M. Hall, and L. Safranyik. 1975. Numbers of *Dendroctonus rufipennis* (Kirby) and *Thanasimus undatulus* Say at pheromone-baited poisoned and unpoisoned trees. J. Entomol. Soc. Br. Columbia 72:20–22.

Greany, P. D., J. H. Tumlinson, D. L. Chambers, and G. M. Boush. 1977. Chemically mediated host finding by *Biosteres (Opius) longicaudatus,* a parasitoid of tephritid fruit fly larvae. J. Chem. Ecol. 3:189–195.

Hassell, M. P. 1968. The behavioral response of a tachinid fly (*Cyzenis albicans* (Fall.)) to its host, the winter moth (*Operophtera brumata* (L.). J. Anim. Ecol. 37:627–639.

Hegdekar, B. M., and A. P. Arthur. 1973. Host hemolymph chemicals that induce oviposition in the parasite *Itoplectis conquisitor* (Hymenoptera: Ichneumonidae). Can. Entomol. 105:787–793.

Hendry, L. B., P. D. Greany, and R. J. Gill. 1973. Kairomone mediated host-finding behavior in the parasitic wasp *Orgilus lepidus*. Entomol. Exp. Appl. 15:471–477.

Henson, R. D., S. B. Vinson, and C. S. Barfield. 1977. Oviposition behavior of *Bracon mellitor* Say, a parasitoid of the boll weevil (*Anthonomus grandis* Boheman). III. Isolation and identification of natural releasers of ovipositor probing. J. Chem. Ecol. 3:151–158.

Jones, R. L., W. J. Lewis, M. C. Bowman, M. Beroza, and B. A. Bierl. 1971. Host-seeking stimulant for parasite of corn earworm: Isolation, identification and synthesis. Science 173:842–843.

Jones, R. L., W. J. Lewis, M. Beroza, B. A. Bierl, and A. N. Sparks. 1973. Host-seeking stimulants (kairomones) for the egg parasite, *Trichogramma evanescens*. Environ. Entomol. 2:593–596.

Kline, L. N., R. F. Schmitz, J. A. Rudinsky, and M. M. Furniss. 1974. Repression of spruce beetle (Coleoptera) attraction by methylcyclohexenone in Idaho. Can. Entomol. 106:485–491.

Lanier, G. N., M. C. Birch, R. F. Schmitz, and M. M. Furniss. 1972. Pheromones of *Ips pini* (Coleoptera: Scolytidae): Variation in response among three populations. Can. Entomol. 104:1917–1923.

Lewis, W. J., R. L. Jones, D. A. Nordlund, and H. R. Gross, Jr. 1975. Kairomones and their use for management of entomophagous insects: II. Mechanisms causing increase in rate of parasitization by *Trichogramma* spp. J. Chem. Ecol. 1:349–360.

Lewis, W. J., M. Beevers, D. A. Nordlund, H. R. Gross, Jr., and K. S. Hagen. 1979. Kairomones and their use for management of entomophagous insects: IX. Investigations of various kairomone-treatment patterns for *Trichogramma* spp. J. Chem. Ecol. 5:673–680.

Mudd, A., and S. A. Corbet. 1973. Mandibular gland secretion of larvae of the stored products pests, *Anagasta kuehniella, Ephestia cautella, Plodia interpunctella* and *Ephestia elutella*. Entomol. Exp. Appl. 16:291–293.

Nelson, D. R. 1978. Long-chain methyl-branched hydrocarbons: Occurrence, biosynthesis, and function. Adv. Insect Physiol. 13:1–33.

Nettles, W. C., Jr., and M. L. Burks. 1975. A substance from *Heliothis virescens* larvae stimulating larviposition by females of the tachnid, *Archytas marmoratus*. J. Insect Physiol. 21:965–978.

Nordlund, D. A., and W. J. Lewis. 1976. Terminology of chemical releasing stimuli in intraspecific and interspecific interactions. J. Chem. Ecol. 2:211–220.

Pittman, G. B. 1973. Further observations on douglure in a *Dendroctonus pseudotsugae* management system. Environ. Entomol. 2:109–112.

Read, D. P., P. P. Feeny, and R. B. Root. 1970. Habitat selection by the aphid parasite *Diaeretiella rapae* (Hymenoptera: Braconidae) and hyperparasite *Charips brassicae* (Hymenoptera: Cynipidae). Can. Entomol. 102:1567–1578.

Tucker, J. E., and D. E. Leonard. 1977. The role of kairomones in host recognition and host acceptance behavior of the parasite *Brachymeria intermedia*. Environ. Entomol. 6:527–531.

Vinson, S. B., R. L. Jones, P. E. Sonnet, B. A. Bierl, and M. Beroza. 1975. Isolation, identification, and synthesis of host-seeking stimulants for *Cardiochiles nigriceps,* a parasitoid of tobacco budworm, Entomol. Exp. Appl. 18:443–450.

Vité, J. P., and D. L. Williamson. 1970. *Thanasimus dubius* prey perception. J. Insect Physiol. 15:233–239.

Weseloh, R. M. 1977. Effects on behavior of *Apanteles melanoscelus* females caused by modification in extraction, storage, and presentation of gypsy moth silk kairomone. J. Chem. Ecol. 3:723–735.

Weseloh, R. M., and B. R. Bartlett. 1971. Influence of chemical characteristics of the secondary scale host selection behavior of the hyperparasite, *Cheiloneurus noxius* (Hymenoptera: Encyrtidae). Ann. Entomol. Soc. Am. 64:1259–1264.

Wood, D. L., L. E. Browne, W. D. Bedard, P. E. Tilden, R. M. Silverstein, and J. O. Rodin. 1968. Response of *Ips confusus* to synthetic sex pheromones in nature. Science 159:1373–1374.

CHAPTER THIRTEEN

SEMIOCHEMICALS IN EVOLUTIONARY TIME

PETER W. PRICE

Department of Biological Sciences
Northern Arizona University
Flagstaff, Arizona

This chapter might be subtitled ''Evolutionary Aspects of B.O.'' (body odor). This translates the subject into one whose importance we can immediately appreciate in our own society although, comparatively speaking, body odor is much more important in the lives of herbivores and their enemies. Also we manipulate human body odor outrageously, whereas such manipulation is not so easily accomplished by insects. However, as we will see, in evolutionary time, body odor and associated odors can change quite dramatically.

Chemical communication may be regarded as language transmitted as coded messages. Some organisms have the key to the code and respond to such messages; others do not and remain oblivious. But a coded message may always be decoded by enemies in evolutionary time, and the benefits of such communication are then weighed against the disadvantages, with natural selection acting as the arbiter. Solution of the code by an enemy may cause strong selective pressure for change in the code, and given an environment that is already heavy with chemical signals, the acquisition of a new dialect may be difficult to achieve. But a change in body odor may result in an extraordinary improvement in fitness, as we shall see.

Thus to understand the evolutionary aspects of body odor, we need to know (1) the ecological setting in which semiochemicals are used, (2) the evolutionary potential of the users, and (3) the evolutionary consequences of their use. These subjects are treated in order and are related particularly to aspects of the life of parasitoids and predators. A final section relates these themes to semiochemical users in populations manipulated for biological control purposes.

THE ENVIRONMENT IN WHICH PARASITOIDS AND PREDATORS SEARCH

There exists a great diversity of chemically mediated information in the four trophic level system commonly found in terrestrial communities (Figure 13.1). Figure 13.1 simplifies the interactions because it concentrates on a single food chain based on a single host plant. The interactions involving only one herbivore species (Sp. 1) are depicted, but these necessarily include other members of the same species (a second member of Sp. 1 is shown), other herbivores on the host plant (Sp. 2), associated plants, and associated herbivores. Interactions at the third trophic level are arranged in a similar way. Associated plants and insects involve plants growing adjacent to or near the host plant and the insects in the food webs based on these plants. Each kind of interaction is described briefly with an example, following the numbers in the figure.

Interactions mediated by chemicals at all trophic levels are considered because the actions of chemicals synthesized by the autotrophic plants may have repercussions throughout the food web of heterotrophs. Thus geographic variation in chemicals within a plant species, or chemical differences between host plants utilized by a single herbivore species, may alter considerably the chemical environment in which the herbivore and its enemies live, thereby modifying the interaction between trophic levels 2 and 3 in which we are most interested.

Interactions Between Trophic Levels 1 and 2

1 The herbivore is attracted to the host plant by volatile chemicals (kairomones) (Kogan 1977, Staedler 1977, Chapter 3). However, although the herbivore may stay to feed, various defensive chemicals (allomones) may influence adversely the capacity of the herbivore to utilize the plant. Such defense is not illustrated by an arrow in Figure 13.1 because these chemicals do not cause a rapid behavioral response by the herbivore. Indeed, it is possible that tannins, which are potent defensive chemicals (Feeny 1968, 1970), may act as cues for host acceptance by generalist feeders on tree leaves (Görnitz 1955). Haukioja and Niemelä (1976, 1977) also found that the geometrid *Oporinia autumnata* gained more weight when fed on detached birch leaves than on those on standing trees, and they grew better on undamaged leaves than on damaged leaves. The mode of defense is not known, but its efficacy is clear. In addition, the plant may contain chemicals that are not effective against specialized herbivores, but render the plant unpalatable to nonspecialists: the qualitative defenses recognized by Feeny (1976), and antifeedants (e.g. Chapman 1974, Munakata 1977).

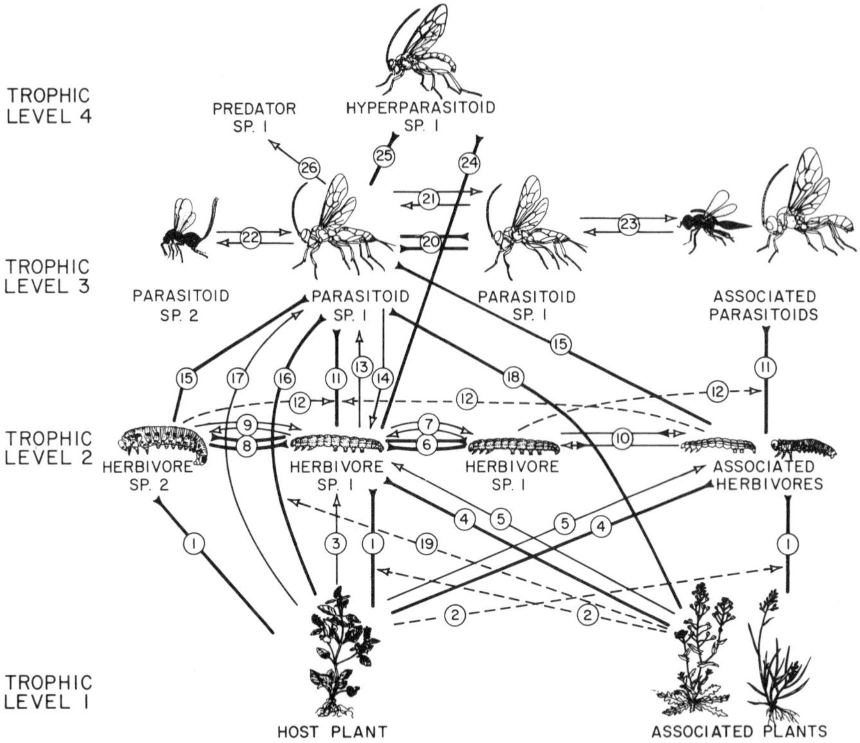

FIGURE 13.1. Interactions in a community of four trophic levels involving semiochemicals. Arrows are placed against the responding organism. Thick solid lines and solid arrows illustrate attraction to a stimulus (e.g. 1, 4, 11, 24). Thin solid lines and open arrows illustrate repulsion (e.g. 3, 13, 17, 26). Thin dashed lines show indirect effects such as interference with another response (e.g. 2, 12, 19). The arrangement of the figure and the numbered responses are discussed in the text.

2 Volatile chemicals from associated plants may influence attraction to the host plant. Tahvanainen and Root (1972) found that when host crucifers were planted with nonfood plants such as tomato and ragweed, fewer flea beetles (*Phyllotreta cruciferae*) colonized the crucifers. There was an "associational resistance" to colonization in the community (see also Atsatt and O'Dowd 1976).

3 The plant produces defensive chemicals in response to feeding at the site of damage. Phenolics are commonly oxidized to quinones, which are common repellent chemicals in plants and arthropods; and glandular hairs when broken may release repellents as in *Primula, Urtica,* and *Laportea* (Rodriguez and Levin 1976).

4 Associated plants may act as attractants to adults by providing nectar, pollen, or chemicals. Lepidoptera in the families Nymphalidae (Danainae and Ithomiinae), Ctenuchidae, and Arctiidae are attracted to plants containing pyrrolizidine alkaloids (Pliske 1975a, b, Pliske et al. 1976).

5 Repellent chemicals from associated plants may influence herbivores directly. In the presence of associated plants, adults of *P. cruciferae* fed less than those in pure stands (Tahvanainen and Root 1972) in addition to the reduced numerical response described in item 2.

Interactions Within Trophic Level 2

6 Herbivore species 1 attracts other individuals of the same species by means of sex pheromones (e.g. Beroza 1970, Shorey and McKelvey 1976, Chapter 11) or by aggregation pheromones (e.g. Wood 1970).

7 Herbivore species 1 repels other individuals of the same species or reduces their fitness in other ways by means of epideictic pheromones as described by Prokopy in Chapter 10. The Mediterranean flour moth (*Anagasta kuehniella*) and the confused flour beetle (*Tribolium confusum*) individuals have deleterious effects on conspecifics (Corbet 1971, Chapman 1926), and aggregated pine looper larvae (*Bupalus piniarius*) show reduced growth and fecundity in the presence of a plentiful food supply (Gruys 1970). Male moths of *Pseudaletia unipuncta* produce a pheromone that inhibits the approach of conspecific males (Hirai et al. 1978). Territories that defend progeny are marked out chemically by fruit flies (Prokopy 1972) and bean weevils (Oshima et al. 1973). Complexity is added to this interaction by the possibility that repellent pheromones are derived from microorganisms on which the herbivores feed. Brand et al. (1976) found that verbenone could be synthesized by a mycangial fungus growing on host plant tissues; and once ingested by the beetles, this chemical may constitute a significant fraction of the pheromone.

8 Herbivore species 1 attracts herbivore species 2 and/or vice versa. Wood (1970) provides several examples of male bark beetle pheromones that attract conspecific females and females from other species.

9 Herbivore species 1 repels herbivore species 2 and/or vice versa. Ipsenol, a component of the pheromone of *Ips paraconfusus,* inhibits attacks by *Ips pini* (Birch and Light 1977), and the sex pheromone of *Rhyacionia buoliana* inhibits the response of male *Rhyacionia frustrana* to conspecific females (Berrisford and Hedder 1978) (see also Steck et al. 1977, Pliske et al. 1976). Verbenone, which may be ingested by bark beetles, acts both as an allomone and as a repellent pheromone (see Brand et al. 1976).

10 Associated herbivores have similar influences as in items 8 and 9.

Interactions Between Trophic Levels 2 and 3

11 The parasitoid or predator is attracted to the herbivore by means of kairomones (e.g. Vinson 1977, Waage 1978, Lewis et al. 1976, Chapters 5, 7, and 8). *Microplitis croceipes* finds larvae of *Heliothis virescens* and *Heliothis zea* via 13-methylhentriacontane located in the larva's frass (Lewis and Jones 1971, Jones et al. 1971). *Cardiochiles nigriceps*, also a parasitoid of *H. virescens*, responds to three chemicals in the mandibular gland secretion of the larvae: 12-methylhentriacontane, 13-methyldotriacontane, and 13-methyltritriacontane. The egg parasitoid *Trichogramma evanescens* was found to utilize kairomones present on scales left by ovipositing moths (Lewis et al. 1972), with tricosane identified as the primary mediator (Jones et al. 1973). The scales from *H. zea* adults stimulate predation on eggs by larvae of *Chrysopa carnea* (Lewis et al. 1977, Nordlund et al. 1977). Predatory clerid beetles are strongly attracted by bark beetle pheromones (see Borden 1977 for review).

12 Any chemical from another herbivore may interfere with the enemies' detection of a host. This appears to be an untested probability.

13 The herbivore is chemically defended against its enemies. Defense takes many forms, such as the osmeterium of *Papilio* sp. (Eisner and Meinwald 1965), regurgitation of resin by sawflies (Prop 1960, Eisner et al. 1974), sequestering from food plants of toxins such as cardiac glycosides (see Duffey 1977, Roeske et al. 1976 for refs.), and synthesis of toxins such as cyanogenic chemicals, as in *Zygaena* spp. (Jones et al. 1962). Alternatively, aphid alarm pheromone may attract attending ants, which remove predators from the host plant (Nault et al. 1976). However, parasitoids are known to inject chemicals that reduce host resistance to internal eggs and larvae (Vinson 1975).

14 The herbivore moves away from or removes chemicals that may act as kairomones. Feeny (1976) describes how the tiger swallowtail larva (*Papilio glaucus*) leaves a leaf and cuts the petiole at its base, thereby removing leaf damage and much body odor. Many larvae feed nocturnally and leave their feeding site during the day when most parasitoids are active. Several defoliators of pine escape attack by the tachinid *Eucarcelia rutilla* in this way (Herrebout 1969a).

15 The herbivore's enemy is attracted principally by another herbivore species, or by herbivore's products such as honeydew. *E. rutilla* is attracted mainly to *B. piniarius* but will attack associated species (Herrebout 1969a). Zoebelein (1956) emphasized the importance of honeydew and showed that honeydew from different herbivore species resulted in different survivorship for adult parasitoids.

Interactions Between Trophic Levels 1 and 3

16 The parasitoid is attracted to the odor of the herbivore's host plant (Vinson 1977, Chapter 4), odors from fermentation products from the host plant (Greany et al. 1977), or plant chemicals released in a kairomone from the herbivore (Hendry et al. 1976). Parasitism by *Diaeretiella rapae* was much higher when a host *Myzus persicae* was on collard (with mustard oils attractive to *D. rapae*) than when it was on beet (with no mustard oils) (Read et al. 1970). Corn contains tricosane, which as pointed out in item 11, is found on the eggs of the corn earworm *H. zea* and is the kairomone utilized by the egg parasitoid *T. evanescens*. Likewise potato contains heptanoic acid, which appears in the frass of the potato tuberworm (*Phthorimaea operculella*), which aids host discovery by *Orgilus lepidus*. The latter examples were studied by Hendry et al. (1976).

17 The parasitoid or predator is repelled or trapped by host plant chemicals. *E. rutilla* during a preoviposition period is repelled by odors from pine on which its host occurs and is attracted to oak. At the end of this period pine odor becomes attractive (Herrebout 1969b, Herrebout and van der Veer 1969). Rabb and Bradley (1968) found that two species of parasitoid were effective on *Manduca sexta* on most solanaceous plants, but they caused no mortality on tobacco because the small wasps were immobilized by the trichome exudate. The overall response of the predatory clerid *Thanasimus dubius* was reduced in the presence of verbenone (Vitánd Williamson 1970), which may be acquired by bark beetles from fungi inoculated into their host plants (Brand et al. 1976).

18 The parasitoid or predator is attracted to associated plants that provide nectar and/or pollen, or more attractive odors. *Itoplectis conquisitor* was much more effective on codling moth (*Laspeyresia pomonella*) and eastern tent caterpillar (*Malacosoma americanum*) when wild flowers were available as a nectar source (Leius 1961, 1967). Herrebout et al. (1969) found that juniper was more attractive than pine to *E. rutilla*, although hosts were not found on juniper.

19 Associated plants interfere with attractive chemicals used by parasitoids (Monteith 1955, 1958a, 1958b, 1960). Most forest trees and plants of the forest floor masked the odor of the larch sawfly (*Pristiphora erichsonii*) and its host plant. This resulted in low levels of parasitism, typically 10–13% compared with up to 86% of hosts in pure larch.

Interactions Within Trophic Level 3

20 Parasitoid species 1 attracts other individuals of the same species by sex pheromones (e.g. Matthews 1974, 1975, Robacker and Hendry 1977, Lewis et al. 1971) and aggregation pheromones (Lewis et al. 1971).

21 Parasitoid species 1 repels other individuals of the same species (e.g. Price 1970, 1972). Females leave a trail that is recognized and avoided by conspecific females. This pheromone is also deposited on host cocoons, forming a chemical defense for the parasitoid's progeny.

22 Parasitoid species 1 repels species 2 or vice versa (Price 1970), and the parasitoids utilizing other hosts on the same plant species. For example, *Pleolophus indistinctus* avoided trails of three other ichneumonid parasitoids tested, although the principal host of one of the species, *Mastrus aciculatus* was probably not a host utilized by *P. indistinctus*.

23 Associated parasitoids have influences similar to those described in item 22. Some species of ichneumonid are known to produce odors that are very pungent, even to the relatively insensitive human nose (Townes 1939). Such odors may influence parasitoids on other plants several decimeters away.

The Fourth Trophic Level

24 The hyperparasitoid is attracted by the presence of the herbivore. The attractive chemicals have not been investigated as far as I know, but chemical cues are likely to be important. For example, Tripp (1961) describes how female *Euceros frigidus* always lay eggs about 10 cm from sawfly larvae, always on a route along which larvae will pass in the next few days. The eventual real host is a hymenopteran parasitoid of sawfly larvae.

25 The hyperparasitoid is attracted by odors from the parasitoid (Read et al. 1970). Females of the hyperparasitoid *Charips brassicae* were attracted to the odor of female *D. rapae* parasitoids and presumably used this cue first in finding parasitized aphids in which to oviposit. Semiochemicals are also strongly implicated in the discovery of *Orgilus obscurator,* a parasitoid on *R. buoliana,* by the cleptoparasitic *Temelucha interruptor* (Arthur et al. 1964), and *Pseudorhyssa sternata,* a cleptoparasitoid of *Rhyssa persuasoria,* is known to respond to secretions of the primary parasitoid (Spradbery 1968).

26 Parasitoids are defended chemically against predators with a pungent odor (Townes 1939), sequestered chemicals (Riechstein et al. 1968), or chemicals injected through the sting. In the last case defense may be more mechanical than chemical, but the painful sting well known in ophionine ichneumonids (Borror and DeLong 1954, Smith 1868) seems to be more potent than mere mechanical injury. Regarding sequestered chemicals, the tachinid *Zenillia adamsoni,* which parasitizes monarch butterfly (*Danaus plexippus*) larvae, acquires from its host cardiac glycosides that were originally synthesized by the milkweed plant on which monarch caterpillars feed.

The great diversity of chemical information transmitted and received, and the many organisms utilizing chemicals in any community, suggest that the acquisition by herbivores and their enemies of a unique, unjammed chemical modality is an important evolutionary step. In several circumstances insects have been found to utilize novel chemicals as defensive chemicals, as cues to host location, and as pheromones. Eisner et al. (1971) found that the grasshopper *Romalea microptera* incorporated in its defense secretion a repellent herbicide derivative. The parasitoid *Bracon mellitor* used a chemical found only in the artificial diet of the boll weevil (*Anthonomus grandis*) as a cue that released ovipositor probing (Vinson et al. 1976). In this case associative learning may have been involved. The southern pine beetle (*Dendroctonus frontalis*) utilizes verbenone as a repellent pheromone, probably obtaining this chemical by feeding on a symbiotic fungus (Brand et al. 1976). Host shifts by herbivores will probably envelop the larvae in a rather different environment of odors, with important ramifications up and down the trophic system. This topic is developed in a later section of this chapter.

This overview of the chemical environment in which enemies of herbivores operate also focuses attention on the interaction between the first and third trophic levels. Changes in plant quality can alter the efficacy of the herbivore's enemies directly (interactions 16 and 17 in Figure 13.1), or they can influence the quality of the herbivores and so alter the response of their enemies indirectly, as discussed later in this chapter. Associated plants cannot be regarded as inert members of the community (cf. interactions 18 and 19), and all should be tested for possible influences on the second and third trophic levels. Thus the study and use of semiochemicals must attempt to involve the whole community. For example, to maximize efficacy of the pest's enemies, use of semiochemicals in integrated pest management should be developed in conjunction with plant breeding. The importance of this integration is developed in more detail later in this chapter.

EVOLUTIONARY POTENTIAL OF PARASITIC SPECIES

The potential for rapid evolution of specialized insect herbivores [which may be regarded as parasites (Price 1977)] and their parasitic enmies needs to be considered whenever aspects of biological control are contemplated. Such rapid evolution during mass rearing, in response to stimuli that become irrelevant or deleterious in the field, is a constant threat to success in biological control. Manipulation of natural populations may result in evolutionary change, for better or for worse, depending on how well we understand and regulate selective factors. Natural populations are likely to be highly variable in space and time because different selective pressures operate; thus the qualities of host and parasitoid and the nature of their interaction are likely to differ geographically, and with time in a single location.

The evolutionary potential of such specialized organisms as insect herbivores and parasitoids derives from their generally small size, high reproductive rate (resulting from short generation times and high fecundity), considerable specificity in host selection, and resultant population structure. Hosts tend to be patchily distributed, and in each colonized patch inbreeding is likely to be common. Divergence between populations will be rapid under different selective regimes. Sympatric speciation may occur relatively frequently through host shifts (Bush 1975a). Chromosomal mutations may become rapidly fixed in small inbreeding populations, again resulting in speciation events (cf. Bush 1975b, Goodpasture and Grissell 1975). Askew (1968) and Gordh (1975) have discussed the characteristics of chalcidoids that result in their rapid evolution. Beaver (1977) has applied similar reasoning to bark and ambrosia beetles, and Price (1977, 1980) to parasitoids in general.

In the comparatively well studied fauna of the British Isles, it is possible to estimate the extent of adaptive radiation of herbivorous parasites and parasitoids relative to that of predators. Parasitic herbivores represent about 35% of the British insect fauna (Table 13.1). Parasitoids constitute more than 27%, made up of 5342 species of parasitic Hymenoptera (about 26% of fauna) and 228 species of Tachinidae (about 1% of fauna). The relative size of families of parasitic species compared to predatory species also indicates the greater adaptive radiation among parasites (Table 13.2). Also, when we consider what is as yet unutilized by insect herbivores and parasitoids, it is clear that the opportunities for further adaptive radiation are still greater than that achieved so far (Figures 13.2 and 13.3).

Thus, when we consider the interaction between insect herbivores and their parasitoids, we must anticipate an evolutionarily dynamic relationship, characterized by rapid and fine tuning of adaptive shifts generated by reciprocal interaction, or coevolution, in which selective forces change rapidly in space and time. Evolutionary time thus refers to change in gene frequencies within a breeding season or within the life of an insectary colony, as well as to major changes over longer spans of time.

In the next section I discuss some of the local and geographic variation to which insect herbivores and their enemies are exposed. This section is speculative and sketchy, since variation in hosts that is relevant to parasites has not received much attention in the literature. Variation relevant to parasitoids serves as the main theme.

TABLE 13.1. Feeding Habits of British Insects, Based on Analysis of the Checklist of British Insects by Kloet and Hincks (1945)[a]

Order	Predators	Nonparasitic Herbivores and Carnivores	Parasites On Plants	Parasites On Animals	Saprophages
Thysanura					23
Protura					17
Collembola					261
Orthoptera		39			
Psocoptera		70			
Phthiraptera				308	
Odonata	42				
Thysanoptera			183		
Hemiptera	123		283	5	
Homoptera			976		
Megaloptera	4				
Neuroptera	54				
Mecoptera	3				
Lepidoptera			2233		
Coleoptera	215	65	909	18	1637
Hymenoptera	170	241	435	5342†	36
Diptera	54	231	922	311	1672
Siphonaptera				47	
Totals	665	646	5941	6031	3646
Percentage of insect fauna	3.9	3.8	35.1*	35.6	21.5

After Price (1977, 1980).

[a] Note the large proportion of parasitic herbivores (*) and the large number of parasitic Hymenoptera † in the fauna. Of the 20,244 insects listed, 16,929 have been classified in this table.

LOCAL AND GEOGRAPHIC VARIATION IN NATURAL POPULATIONS: SPECULATION ON THE EVOLUTIONARY CONSEQUENCES OF SEMIOCHEMICAL USE

The host records for the parasitoid *Hoplismenus morulus* (Ichneumonidae) are as follows: *Nymphalis antiopa*, *Nymphalis californica*, *Vanessa cardui*, *Polygonia comma*, and *Polygonia interrogationis* (Heinrich 1960–1962). This short list is deceptive because it fails to convey the great variation of conditions in which *H. morulus* females must search, and in which populations evolve. The geographical distribution of these five hosts results in populations of *H. morulus* with from

TABLE 13.2. Number of Species in the 10 Largest Families in the British Insect Fauna in the Categories "Predators," "Herbivorous Insects," and "Carnivorous Parasitoids" From Check list by Kloet and Hincks (1945)

Predators		Herbivorous Insects		Carnivorous Parasitoids	
Dytiscidae	110	Cicidomyiidae	629	Ichneumonidae	1938
Sphecidae	104	Curculionidae	509	Braconidae	891
Coccinellidae	45	Aphididae	365	Pteromalidae	649
Corixidae	32	Tenthredinidae	358	Eulophidae	485
Cucujidae	32	Noctuidae	298	Tachinidae	228
Hemerobiidae	29	Chrysomelidae	248	Philopteridae	176
Vespidae	27	Cicadellidae	242	Platygasteridae	147
Asilidae	26	Cynipidae [a]	238	Encyrtidae	144
Anthocoridae	25	Olethreutidae	216	Diapriidae	125
Saldidae	20	Miridae	186	Ceraphronidae	108
Means	45		329		489

After Price (1977).

[a] Contains some carnivorous parasitoids.

one to three host species available (Figure 13.4). Where one host is available, as in northwest North America, we can expect *H. morulus* to evolve as a rather specialized searcher of *N. antiopa* hosts. The host occurs in mesic woodland and may be found on 10 host plant species (Tietz 1972) representing six families. Conditions for searching are favorable and comparatively simple (Figure 13.5). But in the northeast three hosts are available that are known to utilize 21 host plant species in 12 families, although all these occur in mesic woodland. Conditions for searching are much more complex, and adaptive responses to these conditions may be very different from those in the northwest. In southern California conditions are even more complex, since one host occurs in chaparral vegetation, another in mesic woodland, and a third in mesic herbaceous vegetation.

Host plant species utilized by the three butterfly species number 44 in 17 families (Figure 13.5). Thus greater habitat diversity is added to the within-habitat complexity to which *H. morulus* must adapt. *H. morulus* that utilize *N. californica* in chapparal have been recognized as a subspecies (*H. morulus pacificus*), whereas *H. morulus morulus* occurs in mesic woods (Heinrich 1960–1962). We should however expect much more variation between populations throughout the geographic range of the species in physiological and behavioral traits and chemical ecology than is observed in morphology. If the parasitoid were abundant enough, it would be valuable to study variation in searching behavior on a geographic scale.

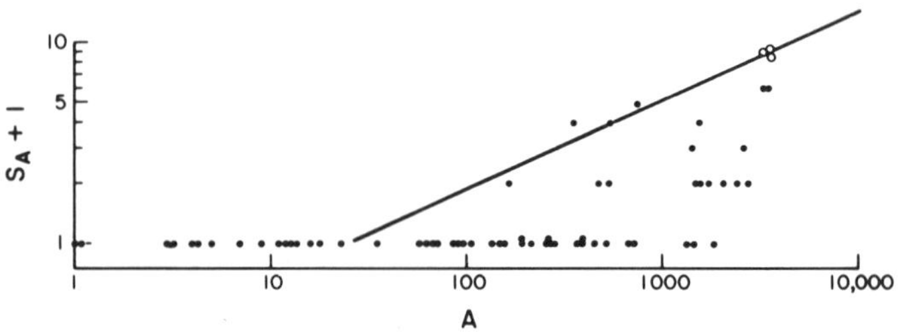

FIGURE 13.2. The relationship between area of host plant species and number of agromyzid herbivore species per host plant species in Britain. Note that many plant species support a number of agromyzid species well below the probable equilibrium number (illustrated by the solid line), suggesting that many resources (or ecological niches) remain unexploited at present. Closed circles are for agromyzids on Umbeliferae and open circles are for the three plant species with largest number of agromyzid species in Britain: two in the Compositae and one in the Ranunculaceae. (After Lawton and Price 1979.)

Many such examples could be used to illustrate the very different conditions to which parasitoid populations of the same species are exposed; the great diversity of B.O. in hosts, and host plants, and associated species. Adaptive shifts in response to these environmental differences are bound to occur, with the result that the adaptive mode of the population is likely to be quite different in each region and even within each habitat. DeBach and Hagen (1964) review the literature on known races of parasitoids used in biological control.

The great morphological variation within the genus *Therion* (Ichneumonidae) has posed problems to systematists for many years (Slobodchikoff 1977). Whether the genus in North America should be considered to be represented by a single species with 10 morphotypes (see Slobodchikoff 1973), or by five distinct species, four intermediate species, and one parthenogenetic clone (Slobodchikoff 1977), is open to debate until more biological characteristics are known, particularly the extent of reproductive isolation between morphotypes or species. Interestingly enough, where Slobodchikoff (1973) examined the behavior of three *Therion* morphotypes he found no differences except in host selection. Two morphotypes responded only to hosts free of secondary setae, and one attacked only hosts with secondary setae present. Such distinct differences probably resulted from divergence of behaviors during a true speciation event. However, a survey of many populations is warranted because great variation may exist in the small, microallopatric populations recognized by Slobodchikoff (1973).

Whatever the true biological status of the genus *Therion,* the considerable morphological continuity between several modal morphotypes probably mirrors similar variation in response to semiochemicals and other characters. Much of this variation is probably selected for by the very different environmental

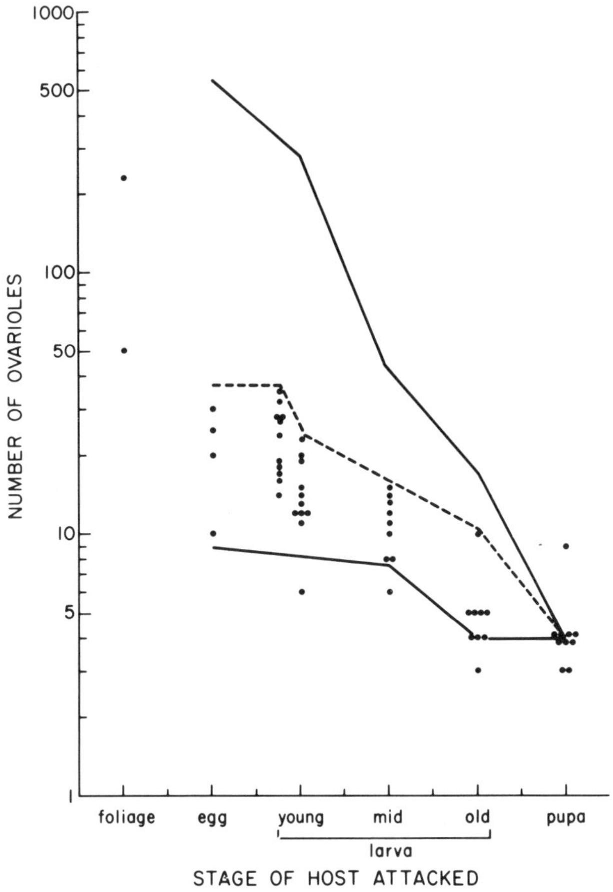

FIGURE 13.3. A resource matrix for Ichneumonidae derived from survivorship curves of lepidopterous hosts indicating the number of ovarioles per ovary required by species to fully exploit the resources available (range is marked by solid lines) and the number actually observed (upper limit is marked by dashed line). The area of the matrix as yet unexploited, indicates the existence of many unutilized niches into which ichneumonids can radiate. (From Price, P. W., 1980, Evolutionary Biology of Parasites. Princeton University Press. Reprinted by permission of Princeton University Press.)

conditions in which populations have existed, as illustrated above for *H. morulus*.

Even a single herbivore host may utilize different plant hosts in different geographic locations, with corresponding changes in body odor of host and plant to which parasitoids must adapt. For example, the white pine weevil (*Pissodes strobi*) is found on three allopatric host plant species: white pine in eastern North

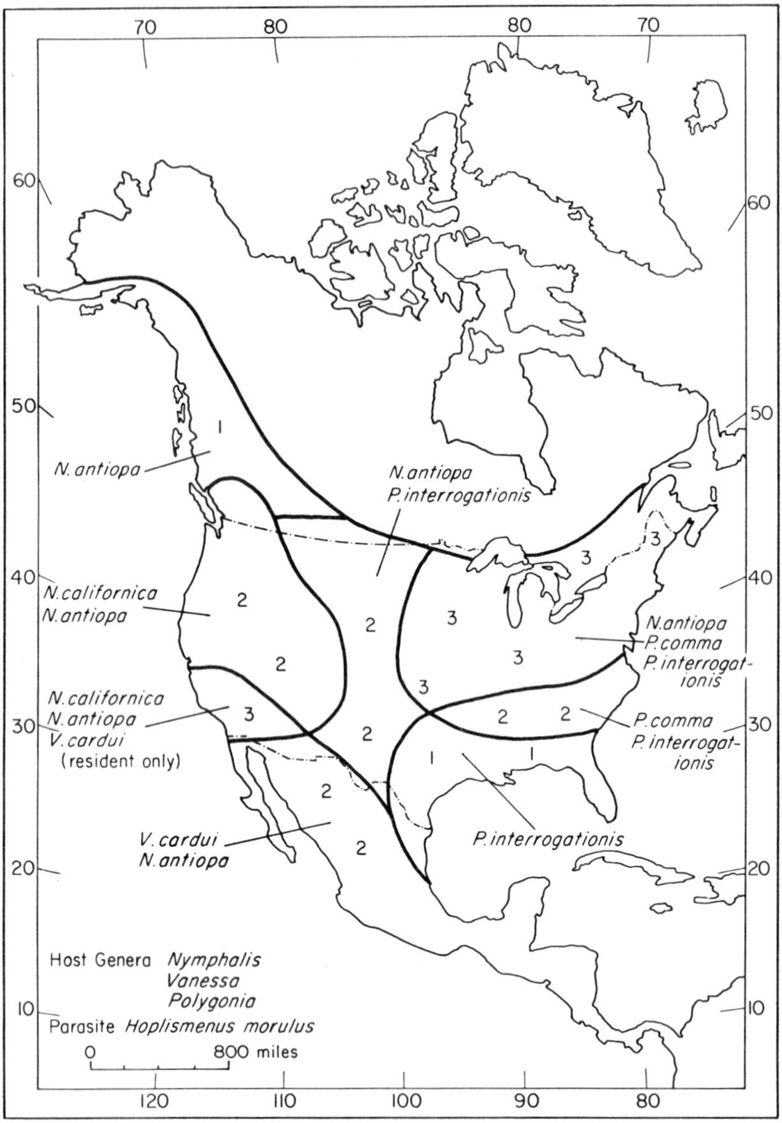

FIGURE 13.4. The distribution of hosts in North America for the parasitoid *Hoplismenus morulus*: hosts from Heinrich (1960–1962), host distributions from Howe (1975).

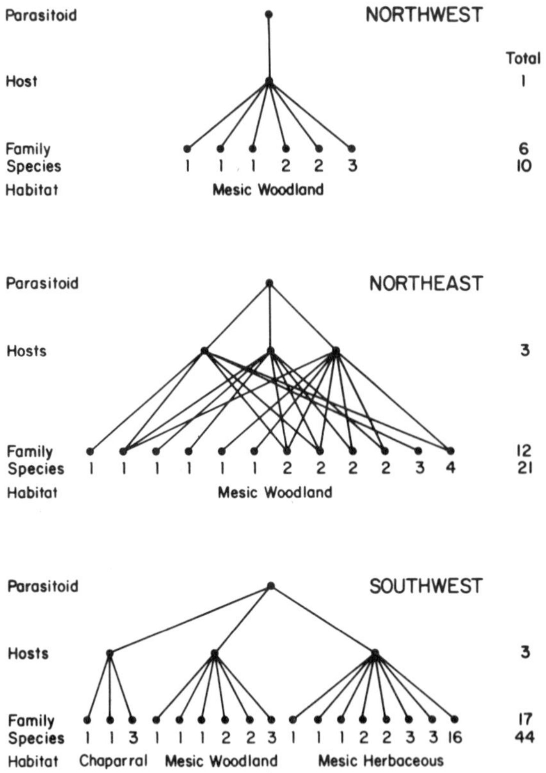

FIGURE 13.5. The spectrum of hosts from which *Hoplismenus morulus* searches and the habitats in which it searches in three geographic locations: northwest, northeast, and southwest North America. The number and species of hosts available in each region are shown in Figure 13.4. Food plants of hosts are from Tietz (1972).

America, Engleman spruce in the west, and Sitka spruce along the western coast (Vandersar et al. 1977). It would be valuable to study parasitoids of the white pine weevil on these three hosts, especially their responses to semiochemicals, and population levels, to see whether parasitoids have tracked the evolution of host utilization by the weevil as envisaged by Vandersar et al. (1977).

We know that where host plants change, it is possible for herbivore semiochemicals to change. The allomone in the osmeterium of *Papilio machaon* differs in the ratio of two constituents, isobutyric acid and 2-methylbutyric acid, when larvae feed on carrot and fennel (ratio is 70–30) and when they feed on parsnip (52–48) (Eisner and Meinwald 1965). The chirality of the precursor in the host tree influences the isomer of verbenol used as an aggregation pheromone by *Ips confusus* (Renwick et al. 1976).

Considering the great variation in chemicals observed in host plant species and within species, such effects are likely to be common. Apple fruits differ considerably in their α-farnesene content, which acts as an attractant and oviposition stimulant for codling moth (Sutherland et al. 1977). *Lotus corniculatus* and *Trifolium repens* show variation in cyanogenic chemical content (Jones 1973), and extensive variation in cardenolide content occurs both within and between milkweed (*Asclepias*) species and other members of the Asclepiadaceae (Roeske et al. 1976). Such differences translate into very significant differences in allomonal content of herbivores, as described by Brower et al. (1968).

The great importance of host plant quality must be recognized when semiochemicals are manipulated in agricultural systems. Unless plant breeders select strains that are compatible with effective parasitoid activity, we may frequently find that different elements in a pest control program are working at odds with each other. For example, the tachinid parasitoid *Lydella grisescens* is much more effective against the European corn borer (*Ostrinia nubilalis*) on one corn hybrid than on another (Table 13.3). This may well result from differences in host plant odor, since the females are attracted first to the plant and then to the insect host (Franklin and Holdaway 1966). If the W22 × A73 hybrid were found to be intrinsically more resistant, a severe net reduction in control might result if this hybrid were used in conjunction with inundative releases of *L. grisescens*. Plant breeders must consider the third trophic level as an important part of the battery of defenses available to plants in evolutionary time.

A shift by a herbivore from one host plant to another immediately changes a diversity of factors important to its survival.

1 There may be a change in plant body odor, which may be used by enemies for detecting hosts.

TABLE 13.3. Effect of Corn Hybrid on the Impact of *Lydella grisescens* on European Corn Borer Populations.

Year	Season	Corn Hybrid	Borers per 100 Plants	Parasites per 100 Borers
1955	Midseason	A322 × A334	92	36.5
		W22 × A73	134	8.2
1957	Midseason	A322 × A334	98	11.8
		W22 × A73	109	0.8
1957	Fall	A322 × A334	129	47.9
		W22 × A73	104	8.0

Data from Franklin and Holdaway (1966).

2 The blend of available kairomones (i.e. herbivore body odor) may change, making the herbivore more or less vulnerable to enemies.

3 There may be a change in food quality, which may increase or decrease fecundity and/or survival. The survivorship curve would change, making it more or less difficult for parasitoids to maintain a population on the herbivore host (Figure 13.6).

4 Physical characters of the host may change which influence the physical protection afforded a herbivore by the host plant. In the milkweed longhorn beetle (*Tetraopes tetrophthalmus*), every character studied differed between a population on the normal host *Asclepias syriaca* and an apparently newly acquired host *Asclepias verticillata* (Price and Willson 1976).

If we imagine that the difference in survivorship of *Pieris rapae* and *Hylemya brassicae* are due to slight differences in resistance in the host plants (see Figure 13.6), we can predict the effect of these differences on success of parasitoids. Parasitoids attacking first instar larvae of *P. rapae* and emerging from the pupae would have probabilities of survival of 0.125 and 0.115, respectively, on brussels sprout and cabbage, a 9% reduction in survival for parasitoids of larvae on cabbage compared to those on brussels sprout. Similar estimates for the differences in survivorship curves for *H. brassicae* would result in 66% reduction in survival (see Table 13.4 for other examples). Thus plant defense, often

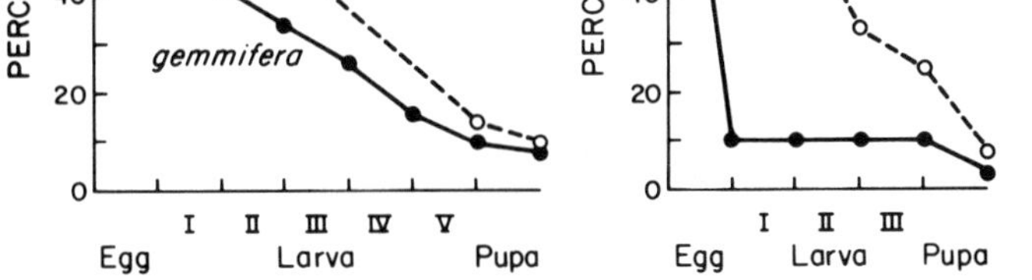

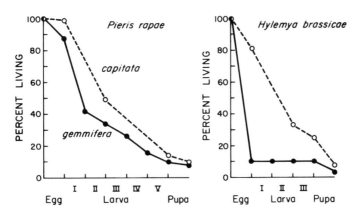

FIGURE 13.6. Survivorship curves. *Left: Pieris rapae* on *Brassica oleracea gemmifera* in England (solid line) and *Brassica oleracea capitata* in Canada (dashed line); *right: Hylemya brassicae* on *Brassicae oleracea capitata* in England (solid line) and in Canada (dashed line). Data are from Dempster (1967) (means of 3 generations), Harcourt (1966) (means of 18 generations), Hughes and Mitchell (1960) (means of 6 generations), and Mukerji (1971) (means of 9 generations), respectively.

TABLE 13.4. The Potential Influence of Host Plant Resistance on Survivorship of Parasitoids Attacking Hosts Whose Survivorships Curves Are Illustrated in Figure 13.6: A Hypothetical Example[a]

Herbivore:	*Pieris rapae*		*Hylemya brassicae*	
Host plant,	*Brassica gemmifera*	*capitata*	*capitata*	*capitata*
Location:	England	Canada	England	Canada
Parasitoid attacks first instar larvae (255 eggs laid)[b]				
Progeny surviving	32	29	77	26
Percentage reduction in survival	9		66	
Parasitoid attacks middle-aged larvae (110 eggs laid[b]				
Progeny surviving	29	26	33	17
Percentage reduction in survival	10		48	

[a] The number of progeny surviving from a female that lays her full complement of eggs is given, plus the relative reduction in survival on the less resistant plant strain. For this example it is assumed that differences in herbivore survival result from differences in plant resistance, and that survivorship of parasitoid larvae in hosts parallels that of the host, as suggested by Price (1975).

[b] Fecundity of parasitoid is estimated from equations for regression lines for ovarioles per ovary against stage of host attacked (Figure 5 in Price 1975) and fecundity against ovariole number (Figure 3 in Price 1975) for Ichneumonidae.

mediated by allelochemics, can have significant influences on the ease with which parasitoids can exploit hosts. In both examples, the host that is more resistant to insect herbivores is more favorable to the parasitoids because it causes early mortality before parasitoids attack. In the *H. brassicae* example the effect is very dramatic. Maternal effects on egg quality could have equivalent influences on egg parasitoids. Plant breeding should play an integral role in manipulation of herbivores and their enemies.

Host shifts can have an enormous impact on the parasitoid fauna. An extreme case is being studied in our department by Paul Gross, who discovered that in one season two closely related coexisting species of *Tildenia,* one leafminer on *Solanum carolinense* and another on *Physalis heterophylla,* are attacked by about 25 species and one species of parasitoid, respectively. Parasitism has reached about 42% of hosts on *S. carolinense* but only 5% of host on *P. heterophylla.*

The factors influencing the number of parasitoid species and their abundance have yet to be determined, but they almost certainly involve differences in semiochemicals. Host shifts may result in evolutionary steps of paramount importance in the exploitation of enemy-free resources (see also Lawton 1978, Lawton and Schröder 1977, Gilbert and Singer 1975). We may frequently aid the herbivore in our breeding programs by selecting plants with more succulent fruits, thicker stems, or perhaps unwittingly for different plant body odor, all tending to reduce access by parasitoids.

Another kind of variation concerns semiochemical differences within species or morphospecies in different parts of their geographic range, apparently independent of host plant differences. Males of the European corn borer from Iowa and New York respond to very different blends of the *E* and *Z* isomers of 11-tetradecenyl acetate (Klun et al. 1973). Attractant pheromones of *I. pini* in California, Idaho, and New York differed even when reared on the same plant host (Lanier et al. 1972). Although these examples are both concerned with pheromones, we should expect similar differences in other semiochemicals. Lanier et al. (1972) did find that the predator *Enoclerus lecontei* much preferred the kairomone of *I. pini* in New York to that produced by beetles from California and Idaho, whereas the parasitoid *Tomicobia tibialis* favored the kairomone from western beetles.

Defense of herbivores against enemies also varies geographically and locally as measured by encapsulation frequency of endoparasitoids. Muldrew (1953) found that percentage of parasitism of the larch sawfly by *Mesoleius tenthredinis* may exceed 70% in Manitoba, but most eggs (>90%) remained unhatched because of encapsulation. In British Columbia, however, no encapsulation was observed. A similar case was observed in *Coccus hesperidium* from Texas, California, and Hawaii, which encapsulated less than 3% of *Metaphycus luteolus* eggs in the Texas population, but almost 30% in the California and Hawaii populations (Bartlett and Ball 1966). Lange and Bronskill (1964) demonstrated that the percentage of encapsulated parasitoids could be influenced by the host's diet and thus may well differ between herbivores of the same species on different host plants (Morris 1976). Morris (1976) has also shown that the percentage of encapsulation of endoparasitoids by *Hyphantria cunea* larvae differs with species of parasitoid, location, and in time at one location. He found that the level of encapsulation correlated well with the genetic quality of the host, as estimated by heat requirements of pupae. There is no doubt that host defense can be a very variable character; this has been particularly well documented by Morris. Allomones may well play a role in herbivore defense, and variability in these substances may well be influenced by chemical variability in the herbivore's host plant.

Many other kinds of variation probably exist, although thus far there seems to be little evidence. For example, responses of parasitoids to deterrent pheromones are likely to differ between individuals at high and low densities. Parasitoid species that encounter much competition or are at the center of distribution of their respective taxa may differ in their response from those that habitually experience little competition or live in areas with low diversity of related species. The parasitoid *Pleolophus basizonus,* which occurs at the center of greatest diversity in its tribe in the eastern Palearctic, has been introduced into North America (Price 1971). In these new surroundings females responded to the allomone of three species with which it competed, but none of these species responded to the allomone of *P. basizonus* (Price 1970). The lack of response would be to the detriment of the indigenous species, since within the host cocoon, larvae of *P. basizonus* were the better competitors.

This documentation of local and geographic variation is important because it illustrates how natural selection is likely to result in great variability in semiochemical production and use in natural populations. Further examples and studies on the mechanisms that result in variation in responses to semiochemicals and their synthesis will be very valuable. Such studies will reveal the evolutionary processes and results concerning semiochemicals.

FITNESS OF SEMIOCHEMICAL USERS

When we manipulate natural enemies of pest species in any way, including by the use of semiochemicals, it is important to keep in mind the influence we have on the fitness of those enemies. If we alter fitness even slightly, evolutionary changes in manipulated populations will result, and these can occur very rapidly in organisms with such high evolutionary potential. Care must be taken to ensure that such changes are beneficial to the integrated management of a pest, rather than detrimental.

Vinson (1977) has reviewed the ways in which semiochemicals may be used in biological control and deals with potential problems with their use in contemporary time. These methods also have implications in evolutionary time, for each method has a particular influence on the fitness of individuals and the evolution of manipulated populations.

Where inundative releases of enemies are used, we should be concerned largely with fitness and evolution during mass rearing because the individuals released are dispensable. When manipulating natural populations, it is important to evaluate the impact of treatment on these populations and the relative fitness of individuals that enter the control area compared with those that remain in the surrounding communities. These two aspects are discussed in order.

One question that is relevant to both inundative releases and manipulation of wild enemies concerns the quality of the crop plant on which the enemies must search. Does the crop plant contain the best possible blend of characters for the greatest total environmental resistance to pest species? This involves innate plant resistance and plant resistance through being adapted to favor the action of the herbivore's enemies. In addition to such qualities as plant odor, defensive chemicals, and precursors of semiochemicals to be produced by herbivores, physical factors that influence search and availability of herbivores must be evaluated. These include the effect of trichomes on searching rate, presence of glandular secretions, and the degree to which herbivores are concealed. All these characters can be manipulated through selection, although I know of breeding programs on only one plant species that have taken the third trophic level into account: namely, experiments on cucumber in glasshouses where a trichome-free variety facilitates search for whitefly by *Encarsia formosa* (Ekbom 1977, Woets and van Lenteren 1976).

Concerning mass production of enemies for inundative release, the major danger is that species evolve adaptations to rearing conditions that become maladaptive after release. Such maladaptive shifts are as likely to involve semiochemicals as any other character, and investigators should monitor for them continually. Quality control has been discussed in detail by Bush (1978), Boller and Chambers (1977), and Mackauer (1976) and is not considered further in this chapter.

When natural populations of the herbivore's enemies are manipulated by the use of semiochemicals, I think two questions need to be answered to gain an evolutionary perspective of the process. (1) What proportion of a natural local population comes under the influence of manipulation? (2) What is the fitness of manipulated individuals versus unmanipulated members of the local population?

If a small proportion of a large population is attracted and held in a crop field by semiochemicals, the quality of the population may not be influenced appreciably, no matter what level of fitness is achieved by the attracted individuals. However, if a large proportion of a population becomes manipulated, rapid evolution may result, and its direction will depend on the fitness of the manipulated individuals versus other members of the local population. If fitness is reduced in responding individuals, selection will favor individuals showing little or no response to certain semiochemicals, to the detriment of the biological control program. Resistance to semiochemicals that are deleterious can evolve just as rapidly as resistance to insecticides or juvenile hormone (e.g. Templeton and Rankin 1978). Should responding individuals realize a greater fitness in the manipulated crop, evolution of the local population may favor the continued use of manipulation, with the potential for improved control over several generations. Selection may work to reduce the threshold for response to attractive semiochemicals, to increase the threshold for response to repellent semiochemicals, and for more or less dispersal (a highly variable character), permitting rapid response to selection (Johnston 1978).

Of course selection influences not only the herbivore's enemies but the herbivore itself. Where enemies are particularly effective, selection will work to reduce cues to host discovery, or to increase defense. Much maneuvering of body odor in evolutionary time probably goes on in nature, with strong selection on herbivores for body odors not yet decoded by enemies.

The dynamic status of management systems involving species with rapid evolutionary potentials cannot be overemphasized. Once an effective system of pest regulation is achieved, a rapid evolutionary shift by the pest may be expected with a resultant decline in control. No mortality-causing agent has purely random effects on a population. The greater the mortality, the more rapidly a population will change. Therefore, as we grow more successful in regulating pests through use of semiochemicals, the more devious and versatile must our range of tactics become.

CONCLUSION

Evolutionary aspects of semiochemicals have been discussed in relation to the environment in which parasitoids and predators search, the evolutionary potential of herbivores and parasitoids, the evolutionary consequences of semiochemical use, and the fitness of organisms in populations manipulated for biological control purposes.

The importance of semiochemical links within and between trophic levels was emphasized. Many interactions between the four trophic levels in a typical terrestrial community are mediated by semiochemicals. Direct links between plants (trophic level 1) and the herbivore's enemies (trophic level 3), and indirect links through the second trophic level, make plant quality an important aspect of community organization. The selection of plant strains that are conducive to the action of predators and parasitoids is a necessary component in any integrated control program, including any attempts at augmentation of natural enemies of herbivores.

Several studies illustrate the existence of local and geographic variation in semiochemical use, showing that natural selection works in different ways in different parts of a species's range. More attention is needed on variation in production of and response to semiochemicals in local populations and throughout the geographic range of a species.

The effects of manipulation of parasitoid and predator populations on the fitness of individuals is considered briefly. Evolution during mass rearing can render a population maladapted for field conditions. When naturally occurring local populations are utilized, it is important to estimate the proportion of individuals in a population that are manipulated, and the fitness of these individuals compared to those that remain unmanipulated. Depending on this relative fitness, selection may work to favor the manipulative technique or to cause it to become less effective. The dynamic status of management systems involving species with great potential for evolutionary change demands a very versatile and devious battery of tactics if it is to be successful.

ACKNOWLEDGMENTS

I am grateful to the following colleagues for helpful comments, useful references, and criticism on this chapter: Carl Bouton, Paul Gross, Richard L. Jones, W. Joe Lewis, Bruce McPheron, Donald A. Nordlund, and Arthur Weis. Financial support was provided through a grant from the National Science Foundation, DEB 78-16152.

REFERENCES

Arthur, A. P., J. E. R. Stainer, and A. L. Turnbull. 1964. The interaction between *Orgilus obscurator* (Nees) (Hymenoptera: Braconidae) and *Temelucha interruptor* (Grav.) (Hymenoptera: Ichneumonidae), parasites of the pine shoot moth, *Rhyacionia buoliana* (Schiff.) (Lepidoptera: Olethreutidae). Can. Entomol. 96:1030–1034.

Askew, R. R. 1968. Considerations on speciation in Chalcidoidea (Hymenoptera). Evolution 22:642–645.

Atsatt, P. R., and D. J. O'Dowd. 1976. Plant defense guilds. Science 193:24–29.

Bartlett, B. R., and J. C. Ball. 1966. The evolution of host suitability in a polyphagous parasite with special reference to the role of parasite egg encapsulation. Ann. Entomol. Soc. Am. 59:42–45.

Beaver, R. A. 1977. Bark and ambrosia beetles in tropical forests. BIOTROP Spec. Publ. No. 2:133–147. (Proc. Symp. Forest Pests and Diseases in Southeast Asia, Bogor, Indonesia).

Berrisford, C. W., and R. L. Hedder. 1978. Suppression of male *Rhyacionia frustrana* response to live females by the sex pheromone of *R. buoliana*. Environ. Entomol. 7:532–533.

Beroza, M. 1970. Chemicals Controlling Insect Behavior. Academic Press, New York.

Birch, M. C., and D. M. Light. 1977. Inhibition of the attractant pheromone response in *Ips pini* and *I. paraconfusus* (Coleoptera: Scolytidae): Field evaluation of ipsenol and linalool. J. Chem. Ecol. 3:257–267.

Boller, E. F., and D. L. Chambers. 1977. Quality aspects of mass-reared insects. P. 219–235. *In* R. L. Ridgway and S. B. Vinson (eds.), Biological Control by Augmentation of Natural Enemies. Plenum Press, New York.

Borden, J. H. 1977. Behavioral responses of Coleoptera to pheromones, allomones, and kairomones. P. 169–198. *In* H. H. Shorey and J. J. McKelvey, Jr. (eds.), Chemical Control of Insect Behavior: Theory and Application. Wiley, New York.

Borror, D. J., and D. M. DeLong. 1954. An Introduction to the Study of Insects. Constable, London.

Brand, J. M., J. W. Bracke, L. N. Britton, A. J. Markovetz, and S. J. Barras. 1976. Bark beetle pheromones: Production of verbenone by a mycangial fungus of *Dendroctonus frontalis*. J. Chem. Ecol. 2:195–199.

Brower, L. P., W. N. Ryerson, L. L. Coppinger, and S. C. Glazier. 1968. Ecological chemistry and the palatability spectrum. Science 161:1349–1351.

Bush, G. L. 1975a. Sympatric speciation in phytophagous parasitic insects. P. 187–206. *In* P. W. Price (ed.), Evolutionary Strategies of Parasitic Insects and Mites. Plenum Press, New York.

Bush, G. L. 1975b. Modes of animal speciation. Annu. Rev. Ecol. Syst. 6:339–364.

Bush, G. L. 1978. Planning a rational quality control program for the screwworm fly. P. 37–47. *In* R. H. Richardson (ed.), The Screwworm Problem: Evolution of Resistance to Biological Control. University of Texas Press, Austin.

Chapman, R. F. 1974. The chemical inhibition of feeding by phytophagous insects: A review. Bull. Entomol. Res. 64:339–363.

Chapman, R. N. 1926. Inhibiting the process of metamorphosis in the confused flour beetle (*Tribolium confusum,* Duval). J. Exp. Zool. 45:292–299.

Corbet, S. A. 1971. Mandibular gland secretion of larvae of the flour moth, *Anagasta kuehniella,* contains an epideictic pheromone and elicits oviposition movements in a hymenopteran parasite. Nature (London) 232:481–484.

DeBach, P., and K. S. Hagen. 1964. Manipulation of entomophagous species. P. 429–458. *In* P. DeBach (ed.), Biological Control of Insect Pests and Weeds. Reinhold, New York.

Dempster, J. P. 1967. The control of *Pieris rapae* with DDT. I. The natural mortality of the young stages of *Pieris*. J. Appl. Ecol. 4:485–500.

Duffey, S. S. 1977. Arthropod allomones: Chemical effronteries and antagonists. Proc. 15th Int. Cong. Entomol. Washington D.C.:323–394.

Eisner, T., L. B. Hendry, D. B. Peakall, and J. Meinwald. 1971. 2,5-Dichlorophenol (from ingested herbicide?) in defensive secretion of grasshopper. Science 171:277–278.

Eisner, T., J. S. Johnessee, J. Carrel, L. B. Hendry, and J. Meinwald. 1974. Defensive use by an insect of a plant resin. Science 184:996–999.

Eisner, T., and Y. C. Meinwald. 1965. Defensive secretion of a caterpillar (*Papilio*). Science 150:1733–1735.

Ekbom, B. S. 1977. Development of a biological control program for greenhouse whiteflies (*Trialurodes vaporarium* Westwood) using its parasite *Encarsia formosa* (Gahan) in Sweden. Z. Angew. Entomol. 84:145–154.

Feeny, P. P. 1968. Effect of oak leaf tannins on larval growth of the winter moth *Operophtera brumata*. J. Insect Physiol. 14:805–817.

Feeny, P. P. 1970. Seasonal changes in oak leaf tannins and nutrients as a cause of spring feeding by winter moth caterpillars. Ecology 51:565–581.

Feeny, P. P. 1976. Plant apparency and chemical defense. P. 1–40. *In* J. W. Wallace and R. L. Mansell (eds.), Biochemical Interaction Between Plants and Insects. Vol. 10, Recent Advances in Phytochemistry, Plenum Press, New York.

Franklin, R. T., and F. G. Holdaway. 1966. A relationship of the plant to parasitism of European corn borer by the tachinid parasite *Lydella grisescens*. J. Econ. Entomol. 59:440–441.

Gilbert, L. E., and M. C. Singer. 1975. Butterfly ecology. Annu. Rev. Ecol. Syst. 6:365–397.

Goodpasture, C., and E. E. Grissell. 1975. A karyological study of nine species of *Torymus* (Hymenoptera: Torymidae). Can. J. Genet. Cytol. 17:413–422.

Gordh, G. 1975. Some evolutionary trends in the Chalcidoidea (Hymenoptera) with particular reference to host preference. J. N.Y. Entomol. Soc. 83:279–280.

Görnitz, K. 1955. Frassauslösende Stoffe für Polyphag an holzgewächsen fressende Raupen. Verh. Dtsch. Ges. Angew. Entomol. E. 13:38–47.

Greany, P. D., J. H. Tumlinson, D. L. Chambers, and G. M. Boush. 1977. Chemically mediated host finding by *Biosteres (Opius) longicaudatus*, a parasitoid of tephritid fruit by larvae. J. Chem. Ecol. 3:189–195.

Gruys, P. 1970. Mutual interference in *Bupalus piniarius* (Lepidoptera, Geometridae). P. 199–207. *In* P. J. den Boer and G. R. Gradwell (eds.), Proc. Adv. Study Inst. Dynamics of Numbers in Populations (Oosterbeek 1970). Centre for Agricultural Publishing and Documentation, Wageningen.

Harcourt, D. G. 1966. Major factors in survival of the immature stages of *Pieris rapae* (L.). Can. Entomol. 98:653–662.

Haukioja, E., and P. Niemelä. 1976. Does birch defend itself actively against herbivores? Rep. Kevo Subarctic Res. Stn. 13:44–47.

Haukioja, E., and P. Niemelä. 1977. Retarded growth of a geometrid larva after mechanical damage to leaves of its host tree. Ann. Zool. Fennici 14:48–52.

Heinrich, G. H. 1960–1962. Synopsis of Nearctic Ichneumoninae Stenopneusticae with particular reference to the northeastern region (Hymenoptera). Parts I–VII. Can. Entomol. Suppl. 15, 18, 21, 23, 26, 27, and 29.

Hendry, L. B., J. K. Wichmann, D. M. Hindenlang, K. M. Weaver, and S. H. Korzeniowski. 1976. Plants—The origin of kairomones utilized by parasitoids of phytophagous insects? J. Chem. Ecol. 2:271–283.

Herrebout, W. M. 1969a. Some aspects of host selection in *Eucarcelia rutilla* Vill. (Diptera: Tachinidae). Neth. J. Zool. 19:1–104.

Herrebout, W. M. 1969b. Habitat selection in *Eucarcelia rutilla* Vill. (Diptera: Tachinidae). II. Experiments with females of known age. Z. Angew. Entomol. 63:336–349.

Herrebout, W. M., A. D. Tates, and J. van der Veer. 1969. Habitat selection in *Eucarcelia rutilla* Vill. (Diptera: Tachinidae). IV. Experiments with gravid females. Z. Angew. Entomol. 64:218–232.

Herrebout, W. M., and J. van der Veer. 1969. Habitat selection in *Eucarcelia rutilla* Vill. (Diptera: Tachinidae). III. Preliminary results of olfactometer experiments with females of known age. Z. Angew. Entomol. 64:55–61.

Hirai, K., H. H. Shorey, and L. K. Gaston. 1978. Competition among courting male moths: Male-to-male inhibitory pheromone. Science 202:644–645.

Howe, W. H. (ed.) 1975. The Butterflies of North America. Doubleday, Garden City, NY.

Hughes, R. D., and B. Mitchell. 1960. The natural mortality of *Erioischia brassicae* (Bouche) (Dipt., Anthomyiidae): Life tables and their interpretation. J. Anim. Ecol. 29:359–374.

Johnston, J. S. 1978. Dispersal behavior and biological control of insects. P. 113–127. *In* R. H. Richardson (ed.), The Screwworm Problem: Evolution of Resistance to Biological Control. University of Texas Press, Austin.

Jones, D. A. 1973. Co-evolution and cyanogenesis. P. 213–242. *In* V. H. Heywood (ed.), Taxonomy and Ecology. Academic Press, New York.

Jones, D. A., J. Parsons, and M. Rothschild. 1962. Release of hydrocyanic acid from crushed tissues of all stages in the life-cycle of species of the Zygaeninae (Lepidoptera). Nature (London) 193:52–53.

Jones, R. L., W. J. Lewis, M. Beroza, B. A. Bierl, and A. N. Sparks. 1973. Host-seeking stimulants (kairomones) for the egg parasite, *Trichogramma evanescens*. Environ. Entomol. 2:593–596.

Jones, R. L., W. J. Lewis, M. C. Bowman, M. Beroza, and B. A. Bierl. 1971. Host-seeking stimulant for parasite of corn earworm: Isolation, identification, and synthesis. Science 173:842–843.

Kloet, G. S., and W. D. Hinks. 1945. A Check List of British Insects. Published by the authors, Stockport.

Klun, J. A., O. L. Chapman, K. C. Mattes, P. W. Wojtkowski, M. Beroza, and P. E. Sonnet. 1973. Insect sex pheromones: Minor amount of opposite geometrical isomer critical to attraction. Science 181:661–663.

Kogan, M. 1977. The role of chemical factors in insect/plant relationships. Proc. 15th Int. Congr. Entomol., Washington, D.C.:211–227.

Lange, R., and J. F. Bronskill. 1964. Reactions of *Musca domestica* L. to parasitism by *Alphaereta pallipes* (Say) with special reference to host diet and parasitoid toxin. Z. Parasitenkd. 25:193–210.

Lanier, G. N., M. C. Birch, R. F. Schmitz, and M. M. Furniss. 1972. Pheromones of *Ips pini* (Coleoptera: Scolytidae): Variation in response among three populations. Can. Entomol. 104:1917–1923.

Lawton, J. H. 1978. Host-plant influences on insect diversity: The effects of space and time. Symp. R. Entomol. Soc. London 9:105–125.

Lawton, J. H., and D. Schröder. 1977. Effects of plant type, size of geographical range and taxonomic isolation on number of insect species associated with British plants. Nature (London) 265:137–140.

Leius, K. 1961. Influence of food on fecundity and longevity of adults of *Itoplectis conquisitor* (Say) (Hymenoptera: Ichneumonidae). Can. Entomol. 93:771–780.

Leius, K. 1967. Influence of wild flowers on parasitism of tent caterpillar and codling moth. Can. Entomol. 99:444–446.

Lewis, W. J., and R. L. Jones. 1971. Substance that stimulates host-seeking by *Microplitis croceipes* (Hymenoptera: Braconidae), a parasite of *Heliothis* species. Ann. Entomol. Soc. Am. 64:471–473.

Lewis, W. J., R. L. Jones, H. R. Gross, Jr., and D. A. Nordlund. 1976. The role of kairomones and other behavioral chemicals in host finding by parasitic insects. Behav. Biol. 16:267–289.

Lewis, W. J., R. L. Jones, and A. N. Sparks. 1972. A host-seeking stimulant for the egg parasite *Trichogramma evanescens:* Its source and a demonstration of its laboratory and field activity. Ann. Entomol. Soc. Am. 65:1087–1089.

Lewis, W. J., D. A. Nordlund, H. R. Gross, Jr., R. L. Jones, and S. L. Jones. 1977. Kairomones and their use for management of entomophagous insects. V. Moth scales as a stimulus for predation of *Heliothis zea* (Boddie) eggs by *Chrysopa carnea* Stephens larvae. J. Chem. Ecol. 3:483–487.

Lewis, W. J., J. W. Snow, and R. L. Jones. 1971. A pheromone trap for studying populations of *Cardiochiles nigriceps*, a parasite of *Heliothis virescens*. J. Econ. Entomol. 64:1417–1421.

Mackauer, M. 1976. Genetic problems in the production of biological control agents. Annu. Rev. Entomol. 21:369–385.

Matthews, R. W. 1974. Biology of Braconidae. Annu. Rev. Entomol. 19:15–32.

Matthews, R. W. 1975. Courtship in parasitic wasps. P. 66–86. *In* P. W. Price (ed.), Evolutionary Strategies of Parasitic Insects and Mites. Plenum Press, New York.

Monteith, L. G. 1955. Host preferences of *Drino bohemica* Mesn. (Diptera: Tachinidae) with particular reference to olfactory responses. Can. Entomol. 87:509–530.

Monteith, L. G. 1958a. Influence of food plant of host on attractiveness of the host to tachinid parasites with notes on pre-imaginal conditioning. Can. Entomol. 90:478–482.

Monteith, L. G. 1958b. Influence of host and its food plant on host-finding by *Drino bohemica* Mesn. (Diptera: Tachinidae) and interaction of other factors. Proc. 10th Int. Congr. Entomol. 2:603–606.

Monteith, L. G. 1960. Influence of plants other than the food plants of their host on host finding by tachinid parasites. Can. Entomol. 92:641–652.

Morris, R. F. 1976. Influence of genetic differences and other variables on the encapsulation of parasites by *Hyphantria cunea*. Can. Entomol. 108:673–684.

Mukerji, M. K. 1971. Major factors in survival of the immature stages of *Hylemya brassicae* (Diptera: Antomyidae) on cabbage. Can. Entomol. 103:717–728.

Muldrew, J. A. 1953. The natural immunity of the larch sawfly (*Pristiphora erichsonii* (Htb.)) to the introduced parasite *Mesoleius tenthredinis* Morley, in Manitoba and Saskatchewan. Can. J. Zool. 31:313–332.

Munakata, K. 1977. Insect feeding deterrents in plants. P. 93–102. *In* H. H. Shorey and J. J. McKelvey, Jr. (eds.), Chemical Control of Insect Behavior: Theory and Application. Wiley, New York.

Nault, L. R., M. E. Montgomery, and W. S. Bowers. 1976. Ant-aphid association: Role of aphid alarm pheromone. Science 192:1349–1351.

Nordlund, D. A., W. J. Lewis, J. W. Todd, and R. B. Chalfant. 1977. Kairomones and their use for management of entomophagous insects. VII. The involvement of various stimuli in the differential response of *Trichogramma pretiosum* Riley to two suitable hosts. J. Chem. Ecol. 3:513–518.

Oshima, K., H. Honda, and I. Yamamoto. 1973. Isolation of an oviposition marker from Azuki bean weevil, *Callosobruchus chinensis*. Agric. Biol. Chem. 37:2679–2680.

Pliske, T. E. 1975a. Attraction of Lepidoptera to plants containing pyrrolizidine alkaloids. Environ. Entomol. 4:455–473.

Pliske, T. E. 1975b. Pollination of pyrrolizidine alkaloid–containing plants by male Lepidoptera. Environ. Entomol. 4:474–479.

Pliske, T. E., J. A. Edgar, and C. C. J. Culvenor. 1976. The chemical basis of attraction of ithomiine butterflies to plants containing pyrrolizidine alkaloids. J. Chem. Ecol. 2:255–262.

Price, P. W. 1970. Trail odors: Recognition by insects parasitic on cocoons. Science 170:546–547.

Price, P. W. 1971. Niche breadth and dominance of parasitic insects sharing the same host species. Ecology 52:587–596.

Price, P. W. 1972. Behavior of the parasitoid *Pleolophus basizonus* (Hymenoptera: Ichneumonidae) in response to changes in host and parasitoid density. Can. Entomol. 104:129–140.

Price, P. W. 1975. Reproductive strategies of parasitoids. P. 87–111. *In* P. W. Price (ed.), Evolutionary Strategies of Parasitic Insects and Mites. Plenum Press, New York.

Price, P. W. 1977. General concepts on the evolutionary biology of parasites. Evolution 31:405–420.

Price, P. W. 1980. Evolutionary Biology of Parasites. Princeton University Press, Princeton, NJ.

Price, P. W., and M. F. Willson. 1976. Some consequences for a parasitic herbivore, the milkweed longhorn beetle, *Tetraopes tetrophthalmus*, of a host-plant shift from *Asclepias syriaca* to *A. verticillata*. Oecologia 25:331–340.

Prokopy, R. J. 1972. Evidence for a marking pheromone deterring repeated oviposition in apple maggot flies. Environ. Entomol. 1:326–332.

Prop, N. 1960. Protection against birds and parasites in some species of tenthredinid larvae. Arch. Neerl. Zool. 13:380–447.

Rabb, R. L., and J. R. Bradley. 1968. The influence of host plants on parasitism of eggs of the tobacco hornworm. J. Econ. Entomol. 61:1249–1252.

Read, D. P., P. P. Feeny, and R. B. Root. 1970. Habitat selection by the aphid parasite *Diaeretiella rapae* (Hymenoptera: Braconidae) and hyperparasite *Charips brassicae* (Hymenoptera: Cynipidae). Can. Entomol. 102:1567–1578.

Reichstein, T., J. von Euw, J. A. Parsons, and M. Rothschild. 1968. Heart poisons in the monarch butterfly. Science 161:861–866.

Renwick, J. A. A., P. R. Hughes, and I. S. Krull. 1976. Selective production of *cis*- and *trans*-verbenol from (−)- and (+)-α-pinene by a bark beetle. Science 191:199–201.

Robacker, D. C., and L. B. Hendry. 1977. Neral and geranial: Components of the sex pheromone of the parasitic wasp *Itoplectis conquisitor*. J. Chem. Ecol. 3:563–577.

Rodriguez, E., and D. A. Levin. 1976. Biochemical parallelisms of repellents and attractants in higher plants and arthropods. P. 214–270. *In* J. W. Wallace and R. L. Mansell (eds.), Biochemical Interaction Between Plants and Insects. Vol. 10, Recent Advances in Phytochemistry. Plenum Press, New York.

Roeske, C. N., J. N. Seiber, L. P. Brower, and C. M. Moffitt. 1976. Milkweed cardenolides and their comparative processing by monarch butterflies (*Danaus plexippus* L.). P. 93–167. *In* J. W. Wallace and R. L. Mansell (eds.), Biochemical Interaction Between Plants and Insects, Vol. 10, Recent Advances in Phytochemistry. Plenum Press, New York.

Shorey, H. H., and J. J. McKelvey, Jr. 1976. Chemical Control of Insect Behavior: Theory and Application. Wiley, New York.

Slobodchikoff, C. N. 1973. Behavioral studies of three morphotypes of *Therion circumflexum* (Hymenoptera: Ichneumonidae). Pan-Pac. Entomol. 49:197–206.

Slobodchikoff, C. N. 1977. Patterns of variation in wasps of the genus *Therion* (Hymenoptera: Ichneumonidae). Univ. California Publ. Entomol. 82:1–65.

Smith, F. 1868. (No title). Proc. Entomol. Soc. London 1868. xxxii.

Spradbery, J. P. 1968. The biology of *Pseudorhyssa sternata* Merrill. (Hym., Ichneumonidae), a cleptoparasite of siricid woodwasps. Bull. Entomol. Res. 59:291–297.

Staedler, E. 1977. Sensory aspects of insect plant interactions. Proc. 15th Int. Congr. Entomol., Washington, D.C.:228–248.

Steck, W., E. W. Underhill, and M. D. Chisholm. 1977. Attraction and inhibition in moth species responding to sex-attractant lures containing *Z*-11-hexadacen-1-yl acetate. J. Chem. Ecol. 3:603–612.

Sutherland, O. R. W., C. H. Wearing, and R. F. N. Hutchins. 1977. Production of α-farnesene, an attractant and oviposition stimulant for codling moth, by developing fruit of ten varieties of apple. J. Chem. Ecol. 3:625–631.

Tahvanainen, J. O., and R. B. Root. 1972. The influence of vegetational diversity on the population ecology of a specialized herbivore, *Phyllotreta cruciferae* (Coleoptera: Chrysomelidae). Oecologia 10:321–346.

Templeton, A. R., and M. A. Rankin. 1978. Genetic revolutions and control of insect populations. P. 83–111. *In* R. H. Richardson (ed.), The Screwworm Problem: Evolution of Resistance to Biological Control. University of Texas Press, Austin.

Tietz, H. M. 1972. An Index to the Described Life Histories, Early Stages and Hosts of the Macrolepidoptera of the Continental United States and Canada, Vols. 1 and 2. Allyn Museum of Entomology, Sarasota, FL.

Townes, H. K. 1939. Protective odors among the Ichneumonidae (Hymenoptera). Bull. Brooklyn Entomol. Soc. 34:29–30.

Tripp, H. A. 1961. The biology of a hyperparasite, *Euceros frigidus* Cress. (Ichneumonidae) and description of the planidial stage. Can. Entomol. 93:40–58.

Vandersar, T. J. D., J. H. Borden, and J. A. McLean. 1977. Host preference of *Pissodes strobi* Peck (Coleoptera: Curculionidae) reared from three native hosts. J. Chem. Ecol. 3:377–389.

Vinson, S. B. 1975. Biochemical coevolution between parasitoids and their hosts. P. 14–48. *In* P. W. Price (ed.), Evolutionary Strategies of Parasitic Insects and Mites. Plenum Press, New York.

Vinson, S. B. 1977. Behavioral chemicals in the augmentation of natural enemies. P. 237–279. *In* R. L. Ridgway and S. B. Vinson (eds.), Biological Control by Augmentation of Natural Enemies. Plenum Press, New York.

Vinson, S. B., R. D. Henson, and C. S. Barfield. 1976. Ovipositional behavior of *Bracon mellitor* Say (Hymenoptera: Braconidae), a parasitoid of boll weevil (*Anthonomus grandis* Boh.). I. Isolation and identification of a synthetic releaser of ovipositor probing. J. Chem. Ecol. 2:431–440.

Vité, J. P., and D. L. Williamson. 1970. *Thanasimus dubius:* Prey perception. J. Insect Physiol. 16:233–239.

Waage, J. K. 1978. Arrestment responses of the parasitoid, *Nemeritis canescens,* to a contact chemical produced by its host, *Plodia interpunctella*. Physiol. Entomol. 3:135–146.

Woets, J., and J. C. van Lenteren. 1976. The parasite-host relationship between *Encarsia formosa* (Hymenoptera: Aphelinidae) and *Trialurodes vaporarium* (Homoptera: Aleyrodidae). VI. The influence of the host plant on the greenhouse whitefly and its parasite *Encarsia formosa*. Proc. 3rd Conf. Biol. Control Glasshouses. OILB/SROP. Antibes, France.

Wood, D. L. 1970. Pheromones of bark beetles. P. 301–316. *In* D. L. Wood, R. M. Silverstein, and M. Nakajima (eds.), Control of Insect Behavior by Natural Products. Academic Press, New York.

Zoebelein, G. 1956. Der Honigtau als Nahrung der Insekten. Z. Angew. Entomol. 38:369–416, 39:129–167.

SECTION V

CONCLUSION

CHAPTER FOURTEEN

SUMMARY OF SIGNIFICANCE AND EMPLOYMENT STRATEGIES FOR SEMIOCHEMICALS

ROBERT D. JACKSON

AR–SEA–USDA
Beltsville, Maryland

W. JOE LEWIS

AR–SEA–USDA
Southern Grain Insects Research Laboratory
Tifton, Georgia

When people first began to culture plants, they began to accumulate information on factors affecting the growth and productivity of plants, especially those they cultivated. The application of this information resulted in the development of agriculture as a viable occupation and way of life, and farming became an integral part of human history.

The operation of the early farms was an art based on knowledge accumulated from years of experience and passed from one generation to the next. The practitioners on these relatively small cultivated areas lacked refined technology and often were unable to produce high yields per unit of area. Nevertheless, their practices were, to a great extent, ecologically sound in that they were rather diversified and utilized, to a considerable degree, a total resource management approach.

Scientific advances, the incorporation of resulting technologies into agricultural practices, and the ever-increasing demand for greater production capacities resulted in specialization in certain commodities and more intensive monoculture cropping.

The innate interrelationships among plants, phytophagous insects, and entomophagous insects were primarily developed under more complex, uncultivated conditions (Figure 14.1). These interrelationships are often adversely affected when plants are grown in less complex monocultures. The superabundant food supplies and otherwise altered ecological conditions prevalent in modern agricultural systems often allow the plant pest populations to explode far beyond the levels present under noncultivated conditions.

The continued employment of cultivated systems will be required if we are to meet our food and fiber needs. However, we must realize that the equilibrium in such systems is very fragile and that certain pest problems can be anticipated. With this realization as a basis, and with an adequate understanding of the interactions and mechanisms involved in both cultivated and uncultivated communities, we will be able to proceed with the development of effective and ecologically harmonious manipulation techniques for managing pest situations.

The vast array of chemical mediators discussed in the preceding chapters are prime candidates as manipulative tools in such programs. Research during the past two decades and the principles set forth in this book provide the concepts necessary for harnessing these valuable pest control tools.

FIGURE 14.1. A predacious coccinellid larva attacking a caterpillar of *Heliothis zea*. Such predators have long been recognized as basic regulating forces in agricultural ecosystems. An understanding of the factors governing the interrelationships of our crops with these phytophagous and entomophagous insects is fundamental to sound pest management practices.

APPROACHES FOR EMPLOYMENT

Let us explore the various avenues and strategies for employment of semiochemicals. The three basic approaches for their utilization are (1) to impede the reproduction and other activities of pest species, (2) to enhance the reproduction and activities of beneficial species, and/or (3) to monitor the status and activities of the pest and/or beneficial species for management decisions. The first approach is a direct control strategy; the second and third approaches could be considered to represent indirect employment.

We outline the various avenues for their use according to classification of the various semiochemicals presented by Nordlund in Chapter 2.

Pheromones

Pheromones, of course, are the chemical cues that mediate interactions among individuals of the same species. The sex pheromones, discussed by Roelofs (Chapter 11), and the epideictic pheromones, discussed by Prokopy (Chapter 10), are the two types of pheromone under consideration. Except for the use of pheromone traps to monitor or detect populations, the approach for employment of pheromones consists primarily of a direct approach to impede the mating and ovipositional activities. Because this direct approach is usually less complex and often is more easily understood, both biologically and chemically, it has received the greatest attention, with by far the bulk of that attention going to sex pheromones.

Sex Pheromones

Monitoring and Detection Traps. The use of sex and aggregation pheromones in traps was one of the first uses of semiochemicals in pest management programs. The development of integrated pest management programs is dependent on a knowledge of the presence and density of insect pest populations. For many programs, the timing of control applications is based on data obtained from traps. Light traps are used for some of these programs, but pheromone traps offer several advantages as discussed by Roelofs (Chapter 10). Although pheromone traps are an inexpensive monitoring tool, the interpretation of the trap data may be complex and should be undertaken only by trained personnel.

In addition to using traps to monitor insect pest densities, investigators use traps to detect the presence of new infestations of pest insects. Pheromone traps are deployed in likely areas of introduction, such as airports, and in habitats offering high potential for invasion. This early warning system allows for the timely deployment of control tactics to eliminate or reduce the immediate impact of the establishment of the invading pest.

Mass Trapping. Mass trapping has been studied intensively for several insect species. These are reviewed by Roelofs and Prokopy (Chapters 9 and 10). The largest mass trapping program to date is that reported by O'Sullivan (1979) for *Ips typographus*. Sweden and Norway have initiated a program to use nearly 1 million traps baited with a total of 100 kg of pheromone at an estimated cost of $23 million annually. The success of the program is yet to be determined, but if the loss of six million trees each year can be substantially reduced, the significance of this technique will be firmly established.

Mating Disruption. Following early work on mating disruption of the cabbage looper (*Trichoplusia ni*) by Gaston et al. (1967), mating disruption has been tried for many insects. The technique is based on the premise that males would be unable to locate females if the environment around the females were permeated with a sex pheromone. Two sex pheromones have been registered for use in mating disruption; these are gossyplure for the pink bollworm (*Pectinophora gossypiella*) and disparlure for the gypsy moth (*Lymantria dispar*). Sex pheromones are known for many more pest species and, no doubt, several more will be registered for use in mating disruption programs. Generally, this technique must be applied as a preventive measure, and it has been successful only when insect populations are relatively low. In some insect species, the sex pheromones consist of both attractant pheromones and mating-stimulant pheromones. In these species, mating-stimulant pheromones rather than attractant pheromones may more effectively disrupt mating, because the males might continuously attempt to mate in the absence of a female and thus would not be responsive to the attractant. The effective use of this technique, as well as the mass trapping technique, requires a thorough understanding of the behavioral sequence involved in the interaction and the various pheromones that mediate each step. The extent to which mating disruption will become established in pest management programs is yet to be determined, but this approach is potentially a very powerful tool for population management.

Epideictic Pheromones

In some insect-plant relationships, there is an optimum population density for the survival of the insects. Among the several species, there exist chemical communication systems (epideictic pheromones) that enable insects to aggregate to levels necessary to exploit a habitat or host, but prevent aggregation of individuals after the high end of the optimum range is reached. One of the best examples of this phenomenon is found in *Dendroctonus* bark beetles. Furniss et al. (1972, 1974) have carried out pest management tests on *Dendroctonus* solely through the use of aggregation disruption. They were able to protect felled Sitka spruce trees from attack for up to 60 days. Hedden and Pitman (1978) used a combination of attractants with aggregation disruption to modify insect behavior to such an extent that the natural defense of the host trees killed the beetles that attacked the trees at densities below the threshold for colonization.

Most of the progress in developing oviposition disruption with epideictic pheromones has been with *Rhagoletis cerasi* in Switzerland by Katsoyannos and Boller (1976, 1979). These investigators achieved substantial reductions in infestations in cherries through application of the pheromones used to mark oviposition sites by the fly. Prokopy (Chapter 10) has reviewed the progress in the development of oviposition disruption and expresses caution against overoptimism in the use of these pheromones as a single approach to control.

The evidence is relatively clear that many phytophagous insect species are able to regulate their population density through the use of epideictic pheromones. Prokopy (Chapter 10) has listed evidence of 33 species of phytophagous insects that release epideictic pheromones that influence their level of host exploitation. As further biological data and information are developed, it is likely that other phytophagous insects will be found to secrete epideictic pheromones, and artificially applied synthetic pheromones can be used to reduce their populations. Much additional information will have to be developed before the real significance of the use of epideictic pheromones in pest management programs can be determined.

The epideictic pheromones have also been shown to be of vital importance in the foraging behavior of beneficial entomophagous insects (Chapter 9). Implementation of this tactic on selected entomophages may be useful. For example, it may be used to regulate the interspecific competition among parasitoids, cleptoparasitoids, and predators. In a field study of parasitoids and predators attacking the eggs of the sugarcane borer (*Diatraea saccharalis*) in Brazil, Teran (1980) shows that rates of attack by the egg parasitoids were very low, but the rates of attack and destruction of eggs by predators were high. However, the rate of predation may not have been adequate to meet the economic threshold. In some situations, it might be necessary to reduce predation to a level low enough to allow the parasitoids to increase to a high population density, to achieve a greater degree of control than would be possible in the presence of heavy predation.

This phenomenon may also be responsible for the failure of parasitoids to become effective in sugarcane fields in Louisiana, where the fire ant is a predator of both parasitized and nonparasitized eggs and larvae of the sugarcane borer. Hensley (1971) has shown that the average level of attack by the parasitoid *Lixophaga diatraeae* on sugarcane borer larvae averages no more than 4%. Following an attack by this parasitoid, the sugarcane borer larvae may feed less, and because of a reduced frass plug, may become more vulnerable to predation than healthy larvae. Differential predation on unhealthy larvae could reduce the frequency and extent of parasitoid outbreaks in some insect populations.

Allelochemics

Allelochemics are the chemical substances that mediate interactions among members of different species. We consider kairomones, which are of benefit to the receiving species, allomones, which are of benefit to the sender, and synomones, which are of mutual benefit. These three types of allelochemic are discussed collectively, each for control of pest insects and for enhancing the performance of beneficial arthropods.

Control of Pest Insects

It is becoming more and more apparent that allelochemics are extremely important factors in the interaction of phytophages and their host plants (Schoonhoven, Chapter 3).

Phytophagous insects use kairomones to locate and recognize host plants, and for oviposition and feeding in a sense analogous to that in which sex pheromones are used in mating interactions. Therefore, the potential exists for using the same strategy for employing kairomones as for employing sex pheromones. Kairomones can be used to directly disrupt the finding and recognition of host plants, for mass trapping, or for indirectly forecasting and monitoring pest species. Such research has received little attention to date because it has been generally assumed that such chemicals did not have sufficient potency to warrant exploitation. However, recent studies indicate that substantial possibilities do exist.

Allomones are of importance to host plants as key mechanisms of defense against phytophages. These chemicals offer great potential as oviposition and feeding disruptants in a manner similar to the use of epideictic pheromones (Prokopy, Chapter 10).

There are various avenues for the employment of these semiochemicals for improving resistant crop breeding programs and for the development of insecticidal baits, as discussed in detail later in this chapter.

Beneficial Arthropods

The use of kairomones to manipulate and enhance interactions of beneficial arthropods with their pest hosts or prey may be considered under two headings.

Importation. First, the employment of knowledge and technology relative to the kairomones and their function may greatly increase the effectiveness of importation programs. As was thoroughly demonstrated in the discussions by Price (Chapter 13), these kairomones are very strategic focal points in the evolutionary strategies of entomophages with their hosts or prey. Consequently, information relative to the involvement of kairomones in the searching behavior could be of critical importance in choosing proper candidates for importation. Also, technology, such as discussed by Greany and Hagen (Chapter 7) and Gross (Chapter 8), can be employed for colonization releases to increase the likelihood of effective establishment.

Augmentation-Manipulation. Kairomones can be used in various ways to improve the flexibility and performance of entomophagous insects in augmentation and manipulation programs. Their greatest potential utility at this time appears to be for aggregating natural predators or retaining released entomophages in target locations. The various approaches and techniques are treated in detail by Gross (Chapter 8). The use of kairomones with this approach appears to be particularly promising when incorporated with food sprays and/or artificially supplied hosts or prey.

Integrated Use of Various Semiochemicals

For simplicity, we have discussed the use of semiochemicals on an individual basis; but in the field environment, insects are simultaneously exposed to several semiochemicals and may react to combinations more strongly than to single chemicals, or in a different way. The utilization of semiochemicals in pest management systems may frequently involve combinations, such as kairomones and synomones or kairomones and pheromones.

Some parasitoids and predators may orient by means of a kairomone emitted by the insect host or prey and by a synomone emitted by a noncrop host. In such circumstances, kairomones applied to a crop may be of limited usefulness in retaining the entomophages in the crop; but the application of both the kairomone and the synomone could markedly increase the retention of those beneficial insects in the crop and subsequently increase the rates of parasitization and predation.

Another scheme might involve the use of at least three semiochemicals. The greater portion of a field might be treated with epideictic pheromones, with small strips of the crop treated with oviposition attractants and/or stimulants, to create a "trap-crop" effect. The concentration of the pest would be exploited by natural enemies, but augmentative releases of reared entomophages, along with application of the pest insect kairomone, could result in very high rates of attack throughout the field with reduced numbers of released entomophages.

Resource Management

As the complex chemical interrelationships among plants, phytophages, and entomophages as described by the authors of the preceding chapters are more fully understood, the total habitat or agricultural ecosystems can be managed to increasingly meet the needs of beneficial organisms and to the detriment of pests. This habitat management may include effective combinations of such practices as intercropping, companion cropping, and encouragement of certain noncrop plants. Certainly, such practices are not new, but their use has been largely based on empirical observations and folklore without a complete understanding of the mechanisms involved. As the mechanisms are elucidated, the practices can be selectively used in totally integrated systems of pest management.

Resistant Crop Varieties

The effect of chemical constituents of plants on insect populations has long been recognized. Dethier (1947) reviewed the chemical attractants and repellents three decades ago, and Painter (1951) recognized that chemicals were, in part, responsible for both nonpreference and antibiosis, two of the three principal mechanisms of host plant resistance. However, Painter (1951) states, "A knowledge of the causes of resistance is highly desirable, but it may or may not be of use in breeding operations." Because plant breeders and entomologists have been able to make progress in the development of resistant crop varieties without knowing the actual causes of resistance, they have not vigorously investigated the chemical basis of resistance. With the development of better and more rapid analytical techniques and an increased understanding of the chemical mediators between plants and insects, greater attention is now being given to the chemical basis of resistance.

In the past, the basis of improved plant resistance to insects has been primarily through increased levels of allomones rather than reduced levels of kairomones. However, these insect-plant interactions consist of complex stimulus-response situations that may involve several phases of host selection. Although a complete understanding of all the interactions involved is not necessary before crop varieities with increased levels of resistance can be developed, an increased understanding of the chemical mediators should greatly enhance the effectiveness of these programs. Also, Callahan (1975) has theorized that plant volatiles are used by insects to detect plants via electromagnetic radiation. Alterations in plant volatiles might have an impact on the termination of long-range insect migration and the subsequent reduction in fitness of insects to survive in the environment. Alteration of the kairomones, as well as the allomones, of crop varieties could have a very profound impact on pest management systems. However, only a limited number of investigators have studied the role of volatile chemicals in the host plant selection process.

Vinson (Chapter 4) points out that plant alleochemics are also important in the host plant selection process of parasitoids and predators. A crop variety may be considered to be resistant because of reduced attraction for phytophagous insects or other qualities; but as pointed out by Price (Chapter 13), such an alteration may affect the ability of parasitoids and predators to attack the phytophagous insects. Consequently, the variety might be more severely damaged than a nonresistant variety. However, the selection of improved crop varieties by plant breeders involves a complex consideration of many factors. Therefore, a better understanding of the interaction between plants and beneficial insects is necessary before plant breeders can effectively incorporate the factor of entomophages into their programs.

Chemical Control

Plant kairomones and phagostimulants can be combined with insecticides to markedly increase their efficiency and to achieve more effective pest management programs, thus reducing the adverse impact on the environment. An example of this is shown by the work of J. A. Onsager, Rangeland Insects Research Laboratory, Bozeman, Montana (personal communication). In research plots, Onsager has been able to control grasshoppers with only 0.03 lb of insecticide/acre in a bait. This control is slightly less than the control achieved by a conventional application of 1.5 lb of insecticide, but the reduced impact on beneficial organisms may result in equal or increased population suppression. In this example, wheat bran was the bait material; but a bait could be manufactured that would have characteristics not available in a "natural" bait. This could be achieved by combining an inert material with kairomones and phagostimulants that would be strongly attractive to the pest grasshoppers. Since most grasshopper infestations are comprised of several species, it may be possible to "tailor make" a bait that would be attractive only to the species of grasshoppers of particular concern in the pest management program. Other species of grasshoppers could be retained in the environment to continue feeding on undesirable vegetation and to provide food for beneficial organisms.

Attractant baits have been used with insecticides in the past, but they are not used extensively now. With an increased understanding of semiochemicals and how they can be utilized to manufacture baits with specific properties, it will be possible to eliminate many of the disadvantages of insecticides while retaining their usefulness in pest management programs. The use of toxicant baits is one of the most underutilized opportunities to reduce the impact of chemical insecticides on the environment; but if the potential of these substances in pest management programs is to be realized, much more effort must be devoted to their development.

CONSTRAINTS TO THE USE OF SEMIOCHEMICALS IN PEST MANAGEMENT PROGRAMS

Several of the chapter authors have considered the various constraints and precautions in the use of semiochemicals. Lewis (Chapter 1) discusses various constraints to the development and use of semiochemicals, and Price (Chapter 13) has given special consideration to the influence of manipulation of natural enemies through the use of semiochemicals on the evolution of the fitness of both pest and beneficial insects. Klassen (1976) has also discussed constraints in the development of several alternative methods of pest control.

The use of semiochemicals to manipulate natural enemies of pest insects could have an impact on the fitness of the natural enemies, but as long as the proportion of manipulated individuals in relation to the entire population is relatively small, there is little likelihood that there will be a significant reduction in the fitness of the natural populations. However, if the use of semiochemicals becomes so widespread that a large portion of the population is affected, careful consideration must be given to the influence being exerted on insect populations. Also, insects may become resistant to manipulation by semiochemicals just as they have become resistant to insecticides. This is especially true if the semiochemicals bring about mortality of large proportions of insect populations. These restraints will not necessarily reduce the significance of semiochemicals in pest management programs; but as semiochemicals become successfully integrated into pest management programs, we must have sufficient versatility of action to avoid any deleterious effects. This versatility of action, however, must be based on a thorough understanding of biological systems.

Indeed, many of the difficulties we have experienced in our attempts to regulate pest insect species have been due to a lack of thorough understanding of the biological systems. At the dawn of the insecticide age following World War II, it appeared that we would, for the first time in history, be able to achieve a satisfactory level of insect control and to reduce the severe losses being caused by insects to the human food supply. This rapid and extensive development of organic insecticides was made possible and brought about largely by the rapid development of the chemical industry throughout the world. During the years from World War II to the present, investments in the chemical industry were very large. Only a small fraction of this investment was devoted to developing methods to control insects, but much of the chemical technology that accumulated was applicable to the chemical insecticide field.

Thus the development of organic insecticides was due not to an increase in biological information, but to the development of the chemical industry and the concurrent increase in chemical technology. However, after organic insecticides came into general use, some of the biological implications of their use, such as insecticide resistance and adverse impacts on other organisms in the environment, became apparent. Investigations into alternative methods of insect control were then started or renewed with increased vigor. Many alternative approaches to insect population suppression are now being studied (Knipling 1979), but all alternative approaches require an increased level of understanding of the biological phenomena.

Scientists and the general public now realize that insect control must be based on a thorough understanding of biological systems, and this point of view is reflected in the development of the integrated pest management concept of pest control. However, integrated pest management programs are often constrained by lack of complete biological information. The reasons for this lack of biological technology are diverse but are primarily related to the cost of investment in biological research. The return on investments in biological research is often unclear, difficult to document, and difficult to protect through proprietary protection laws. For these reasons, biological information is developed mainly by researchers supported by public funds; and because of the strong competition for the use of public funds, research on biological phenomena may be limited. This is especially true when long-term studies are required before any return on the investment will be evident. The use of semiochemicals in pest management programs is usually very complex, involving a high level of both chemical and biological technology. Thus the development of semiochemicals has required such a long-term research investment.

TERMINOLOGY

Wenk (1979) points out that scientists must communicate to the public concerning the need for research activities and the potential importance to society of such investigations. This is especially important for biological scientists; and if support for the development of semiochemicals is to be obtained from the public sector, the need for such development must be generally understood. Therefore, scientists cannot continue to limit their communications to discussions with one another, as they have often been prone to do; they must also communicate with the general public.

Nordlund (Chapter 2) discussed in detail a system of terminology for chemicals that mediate various interactions. Such a precise terminology is important for effective conveyance of ideas among scientists and should be implemented when it is of value. However, we must also be careful not to allow the scientific terms to become a divisive jargon that sets the scientist apart from the public—the public that provides the financial support for research. For this reason, scientists engaged in research on chemical ecology, and other fields as well, should constantly test the terminology they use to determine whether it does serve a useful purpose, properly define terms that may not be familiar to those being addressed, and use examples to illustrate the meanings of the definitions. These measures will help ensure that the terminology used does not limit the development of science by alienating the public.

CONCLUSION

We are much further along today in the use and exploitation of chemical cues to achieve insect pest management than Glover (1856) was when he attempted to use light to control cotton insects by building bonfires around the fields. However, when scientists view our attempts to use semiochemicals one hundred years from now, we anticipate that insect population management will have become so sophisticated that our use of semiochemicals in pest management programs in 1980 will appear nearly as elementary as Glover's bonfires appear to us today. The implementation of semiochemicals in pest management programs will not be rapid; rather, it will be a gradual process. The complex chemical and biological technology required for the use of semiochemicals must be developed for each pest species. This will often require understanding the chemical communication system not only of a single pest species but also of several associated species. The action of semiochemicals is very specific; semiochemicals are so fundamental to the survival of insect species that they offer nearly limitless potential to increase the effectiveness of pest management programs. As Shorey (1977) says:

> An insect does not think; it reacts. The reactions are usually triggered by external stimuli and modified by a host of environmental variables, internal variables, and some rudimentary learning. The reactions are often highly stereotyped and cause the insects to perform appropriate behaviors that enhance species survival when appropriate stimuli are encountered. Much of the sensory world of the insect involved in stimulation or inhibition of such behaviors as mating, feeding, and egg-laying is chemical. The reactions of the insects to these chemicals are so predictable that if man could learn enough about the attendant behaviors, he could literally make the insects jump through a hoop.

This potential is so great that almost without exception, future pest management programs will utilize semiochemicals; and often, multiple approaches in the use of semiochemicals will be integrated into programs for each species.

However, before this somewhat rosy prediction can be realized, much more scientific talent must be dedicated to understanding the chemical interactions of insects. As detailed and extensive as the preceding reviews are, they only chronicle the beginnings of the work that will be required before the real significance of semiochemicals in pest management programs can be evaluated.

REFERENCES

Callahan, P. S. 1975. Insect antenna with special reference to the mechanism of scent detection and the evolution of the sensilla. Int. J. Morphol. Embryol. 4:381–430.

Dethier, V. G. 1947. Chemical Insect Attractants and Repellents. Blakiston's, Philadelphia.

Furniss, M. M., L. N. Kline, R. F. Schmitz, and J. A. Rudinsky. 1972. Tests of three pheromones to induce or disrupt aggregation of Douglas fir beetles on live trees. Ann. Entomol. Soc. Am. 65:1227–1232.

Furniss, M. M., G. E. Daterman, L. N. Kline, M. D. McGregor, G. C. Trostle, L. F. Dettinger, and J. A. Rudinsky. 1974. Effectiveness of the Douglas fir beetle anti-aggregation pheromone, methylcyclohexenone, at three concentrations and spacings around felled trees. Can. Entomol. 106:381–392.

Gaston, L. K., H. H. Shorey, and C. A. Saario. 1967. Insect population control by use of sex pheromones to inhibit orientation between sexes. Nature (London) 213:1155.

Glover, T. 1856. Insects frequenting the cotton-plant. *In* Report of the Commissioner of Patents for the Year 1855. Washington, D.C. P. 64–69.

Hedden, R. L., and G. B. Pitman. 1978. Attack density regulation: A new approach to the use of pheromones in Douglas fir beetle population management. J. Econ. Entomol. 71:633–637.

Hensley, S. E. 1971. Management of sugarcane borer populations in Louisiana—A decade of change. Entomophaga 16:133–146.

Katsoyannos, B. I., and E. F. Boller. 1976. First field application of oviposition-deterring pheromone of European cherry fruit fly, *Rhagoletis cerasi*. Environ. Entomol. 5:151–152.

Katsoyannos, B. I., and E. F. Boller. 1979. Second field application of oviposition-deterring pheromone of the European cherry fruit fly, *Rhagoletis cerasi*. Z. Angew. Entomol. (in press).

Klassen, W. 1976. A look forward in pest management. Proc. Tall Timbers Conf. Ecological Animal Control by Habitat Management. 6:173–193.

Knipling, E. F. 1979. The Principles of Insect Population Suppression and Management. U.S. Department of Agriculture, Agriculture Handbook No. 512.

O'Sullivan, D. A. 1979. Pheromone lures help control bark beetles. C&E News, July 30:10–13.

Painter, R. H. 1951. Insect Resistance in Crop Plants. University Press of Kansas, Lawrence.

Shorey, H. H. 1977. Interaction of insects with their chemical environment. P. 1–5. *In* H. H. Shorey and J. J. McKelvey, Jr. (eds.), Chemical Control of Insect Behavior. Wiley, New York.

Teran, F. O. 1980. Natural control of *Diatraea saccharalis* (Fabr. 1794) eggs in sugarcane fields of Sao Paulo. Proc. XVII Congr., International Society of Sugarcane Technologists, Manila, Philippines (in press).

Wenk, E., Jr. 1979. Margins for Survival. Pergamon Press, New York.

INDEX